Structural Integrity

Volume 9

The *Structural Integrity* book series is a high level academic and professional series publishing research on all areas of Structural Integrity. It promotes and expedites the dissemination of new research results and tutorial views in the structural integrity field.

The Series publishes research monographs, professional books, handbooks, edited volumes and textbooks with worldwide distribution to engineers, researchers, educators, professionals and libraries.

Topics of interested include but are not limited to:

- Structural integrity
- Structural durability
- Degradation and conservation of materials and structures
- Dynamic and seismic structural analysis
- Fatigue and fracture of materials and structures
- Risk analysis and safety of materials and structural mechanics
- Fracture Mechanics
- Damage mechanics
- Analytical and numerical simulation of materials and structures
- Computational mechanics
- Structural design methodology
- Experimental methods applied to structural integrity
- Multiaxial fatigue and complex loading effects of materials and structures
- Fatigue corrosion analysis
- Scale effects in the fatigue analysis of materials and structures
- Fatigue structural integrity
- Structural integrity in railway and highway systems
- Sustainable structural design
- Structural loads characterization
- Structural health monitoring
- Adhesives connections integrity
- Rock and soil structural integrity.

Springer and the Series Editors welcome book ideas from authors. Potential authors who wish to submit a book proposal should contact Dr. Mayra Castro, Senior Editor, Springer (Heidelberg), e-mail: mayra.castro@springer.com

More information about this series at http://www.springer.com/series/15775

Oleksandra Datsyshyn · Volodymyr Panasyuk

Structural Integrity Assessment of Engineering Components Under Cyclic Contact

Springer

Oleksandra Datsyshyn
Karpenko Physico-Mechanical Institute,
National Academy of Sciences of Ukraine
Lviv, Ukraine

Volodymyr Panasyuk
Karpenko Physico-Mechanical Institute,
National Academy of Sciences of Ukraine
Lviv, Ukraine

ISSN 2522-560X ISSN 2522-5618 (electronic)
Structural Integrity
ISBN 978-3-030-23071-5 ISBN 978-3-030-23069-2 (eBook)
https://doi.org/10.1007/978-3-030-23069-2

This Springer imprint is published by the registered company Springer Nature Switzerland AG
The registered company address is: Gewerbestrasse 11, 6330 Cham, Switzerland

Preface

Solids often operate in contact in the units of machines and structures. Indeed, the elements of movable and immovable joints of machines and structures operate under the conditions of cyclic contact either due to their functional purposes or due to the influence of the operational factors (e.g., vibration). The defects of the contact surfaces or even the fracture of joints under cyclic contacts may lead to a decrease in the efficiency or even to the loss of the serviceability of the entire machine or structure. The problem of guaranteeing contact strength, durability, and wear resistance of the mechanical components is an important scientific and engineering problem. Within the framework of this problem, it is important to study the nature and causes of the typical defects (damages) appearing in tribojoints and give a theoretical description of the processes of their formation.

One of the most common types of mechanical damages formed in solids in the zone of their cyclic contact is cracking. The cracks appearing and propagating in the surface layers of contact zones of the bodies may form pits, pitting, spall, spalling, gaps, checks, crumbling, etc., on the surfaces of the bodies and, thus, cause the loss of the serviceability of contact surfaces. In some cases, the cracks formed near the surfaces initiate the process of propagation of the main crack and lead to the complete destruction of the products (parts). Under the conditions of cyclic contact of deformable bodies, the processes of initiation and propagation of cracks in the contact zone are observed mainly in cases where the contact interaction is realized in the form of rolling, fretting fatigue, pulsed (pulsating) contact, friction fatigue, etc.

The phenomenon of contact rolling fatigue is typical of tribosystems (tribojoints) in the form of wheel–rail systems, backup and working rolls of the rolling mills, various types of gearings, ball and roller bearings, etc.

The phenomenon of fretting fatigue is realized in the units of machines and structural joints, whose surface elements suffer insignificant fluctuating mutual sliding caused mainly by the cyclic loads or in-service vibrations. This is especially pronounced for joints in the bodies and casings of various means of transportation (cars, aircraft, spacecraft, and ships) and numerous industrial structures (turbines of the nuclear power plants, gas turbine engines, oil platforms, bridges, and pipelines), as well as for various spline, bolt, and key connections. It is known that the cracks appearing in the contact zone of elements of the units and joints suffering fretting fatigue may lead to a severalfold decrease in the durability of machines, which can be very dangerous.

The fracture of friction surfaces is called friction fatigue. It is observed, e.g., in sliding bearings, friction clutches, braking elements (brake shoes and disks), etc.

The contact interaction in the form of pulsed contact takes place under the conditions of cyclic compression of two bodies (or their parts) by normal (or, in general, inclined to the contact surface) forces with possible disconnection of the contacting bodies. This interaction leads to the appearance of arc or annular fatigue cracks in the contact zone and is often accompanied by the phenomenon of fretting corrosion in this zone.

All indicated contact interactions are responsible for the contact fatigue of the material, i.e., in other words, the process of preparation and propagation of fracture in the surface layers of the materials of joined parts whose surfaces operate under the action of variable cyclic contact loads.

The problem of contact strength and durability of various mechanical systems is an urgent scientific and engineering problem. It is extensively studied by the researchers and engineers in scientific centers of many countries.

In the proposed monograph, we present both the results of investigations of the authors and the results of available literature publications in the field of fracture of the surface layers of bodies operating under the conditions of cyclic contact (including tribojoints). We propose a new procedure (i.e., the computational models and algorithms) aimed at the computer investigation of the paths of crack propagation and the specific features of formation of typical contact fatigue damages (pitting, spalling, gaps, crumbling, squats ("dark spots"), checks, cavities etc.) in the contact area. We present examples of evaluation of the residual service lifetime of tribojoints, including the rolling (wheel–rail systems, backing and working rolls of the rolling mills) and fretting (shelf-blade gas turbine engines (GTE)) couples according to the criteria of formation of typical contact fatigue damages in these couples. We also formulate some recommendations concerning the possibility of optimization of the operating parameters of joints (contact loads, friction/lubrication in contact, cyclic crack growth resistance of the materials in transverse shear and normal opening, etc.).

The monograph consists of five chapters.

Chapter 1 represents an overview of the methods used for the evaluation of the contact durability of tribojoints, including, in particular, rolling and fretting couples. We analyze the results of investigations in the related scientific fields, which can be regarded as basic for the proposed methods, namely, in the field of mechanics of fatigue fracture of bodies in cyclic contact and in the field of contact problems of the mathematical theory of cracks. We also present a survey of the well-known scientific and engineering publications dealing with the analyzed problems.

Chapter 2 contains the key elements of the theory of fatigue fracture, namely, the main stages and fatigue characteristics of the materials; the criteria and diagrams of fatigue fracture, and the characteristics of fatigue crack growth resistance of the materials. We present a brief description of the specific features of cyclic contact interaction accompanied by the crack initiation in the nearsurface contact zone: rolling, rolling with sliding, fretting fatigue, pulsed contact, friction fatigue, etc. We also describe typical contact fatigue damages formed by cracks on the working surfaces of elements of the tribojoints. In the same chapter, we propose a new computational model for the investigation of the fracture processes and evaluate the contact residual lifetime of the bodies (elements of tribojoints) under the conditions of their cyclic interaction. The model is based on the step-by-step determination of curvilinear paths in the zone of cyclic contact of the elements of tribojoints depending on operating parameters and the mechanisms of crack propagation: transverse shear (mode II) or normal opening (mode I) in the process of deformation of the material). The algorithms used for the construction of the paths of crack propagation take into account the changes in the stress–strain state caused both by crack elongation and the motion of the counterbody (variations of the load) in the contact cycle and also possible changes in the mechanism of fracture caused by crack propagation and friction between the crack faces. The proposed computational model is developed in Chaps. 4 and 5 for two separate types of contact interaction of the bodies: rolling with sliding and fretting fatigue.

Chapter 3 is devoted to the mathematical method of modeling the deformation and fracture process for elements of moving and fixed joints (tribojoints). We propose singular integral equations (SIE) for the elastic half-plane weakened by a system of curvilinear cracks under the action of various model contact loads applied to the boundary of the half-plane. We briefly describe the Gauss–Chebyshev method of mechanical quadratures that enables one to efficiently construct the numerical solutions of these SIE. We also deduce the relations for the stress intensity factors (SIF) at the crack tips can be expressed via the solutions of SIE for inside and edge curvilinear cracks in the half-plane. In this chapter, we present both known results available from the literature and new results.

Chapter 4 contains the results of investigations of the fracture processes (paths of crack propagation) for bodies under the conditions of rolling (rolling with sliding) contact. These results were obtained by using the model proposed in Chap. 2, the solutions of SIE for the corresponding problems, and also the results of evaluation of lifetime under the conditions of contact fatigue. We study and describe the specific features of formation of the typical defects usually observed in rolling bodies, such as pitting, spalling, squat ("dark spot"), checks, crumbling depending on the operating parameters of the analyzed rolling pair and the characteristics of cyclic crack growth resistance of the corresponding materials. In particular, we present a theoretical confirmation of the well-known Wei hypothesis that the main cause of pitting formation is the presence of oil in the contact zone (edge cracks wedged out by pressure of any kind propagate toward the contact surface and lead to its pitting). We also show under what conditions (operating conditions and the characteristics of crack growth resistance) a subsurface crack may lead to the formation of the following defects: spalling, gaps, very long subsurface cracks that do not appear on the rolling surface, and spontaneous fracture of elements caused by the crack propagation into the bulk of the material.

We present examples of evaluation of the residual contact lifetime of rail, wheel, and roll steels according to the criteria of pitting and spalling formation. The contact fatigue curves (i.e., the dependences of the number N_f of rolling cycles on the maximum value of contact pressure p_0) are plotted by using the criteria of pitting or spalling formation on the rolling surfaces. We also give some recommendations useful for engineering practice.

Chapter 5 deals with the investigation of contact interaction of two bodies under the conditions of fretting fatigue. We study the fracture processes in the material (and construct the paths of crack propagation) in the zone of cyclic contact of two bodies under the conditions of fretting fatigue, in particular, depending on the friction coefficient and stick/slip conditions between the bodies, the form of the base of counterbody (the type of modeling contact loading), etc. We also present some examples of evaluation of the residual service life for turbine blades of the GTE (gas turbine engine) made of titanium alloy.

In addition, Chaps. 4 and 5 contain a large body of data on the dependences of the SIF range in a contact cycle $\Delta K_{II}(\beta)$ on the angle of orientation β of the edge crack for the operating parameters of the wheel–rail system. Note that the parameter $\Delta K_{II}(\beta)$ controls the process of crack propagation by the mechanism of transverse shear (mode II) in the contact zone. We also establish the existence of a characteristic angle $\beta = \beta^*$ for the propagation of shear edge cracks and formulate the conditions required for its determination. It is shown that cracks initiated at this angle are both independent contact fatigue defects and form a base for the initiation of typical contact fatigue defects, such as checks, squats, pitting, etc. Similar conditions are also proposed for the evaluation of the angle that forms a "fretting tongue" in the elements of fretting pair.

On the basis of the accumulated results, we make the necessary conclusions and give some practical recommendations.

Lviv, Ukraine

Oleksandra Datsyshyn
Volodymyr Panasyuk

Acknowledgements

The monograph is written on the basis of the results of joint investigations of the authors and other researchers from the Karpenko Physicomechanical Institute of the National Academy of Sciences of Ukraine: H. P. Marchenko, A. Yu. Glazov, A. B. Levus, R. Ye. Pryshlyak, V. M. Kadyra, M. M. Kopylets, and R. B. Shchur. The authors are also grateful to O. A. Kravchuk for her assistance in preparing this monograph for publication.

The authors also express their deep gratitude to Prof. M. P. Savruk for the discussion of the results and valuable comments.

Contents

Chapter 1
Contact Lifetime Estimation Methods for Tribojoint Elements. A Survey

Abstract In this chapter, we give a survey of the available literature dealing with the methods aimed at the evaluation of the contact lifetime of elements of the rolling and fretting couples. We analyze the results of investigations in the scientific directions that can be regarded as basic for the proposed methods, namely, in the fatigue fracture mechanics and in the theory of contact problems of the mathematical theory of cracks. We also consider the applications of the contact fatigue approaches to the problems of the contact lifetime evaluation of the tribojoint elements.

1.1 Fatigue Fracture Mechanics

The phenomenon of fracture of structural element after a certain number of repeated loads much lower than a single fracture load was first called fatigue by J.V. Poncelet in his book "Introduction to Industrial Mechanics" (1837–1839) [205, 206]. In particular, this phenomenon manifests itself in the decrease in the lifetime of steel articles under repeated variable loads. Clearly, in the course of industrial development, it was necessary to guarantee the required serviceability of products and prevent their premature sudden failures and, hence, the accidents caused by these failures. The indicated period of time was characterized by the rapid development of rail transport, which was accompanied by the breaks of rails, wheels, and axles under the influence of variable loads. Therefore, for almost the entire nineteenth century, the investigations of the phenomenon of fatigue of metals (at that time, carbon steels) were concentrated on the problems of railway transport [206]. At the end of the 19th century and the beginning of the 20th century, the predominant role in the stimulation of the investigations of fatigue of materials was played by the automobile industry. Somewhat later, since the 1920s, the leading role passed to the aircraft construction. The approaches used at that time in the science on strength, fatigue and fracture of materials are now called classical [149].

The classical approaches are based on the assumption that an element of a deformed solid subjected to loading can be either in the continuous state or in the destroyed state. Moreover, the transition of the material (body) from the continuous to destroyed state, i.e., the fracture process occurs instantaneously when the

O. Datsyshyn and V. Panasyuk, *Structural Integrity Assessment of Engineering Components Under Cyclic Contact*, Structural Integrity 9,
https://doi.org/10.1007/978-3-030-23069-2_1

characteristics of the stress-strain state reach a certain critical value typical of a given material. If the characteristics of stress-strain state in the material do not reach their critical values, the fracture processes do not occur and the body preserves its integrity (strength). The classical ideas concerning the fracture processes were used to establish the corresponding characteristics of strength of structural materials and develop the methods for their determination. They were extensively used in the engineering practice and formed the basis of the well-known phenomenological hypotheses on strength [78, 168]. However, the classical approaches do not explain the mechanism of fracture itself, do not reveal the factor accelerating or decelerating this process, and do not enable one to formulate the principles of control over the mechanical characteristics of materials and, in particular, their fracture strength. Hence, it was necessary to develop new approaches.

The main idea of the nonclassical approaches is connected with the introduction of a certain intermediate state of the analyzed material between its continuous and destroyed states [149], which should be taken into account in the evaluation of its determining its strength and durability, especially in the presence of crack-like defects. As an important feature of the regions of strained material characterized by the formation of its intermediate state (process zones), we can mention the fact that, in these zones, the material is always deformed beyond the limits of elasticity and, just in these zones, we observe the most intense realization of the local plastic yield, interaction with the ambient medium, diffusion processes, and other phenomena preceding the local fracture.

The analysis of intermediate states within the framework of continuum mechanics led to the creation of new concepts and computational models. These concepts and models formed a basis of the new branch of science of strength, namely, of the mechanics of quasibrittle fracture of materials. The investigations in this direction were originated in the 1920–1930s in the works by A.A. Griffith, J.N. Taylor, E.O. Orowan, and somewhat later by Irwin [71, 82, 83, 97] (see also the bibliography in [149]).

The results of investigations in the mechanics of fatigue fracture (according to the nonclassical approaches) are presented in the generalizing papers and monographs by O.E. Andreikiv, V.S. Ivanova, A.Ya. Krasovs'kyi, N.A. Makhutov, H.M. Nykyforchyn, I.A. Oding, O.P. Ostash, V.V. Panasyuk, H.S. Pysarenko, V.I. Pokhmurs'kyi, O.M. Romaniv, M.M. Stadnyk, S.V. Serensen, V.F. Terent'ev, V.T. Troshchenko, G.P. Cherepanov, S.Ya. Yarema, A. Carpinteri, S. Kocianda, K. Miller, P. Paris, J. Rice, R. Ricci, D. Taylor, T. Yokobori and other researchers [12, 23, 26, 99, 149, 150, 171, 174, 203, 207].

One of the most important results of investigation of the fatigue fracture of structural materials is the development of standard testing methods, the representation of results of these tests in the form of fatigue fracture diagrams (FFD), and the determination of the characteristics of cyclic crack growth resistance of materials [6, 148, 170]. The specific features of these tests and the history of investigations of this problem were discussed in [148, 174] and in the works by Yarema [216–218] for the case of fracture realized according by the mode I (normal opening) mechanism. The analysis of the case where the fracture process at the crack tip (along its front) is realized according to the mechanism of longitudinal shear was also quite successful and gave results in the form of methodical recommendations for testing (see, e.g., [100]). However, up to now, there are no proper standard testing methods in the case of transverse shear at the crack tip, i.e., in the case of shear realized perpendicularly to the crack front. In this case, only separate procedures are available [19, 137]. Thus, the available literature contains fairly rich data on the characteristics of cyclic crack resistance of materials in normal opening and in the case of longitudinal shear. At the same time, the results of investigation of transverse shear are quite poor. The current state of the problem of evaluation of the characteristics of cyclic crack growth resistance in transverse shear and the description of the corresponding FFD can be found in the survey [122].

Among the indicated investigations, we can also mention the works devoted to the nearthreshold sections of the fatigue fracture diagrams of materials in the transition between the stage of macrocrack initiation and the stage of its propagation. These works deal with the determination of the threshold values of the stress intensity factors (SIF) K_{th} or their ranges ΔK_{th} below which the crack does not grow in the analyzed material. As for its importance and physical nature, the threshold value of the SIF is compared [148, 174] with the endurance limit, which is one of the main characteristics specifying the resistance of structural materials to fatigue fracture within the framework of the classical approaches. On the one hand, this characteristic of cyclic crack growth resistance of the material is especially important for the estimation of the lifetime of structural elements and, on the other hand, it is very sensitive to the structure of the material, possible closure of the crack faces at the crack tip, ambient temperature, frequency of loading, etc. The determination of the threshold values of the SIF requires both the investigation of growth of the macrocrack and the analysis of the stages of formation of so-called "short cracks" and their initiation because "there is no distinct boundary between the stages" [55, 148, 174, 217]. "The duration of the period of macrocrack initiation varies within broad limits. According to different estimates, it constitutes 20–90% of the total lifetime depending on the material, sizes and shape of the body (and, in particular, on the presence of stress concentrators), the character of loading, the properties of the medium, and other operating conditions, as well as on the accepted criteria of separation into periods" ([174], Sect. 4). A significant contribution to these complex investigations was made by O.E. Andreikiv, V.S. Ivanova, Ya.L. Ivanyts'kyi, H.M. Nykyforchyn, O.P. Ostash, V.V. Panasyuk, O.M. Romaniv, V.T. Troshchenko, S. Shtayura, S.Ya. Yarema, M.W. Brown, K.J. Miller, Y. Murakami, J.R. Rice, R.O. Ricci, D. Taylor and other researchers (see the bibliography in [148, 149, 171, 174, 203]).

The analytic description of fatigue fracture diagrams was an important achievement in the mechanics of fatigue fracture of solids. A formula for the description of these diagrams was first proposed by Paris [163, 164]. Later, it was somewhat modified by Yarema [174]. Cherepanov [26], Andreikiv [5], Yarema and Mykytyshyn [219] and other researchers proposed the relations for the description of the complete FFD. These formulas can be found in [150, 174].

1.2 Methods of Mathematical Theory of Cracks

The information about the stress-strain state or the stress intensity factors in bodies with cracks is used as basic for modeling the process of propagation of the cracks in the course of cyclic contact of the bodies. There are numerous publications devoted to the solution of the main problems of the theory of elasticity for cracked bodies (in which a deformed body is subjected to the action of a system of forces or displacements). These are, in particular, the monographs by O.E. Andreikiv, L.T. Berezhnyts'kyi, H.S. Vasyl'chenko, O.P. Datsyshyn, L.V. Ershov, A.O. Kamins'kyi, L.M. Kachanov, H.S. Kit, P.F. Koshelev, A.M. Lin'kov, N.A. Makhutov, E.M. Morozov, G.P. Nikishkov, V.V. Panasyuk, P. Paris, V.Z. Parton, M.P. Savruk, J. Sih, L.I. Slepyan, M.M. Stadnyk, M.P. Stashchuk, V.M. Finkel', M.V. Khai, G.P. Cherepanov, M.H. Aliabadi, T.L. Anderson, A.J. Carlson, E.E. Gdoutos, B.M. Lowengrab, A. Neimitz, D.P. Rooke, I.N. Sneddon, J. Williams, H.R. Wu and others. The data on the stress intensity factors near the crack tips were systematized in the handbooks by Savruk [180], Aliabadi [2], Murakami [134], Rooke and Cartwright [175], Sih [190], Tada et al. [200]. In all these works, the authors mainly analyzed the cases of rectilinear cracks whose faces are not in contact.

The problems of the theory of elasticity for cracked bodies were solved by different methods. The surveys of these methods can be found in the works [150, 165, 180, 200].

In the case of two-dimensional problems for bodies with cracks, the methods of finite elements [132, 188], boundary elements [126], Green functions [214], and singular integral equations [161, 181, 184] are proved to be most efficient and, hence, are used most extensively. At the same time, the methods of conformal images, boundary collocations, bulk forces, asymptotic methods, etc. are now used less commonly.

Note that, in the two-dimensional problems of the theory of elasticity, the researchers often apply various representations of the general solution in the form of certain functions of complex variable (complex potentials). In this case, the so-called Kolosov–Muskhelishvili representations are used especially extensively [112, 139]. If we express the Kolosov–Muskhelishvili complex potentials in the form of Cauchy-type integrals of the unknown derivatives of discontinuities of displacements along the crack contours, then, as shown in [31, 32, 160–162, 182, 183], the problems posed for systems of arbitrarily located rectilinear cracks in infinite and semiinfinite planes, bands, and domains bounded by circles can be reduced to systems of singular integral equations (SIE), which can be efficiently solved (numerically) by the methods of mechanical quadratures. In the monographs by Savruk [181, 184], basic problems of the two-dimensional theory of elasticity for simply connected and multiply connected bodies weakened by curvilinear cracks were reduced to SIE. Similarly, basic problems of the theory of elasticity for a half-plane weakened by a system of curvilinear cracks [29, 30, 96, 209] and the contact problems of action of a rigid punch upon the half-plane with curvilinear cracks [155, 157, 185] were also reduced to systems of SIE. In the present work, we extend the method of SIE to some new contact problems for a half-plane with cracks and use it for the construction of the paths of crack propagation in the zone of cyclic contact between two bodies.

The contact problems in the case where the faces of rectilinear cracks are closed under the action of a system of loads in an infinite elastic plane were solved in [84, 95, 204], in the case of a crack located along a circular arc, they where solved in [85, 145], and, in the case of a crack in the form of broken line, in [221]. A more detailed survey of these investigations can be found in [110, 181]. In the major part of the cited works, the problem of partial contact or contact along the entire crack length was studied by using different methods with regard for the friction between the crack faces. Savruk extended the method of SIE to plane problems for bodies with curvilinear cracks whose faces are under the conditions of smooth contact or sticking [181] and illustrated this approach by numerical results obtained for the SIF at the crack tips along a circular arc, parabola, or semiellipse. At present, the problem of construction of the solution of the problem for curvilinear cracks under the general conditions of contact of the crack faces (sliding with friction, sticking, and opening displacement) remains urgent. For the problems of cyclic contact, it is important to take into account the history of loading. An attempt to formulate and solve contact problems with the history of loading was made in [3, 4].

1.3 Model Schemes of Contact Interactions. Contact of Crack Faces

As a rule, complex stressed states are formed in the elements of friction couples. Its determination with regard for the presence of cracks is connected with serious mathematical difficulties. Therefore, in the construction of computational models, it is customary to use certain simplifications. Thus, within the framework of two-dimensional models, damaged bodies are modeled by elastic half-planes weakened by cracks and the influence of the counterbody is replaced by the action of normal $p(x)$ and tangential $q(x)$ forces distributed in a certain way along the boundary of half-plane. These forces describe the shapes and sizes of the counterbody, the type of contact interaction, the mechanical characteristics of materials and contact surfaces, etc. In the case of contact interaction and rolling or rolling with sliding, the most widespread model loads are concentrated forces and elliptic (Hertzian) distributions of contact pressure $p(x)$. The tangential forces $q(x)$ are, as a rule, connected with the normal forces $p(x)$ by the Amonton–Coulomb law via the coefficient of contact friction f, which enables us to construct a fairly adequate model of the conditions of sliding between the bodies. In the case of contact interaction of fretting fatigue, the role of model loading is played not only by the elliptic distribution of contact pressure but also by the uniform distribution and, under conditions of sticking, by the Cattaneo–Mindlin distribution [24, 101].

Among the configurations of domains used to model the problems of contact interaction, the model of a half-plane with arbitrarily located edge or subsurface cracks is used especially extensively.

The engineering practice and the results of experimental investigations demonstrate [19, 64, 103, 131, 137, 223] that, under the conditions of cyclic contact, the initial macrocracks in the zone of compression propagate almost rectilinearly. In the literature, there are numerous publications devoted to the solution of two-dimensional problems of the theory of elasticity for a half-plane weakened by a single rectilinear crack or a system of cracks of this kind. Their detailed survey and solutions can be found in [110, 161, 181]. We also note also that various brief versions of these results are presented in the handbooks [2, 134, 175, 180, 190, 200]. However, these cases mainly deal with the loads under which the crack faces are not in contact. In fact, in the contact cycles (displacements or changes in the model contact load), the crack faces usually come in contact, which changes during the cycle and this requires the solution of complex contact problems.

Consider two-dimensional problems for a half-plane weakened by a rectilinear subsurface (inside) crack. For the case where a concentrated force moves along the boundary of the half-plane, the SIF was found in the works [25, 73, 88, 113, 114, 146, 186, 187, 196, etc.] depending on the location of the load. At the same time, in the case where the Hertzian pressure plays the role of contact loading, the SIF was determined in [7, 25, 86, 109, 121, 178, etc.]. It should be emphasized that Sheppard et al. [25, 186, 187] reduced the posed problems to systems of singular integral equations in the real form (which has a cumbersome structure) and solved these systems numerically. In this way, the values of the SIF K_{II} and ΔK_{II} were established and the approximate maps of contact of the crack faces were constructed in [186, 187]. Similar results were obtained by Komvopoulos and Cho [113, 114] by the finite-element method. The analysis of these results and the survey of the cited works can be found in [67, 128]. In recent publications [38, 60], the SIF K_{II} and ΔK_{II} for the subsurface crack in a contact cycle were found by the method of SIE in the complex form with regard for the slip–stick conditions between the faces.

There are much more works devoted to the investigation of inclined rectilinear edge cracks in the half-plane as compared with the case of subsurface cracks. Thus, in particular, in the case of action of concentrated forces on the boundary of the half-plane with edge crack perpendicular to the boundary; Rooke and Jones [176] obtained an approximate analytic solution. Numerous authors use this solution as the Green function in problems with more complicated loads. The cases of Hertzian distributions of contact pressure with the corresponding tangential component for cracks inclined to the boundary (as a rule at a certain angle typical of the given rolling couple) under different contact conditions of the crack faces were considered in [1, 9, 16–18, 21, 47, 50, 51, 58, 63, 65, 76, 77, 108, 109, 120, 142, 173, 215, 221, etc.] (see also the surveys [51, 76, 77]).

Thus, Keer and Bryant [108] solved the problem in which the faces of an edge crack contact without friction (conditions of smooth contact) or with an insignificant friction over the entire crack length. By using the dislocation approach (the dislocations are regarded as jumps of displacements along the crack contour), they reduced the original problem to a system of two real singular integral equations. The contact problems for the conditions of partial smooth contact of the faces of an inclined edge crack subjected to the action of a Hertzian load moving along the edge of the half-plane were solved in [47, 50, 51, 123, 124]. These problems were reduced to complex SIE. The values of the SIF and the boundaries of the contact sections of the faces were established.

The most extensive investigations were devoted to the solution of the problems dealing with the case where the crack faces are in contact with friction over the entire crack length. Different methods were used for this purpose and the finite-element method was applied especially extensively [17, 18, 173]. Note that the methods of bulk forces [142], boundary elements [1], weight functions [9, 74], and SIE [38, 58, 59, 63] were also used in this case. The maximal values of the SIF K_{II} and their range ΔK_{II} in a cycle of displacements of the contact load established in these investigations were also used to estimate the contact life of the elements of some rolling couples.

We now mention an important work by Bower [21] in which, for the same set of parameters (orientation and length of the crack, the value of the friction coefficient in contact between the bodies and the crack faces), the values of the SIF were determined and the maps of contact between the crack faces were constructed with regard for the following conditions of contact of the faces: restraint (sticking), sliding with friction, and opening displacement of the faces. As in [108], the method of SIE in the real form was used in the cited work. An approximate solution by the finite-element method with an attempt to take into account the sticking of crack faces was obtained in [215]. More complete investigations of the SIF K_{II} and ΔK_{II} in a single contact cycle under the conditions of sliding and sticking of the crack faces were carried out in [39, 52, 53], in particular in [49, 54] the nearsurface compression residual stresses were taken into account.

Finally, we also mention that the survey and analysis of similar results obtained in the other publications is presented in [51, 52].

All mentioned works were devoted to the analysis of the influence of various in-service factors on the variations of the SIF in the course of motion of the contact load along the edge of the half-plane, namely, friction between the rolling bodies and crack faces, crack length and orientation, size of the section of contact between the rolling bodies, and pressure of oil or different working media in the crack.

Note that the solutions obtained by Keer and Bryant [109], Bower [21], and Sheppard et al. [186, 187] are used as reference solutions up to now. In the works published later, the solutions were mainly obtained by the methods of finite or boundary elements. The results available from the literature are not always in good agreement. The accuracy of these solutions is important for the proper choice of configuration of the initial geometric-force scheme (half-plane–crack–load) used for the construction of the paths of crack propagation.

We now also mention the works [104, 136] in which the SIF and the kinetics of contact of the faces of plane crack (inside or edge; circular or elliptic) in a half-space under the action of a moving Hertzian load (cylinder or sphere) were studied in detail. These results were obtained by Y. Murakami, M. Kaneta, H. Yatsuzuka et al. by the method of bulk forces. They are important for the comparison with the data obtained by analyzing two-dimensional problems.

The progress attained in the mechanics of fatigue fracture and in the mathematical theory of cracks formed a basis for the investigation of the fracture of deformed solids in the case of their cyclic contact.

1.4 Contact Interactions in Tribojoints. Phenomenological Models

An assumption that the fatigue processes running in the course of rolling, fretting fatigue, friction fatigue, and impulsive contact have common symptoms and regularities was made already in [101, 111, 167]. Thus, in particular, the experimental data about these contact interactions are presented in the monograph [167] and some theoretical results in this field can be found in [101]. At present, this concept is extensively developed.

The well-known publication by Way "Pitting due to rolling contact" [213] can be regarded as the first work devoted to the study of the nature of pitting. In the cited work, the investigations were carried out on bearings and disks made of bearing steel. The work by Lundberg and Palmgren [127] concerning the phenomenon of spalling and contact fatigue durability of the bearings is also worth mentioning.

In the 1950–70s, Kostetskii [115] performed precision experimental investigations of contact fatigue and wear via the formation of varioliform pitting and spalling on rolling bearings. The results of studies of the formation of contact fatigue defects in the case of fretting fatigue contact interaction were generalized by Waterhouse in his monograph "Fretting-corrosion" [212]. The surveys of investigations carried out in this field in the 1960–80s were presented in [111, 193–195].

In the 1970–90s, the number of publications dealing with the problems of contact fatigue rapidly increased. The investigations were performed in different directions adapted to the analysis of specific tribojoints.

The investigations of the contact strength and durability of elements of wheel-rail systems are performed especially extensively. Thus, in particular, we can mention a series of large survey papers [22, 68, 98, 125, 147, 191, 224]. Noticeable contributions to these investigations were made by R. Allen, P. Anderson, J. Beynon, M.W. Brown, P.E. Bold, P. Clayton, G. Donzella, M.F. Frolish, D.I. Fletcher, M. Kaneta, A. Kapoor, A. Mazzu, Y. Murakami, M. Sato, K. Savley, X. Su, W.R. Tyfour and many other researchers (see [11, 13, 27, 75, 102, 179, 189, 198, 211]). A large body of investigations of the problems of crack propagation and contact endurance of the rolls of rolling mills and roll steels was carried out in [79, 106, 130, 140, 143, 144, 169, 201, 222]. The results obtained for the bodies of rolling bearings and bearing steels were presented in [14, 107, 118, 127, 137, 202, 213]. For the elements of gearings and the corresponding steels, see [8, 70, 80, 94, 115, 137].

On the basis of the accumulated experience, the corresponding standards were formulated in [19, 64, 79, 137, 211, etc.]. According to these requirements, the basic investigations of contact fatigue of the elements of rolling couples were performed mainly on double-disk machines with diameters more than 40 mm and thicknesses over 10 mm made of identical or different materials under the following conditions: dry friction, moisture (water) or working oil in the contact zone under various contact loads. The researchers studied the processes of crack initiation and growth in the nearsurface zone of rolling bodies and their lifetime, i.e., the number of rolling cycles up to the appearance (recording) of crumbling particle(s) in the contact zone. Note that the study of crack growth shows that, after initiation in the nearsurface contact zone, the cracks propagate by the shear mechanism almost rectilinearly (in their planes) for a certain (quite long) period of time and then pass to the mechanism of normal opening. A large body of experimental data on the specific features of fracture (formation of defects), contact lifetime of the main types of rolling couples (wheel-rail, rolls of rolling mills, and gearings), and the characteristics of cyclic crack growth resistance of materials was obtained by the Japanese (Y. Murakami, M. Kaneta, M. Ishida, G. Yatsuzuki, et al.) and English (K. Johnson, K. Miller, D.I. Fletcher, M.W. Brown, J. Beynon, M.F. Frolish, A. Kapoor, R.A. Smith, et al.) researchers.

It was clear that the necessity of systematization and formulation of the criteria of contact fatigue of the structural materials of tribojoints, as well as the generalization of the accumulated experimental data required the creation of the theoretical foundations of the mechanics of fatigue contact fracture.

In the 1970–90s, the first theoretical models were formulated in the mechanics of fatigue contact fracture including, in particular, the cases of fretting fatigue [91, 176, 208] and rolling [21, 105, 108]. These models and the models proposed later [16, 18, 69, 79, 81, 119, 137, 138, 172, 173] were based on the concepts of fatigue fracture mechanics, solutions of the contact problems of the theory of elasticity for cracked bodies, and the evaluation of the determining operating parameters (residual stresses, medium, distributions of contact forces and friction between the contacting bodies, loading frequency, and temperature) for various types of tribojoints. As the criteria of local fracture, it is customary to use the force or energy criteria of fracture mechanics.

The last years are characterized by the appearance of numerous new investigations and results, especially for the elements of wheel-rail systems. These results are described in detail in the surveys [66, 67, 76, 129, 197, 223]. We also mention new results on the process of crack propagation under the conditions of fretting fatigue contact interaction [10, 87, 92, 166] presented at the Sixth and Seventh International Symposia on Fretting Fatigue (ISFF6, 2010; ISFF7, 2013). The methods of standard testing for the lifetime of elements of the fretting couples were also developed [141].

1.5 Pitting and Spalling

The mechanisms of formation of the most widespread contact fatigue defects, namely, pitting and spalling, were proposed by the researchers on the basis of the outlined models and the results of new experimental investigations.

Thus, some mechanisms of formation of varioliform pitting in rolling bodies were described in [21, 103, 108, 137]. Indeed, in [108], it was assumed that, in the case of unidirectional rolling, the pits form, as a rule, due to the action of contact forces between the bodies. In other words, as a result of action of the cyclic shear stresses in the contact zone, the initial edge cracks propagate by the mechanism of transverse shear mainly rectilinearly with subsequent possible bending of their paths under the influence of tensile stresses and crack propagation by the opening mechanism. The role of liquid (oil) in the contact zone is mainly connected with the decrease in the level of friction between the crack faces. At the same time, the possibility of action of the oil pressure onto the crack faces does not lead to the crack growth toward the surface of the rolling body with subsequent crumbling. In [21, 103, 137], it was assumed that macrocracks can be initiated on the surfaces of both driven and driving bodies of the rolling couple, whereas pitting occurs solely on the surface of the driven body in the presence of liquid and in the case where a shear crack is inclined at an acute angle to the rolling surface in the direction of motion of the counterbody. Under these conditions, liquids penetrate into the crack. As in [108], we now assume that the initial macrocrack first propagates almost rectilinearly by the mechanism of transverse shear. The length of this shear macrocrack can be quite large [137] and comparable with the size (length) of the contact zone. Later, under the pressure of oil upon the crack faces, the mechanism of crack propagation transforms into the normal opening mechanism and the crack turns toward the surface, which leads to its crumbling.

Note that the authors of the works cited above restricted themselves solely to the analysis only the stress intensity factors and the angles of initial deviation θ_0^* [149, 161] for rectilinear (or plane) edge cracks. On the basis on these data, they predict the path of second (curvilinear) stage of crack growth leading (or not leading to the crumbling of the surfaces of rolling bodies. These data were also used in [19, 21, 103] to estimate the lifetime of the nearsurface layers of rolling bodies. However, the indicated predictions of the shapes and sizes of defects and the lifetime of the body are not properly substantiated and can be erroneous.

Some models of spalling formation under the conditions of contact rolling fatigue based on the large body of engineering data and the results of laboratory investigations were presented in [79, 106, 140, 143, 144, 201]. In these works, it is assumed that the cracks of contact rolling fatigue are initiated both on and under the surface. However, in the absence of noticeable subsurface defects, there is no common opinion about the causes of crack initiation and the mechanisms of its propagation. Thus, in [140], it was assumed that subsurface cracks are initiated at the points of maximal shear stresses. At the same time, in [201], it was assumed that microcracks can be initiated in the lower part of the quenched working layer. On the contrary, in [106, 143], the authors made a conclusion that, in practice, the major part of failures caused by spalling occur as a result of crack initiation on the surface caused by the in-service incidents and inhomogeneous contact pressures, thermal shocks, etc.

In [79, 106, 137, 144], the authors made an attempt to explain the propagation of nearsurface cracks along a circle for great distances parallel to the surface of the backup roll or their penetration into the bulk of the material. In particular, a model of fracture mechanics for the contact rolling fatigue was proposed in [79] in order to understand the mechanisms of fracture caused by spalling of the rolls. The aim of the present work is to give a quantitative representation of the stress intensity factors K_{I} and K_{II} at the tips of surface cracks inclined to the contact surface and compare these quantities with the threshold values of the characteristics of cyclic crack-growth resistance $K_{\mathrm{I}th}$ and $K_{\mathrm{II}th}$ in the investigated materials. The model is formulated in terms of the crack length and the directions of its growth and enables one to predict the contact rolling fatigue of backup rolls.

1.6 Nearsurface Cracks Growth Paths. Estimation of Contact Lifetime

Under the conditions of cyclic contact of bodies and complex stressed state in the contact zone, the cracks propagate along curvilinear paths. In this case, either a single crack (or a system of cracks) grows and forms different types of defects of the surface layer of the elements of tribojoints: varioliform and V-like pitting, spalling, cracking, checks, darkening and settling of the surface caused by the subsurface crack branching, i.e., squats, cavities, etc. To estimate the lifetime of contacting bodies, it is necessary to construct the paths of propagation of nearsurface cracks specifying the shapes and sizes of contact-fatigue defects. The number of cycles prior to their formation determines the contact fatigue lifetime of the body (element of the joint). It is also very important to establish the ranges of operating parameters and the characteristics of crack growth resistance of the material that may lead to the crack growth into the bulk of the material and create a hazard of failure for the product (article). The indicated fracture processes in contacting bodies are the contemporary subjects of inquiry in the field of fracture mechanics and require the development of adequate computational models. The presented analysis of the available publications demonstrates that this (main) aspect of the problem of estimation of the contact lifetime of tribojoints has not been properly investigated yet.

In [70, 72, 220], a group of Slovenian researchers (J. Flašker, G. Fajdiga, S. Glodež, B. Zafosnyak, Z. Ren, et al.) developed a theoretical-and-experimental methodology for the construction of the paths of propagating curvilinear cracks, which form microscopic pits in elements of the gearings under the action of contact loads and oil pressure upon the crack faces. The path is constructed step-by-step from the spline of rectilinear segments by using the finite-element method and the criterion of maximum rate of energy release. In a similar way, but by the method of SIE, Goshima

[81] constructed the paths of formation of the pits for the nearsurface zone of the bearing materials. However, in the analyzed cases, the friction between the crack faces is almost not taken into account. At the same time, it strongly affects both the angle of initial orientation of the edge crack and the contact lifetime of elements of the rolling couple. The characteristics of cyclic crack growth resistance of materials in shear are also not taken into account. In the cited works, the authors obtained some contradictory results; in particular, that the pits may be formed without wedging of the crack faces (without pressure of oil upon the crack faces).

A new approach to the solution of this problem was proposed by Datsyshyn et al. [28, 37, 56–58, 60, 153, 158]. Within the framework of the fracture mechanics of materials, Datsyshyn [34–37, 60] formulated a computational model for the investigation of fracture processes and prediction of the residual lifetime of solids (elements of tribojoints) under the conditions of cyclic contact. This model is based on the step-by-step evaluation of the paths of crack propagation by using the solutions of singular integral equations of the two-dimensional contact problems of the theory of elasticity for bodies with curvilinear cracks and the criteria of local fracture in the complex stressed state with regard for the characteristics of the cyclic crack growth resistance of materials in shear and normal opening and the operating parameters of tribojoints. The algorithms used for the construction of the paths of crack growth in the contact zone take into account the variations of the stress-strain state caused both by crack elongation and by the motion of counterbody (variable loading) in a contact cycle, as well as possible changes in the mechanism fracture (from transverse shear to normal opening) in the course of crack propagation and friction between the crack faces. In the present work, within the framework of this model, we suggest a procedure for the estimation of the residual lifetime of rolling and fretting couples according to the formation of contact fatigue defects.

As a specific feature of this model, we can mention the separation of the stages of crack growth in the contact zone by the mode II and mode I mechanisms. In the present work, we briefly describe the results of our preliminary investigations of the curvilinear paths of propagation of initially rectilinear nearsurface cracks of fixed length, which develop according to the mode I mechanism and form typical contact fatigue defects, such as pitting [28, 45, 48, 57, 61, 158], spalling [42–44], squat [154, 156], checks, cracking [46] in rolling bodies, and the growth of edge cracks in elements of fretting couples under the conditions of sliding or sliding/sticking between them [33, 36, 37, 40, 41, 60, 62, 151, 159]. In our monograph, we also present recent results on the growth of edge and subsurface cracks by the mode I and mode II mechanisms and, as a synthesis of both approaches (stages), model the process of formation of typical contact fatigue defects: pitting [38, 58, 60, 63, 152, 153] and spalling [38, 60, 152, 200]. Finally, we give examples of evaluation of the residual lifetime in both stages according to the criteria of formation of these defects.

1.7 Contact Fatigue and Wear

It is necessary to emphasize the following fact: The processes of formation of contact fatigue defects in the course of cyclic contact of solids are closely connected with the processes of wear. According to GOST 27674–88, wear is a **process** of fracture of the material with separation of its fragments (wear particles) from the surface and (or) the accumulation of residual strains in the course of friction, **which** manifests itself in the gradual changes in the sizes and/or shape of the analyzed body. At present, it is customary to distinguish [89, 117, 133] the following main types of wear: mechanical, fatigue, abrasive, hydroabrasive, electro- and hydro-erosive, cavitation, oxidation, stress corrosion, wear caused by sticking, and wear in the course of fretting. For almost all types of wear, an important role is played by the mechanical influence on the surface as a result of which small (comparative with roughness) or somewhat larger fragments of the material (wear particles) separate from its surface. In the opinion of Chichinadze, Kragel'skii, Morozov and Holmberg [89, 93, 117, 133], in all cases connected with the processes of formation and separation of material fragments (fracture processes), it is possible to use the approaches of fracture mechanics. It is worth noting that, according to the sizes of wear particles, it is customary to distinguish [93] the processes of microwear (10^{-6} m...10^{-4} m) and macrowear (10^{-4} m...10^{-2} m). The works [20, 74, 90, 93, 115–117, 120, 135, 177, 210] are devoted to the experimental and theoretical investigations in which the process of wear is studied from the viewpoint of crack initiation and propagation on the micro- or macrolevels. In [111, 116–118, 133, 193, 199], one can find the surveys of the models of wear based on the application of the approaches of fracture mechanics to the problem of wear. These works are mainly based on the introduction of the characteristics of crack growth resistance of the materials in the relations for the rate and intensity of wear and the reliability of elements of the tribojoints. Note that, for the last three decades (since 1985), the direction of investigations on the boundary of tribology and fatigue fracture mechanics is called tribofatics [15, 192]. Thus, the results described in the present monograph can be regarded as investigations of macrowear processes with a contribution to tribofatics.

References

1. Akama M, Mori T (2002) Boundary element analysis of surface initiated rolling contact fatigue cracks in wheel/rail contact systems. Wear 253:35–41
2. Aliabadi MH (1996) Database of stress intensity factors. Computational Mechanics Publications, Southampton, UK
3. Andreev AV, Gol'dshtein RV, Zhitnikov YM (1999) Ravnovesiye krivolineinykh razrezov s uchetom obrazovaniya oblastey naleganiya, skolzheniya i scepleniya beregov treshchiny (Balance of curvilinear cuts with regard for overlapping, slip, and sticking of crack faces). Preprint No. 643. Russian Academy of Sciences, Institute for Problems of Mechanics, Moscow
4. Andreev AV, Gol'dshtein RV, Zhitnikov YM (2001) Evolyuciya ravnovesnogo sostoyaniya gladkikh krivolineinykh treshchin so vzaimodeistvuyushchimi s treniyem poverkhnostyami

v processe nagruzheniya (Evolution of the equilibrium state of smooth curvilinear cracks with surfaces interacting with friction under loading). Preprint No. 676. Russian Academy of Sciences, Institute for Problems of Mechanics, Moscow
5. Andreikiv AE, Darchuk AI (1992) Ustalostnoye razrusheniye i dolgovechnost‘ konstrukciy (Fatigue fracture and durability of structures). Naukova Dumka, Kiev
6. ASTM Designation E647-00 (2000) Standard test method for measurement of fatigue crack growth rates
7. Bastias PC, Hang GT, Rubin CA (1989) Finite element modeling of subsurface mode II cracks under contact loads. Eng Fract Mech 33:143–152
8. Batista AC, Dias AM, Lebrun JL et al (2000) Contact fatigue of automotive gears: evolution and effects of residual stresses introduced by surface treatments. Fatigue Fract Eng Mater Struct 23, 217–228
9. Beghini M, Bertiny L, Fontanari V (2005) Parametric study of oblique edge cracks under cyclic contact loading. Fatigue Fract Eng Mater Struct 28(1/2):31–40
10. Bellecave J, Pommier S, Nadot Y, Meriaux J, Araújo JA (2014) T-stress based short crack growth model for fretting fatigue. Tribol Int 76:23–34
11. Beynon JH, Brown MW, Kapoor A (1999) Initiation, growth and branching cracks in railway track. In: Beynon JH, Brown MW, Lindley TC et al (eds) Engineering against fatigue. A.A. Balkema Publisher, Rotterdam, pp 461–472
12. Beynon JH, Brown MW, Lindley TC, Smith RA, Tomkins B (eds) (1999) Engineering against fatigue. A. A. Balkema Publisher, Rotterdam
13. Beynon JH, Garnham JE, Sewley KJ (1996) Rolling contact fatigue of three pearlitic rail steels. Wear 192:94–111
14. Bhargava V, Hahn GT, Rubin CA (1986) Analysis of cyclic crack growth in high strength roller bearings. Theor Appl Fract Mech 5(1):31–38
15. Bogdanovich AV (ed) (1996) Slovo o tribofatike (A Treatise on Tribofatigue). Remiko, Gomel
16. Bogdanski S (2002) The behaviour of kinked cracks in contact. In: Proceedings of 14th bienniel conference on fracture “Fracture mechanics beyond” (ECF 14), vol I/III. EMAS Publishing, pp 289–296
17. Bogdanski S, Olzak M, Stupnicki J (1996) Numerical stress analysis of rail rolling contact fatigue cracks. Wear 191:4–24
18. Bogdanski S, Trajer M (2005) A dimensionless multi-size finite element model of a rolling contact fatigue crack. Wear 258:1265–1272
19. Bold PE, Brown MW, Allen RJ (1991) Shear mode crack growth and rolling contact fatigue. Wear 144:307–317
20. Bowden FP, Tabor D (1964) The friction and lubrication of solids. Clarendon Press, Oxford
21. Bower AF (1988) The influence of crack face friction and trapped fluid on surface initiated rolling contact fatigue cracks. J Tribol Trans ASME 110(4): 704–711
22. Cannon DF, Edel KO, Grassie SL, Sawley K (2003) Rail defects: an overview. Fatigue Fract Eng Mater Struct 26(10):865–886
23. Carpinteri A (ed) (1994) Handbook of fracture crack propagation in metallic structures (in 2 volumes). Elsevier, Amsterdam
24. Cattaneo C (1938) Sur contatto di due corpi elatici: distribuzione locate degli sforzi, Rend. dell Academia nazionale dei lincei, 27(Ser 6): 342, 434, 474
25. Chang F-K, Comninou M, Sheppard S, Barber JR (1984) The subsurface crack under conditions of slip and stick caused by a surface normal force. J Appl Mech 51:311–316
26. Cherepanov GP (1974) Mekhanika khrupkogo razrusheniya (Mechanics of brittle fracture). Nauka, Moscow
27. Clayton P, Su X (1996) Surface initiated fatigue of pearlitic and bainitic steels under water lubricated rolling/sliding contact. Wear 200:63–73
28. Datsishin OP, Marchenko GP, Panasyuk VV (1993) Theory of crack growth in rolling contact. Mater Sci 29(4):373–383
29. Datsyshin AP, Marchenko GP (1985) An edge curvilinear crack in an elastic half plane. Sov Mater Sci 21(1):66–70

30. Datsyshin AP, Marchenko GP (1985) Interaction of curvilinear cracks with the boundary of an elastic half plane. Sov Mater Sci 20(5):466–473
31. Datsyshin AP, Savruk MP (1973) A system of arbitrarily oriented cracks in elastic solids. J Appl Math Mech 37(2):306–313
32. Datsyshin AP, Savruk MP (1974) Integral equations of the plane problem of crack theory. J Appl Math Mech 38(4):677–686
33. Datsyshyn OP (1999) Durability and fracture calculate model for structural materials under fretting fatigue. Naukovyy Visnyk Ukrauinskogo Derzhavnogo Lisotekhnichnogo Universytetu 9:139–149
34. Datsyshyn OP (2000) Fatigue fracture of calculation model of solids under their contact interaction. Naukovi Notatky Luts'kogo Derzhavnogo Universytetu 7:74–78
35. Datsyshyn OP (2009) Modeling of the development of typical contact fatigue defects in rolling bodies. In: Panasyuk VV (ed) Mekhanika ruynuvannya materialiv ta mitsnist' konstruktsiy (Fracture Mechanics of Materials and Strength of Structures). Proceedings of the 4th international scientific conference. Lviv, pp 49–54
36. Datsyshyn OP (2011) Modeling of the initiation of contact fatigue damages and estimation of the durability of elements of tribological conjunctions. Mater Sci 47(2):188–200
37. Datsyshyn OP (2005) Service life and fracture of solid bodies under the conditions of cyclic contact interaction. Mater Sci 41(6):709–733
38. Datsyshyn OP, Glazov AY (2013) Rolling bodies durability evaluation by formation of typical contact fatigue damages—pitting and spalling. Visnyk Ternopil'skogo Natsional'nogo Tekhnichnogo Universytetu, vol 3, pp 75–87
39. Datsyshyn OP, Hlazov AY, Levus AB (2014) Specific features of contact of the faces of an edge crack under moving hertzian loads. Mater Sci 49(5):589–601
40. Datsyshyn OP, Kadyra VM (2006) Development of edge cracks under fretting fatigue under stick–slip conditions in contact between the bodies, vol 3. Mashynoznavstvo, pp 9–15
41. Datsyshyn OP, Kalakhan OS, Kadyra VM, Shchur RB (2004) Pitting formation under the conditions of fretting fatigue. Mater Sci 40(2):159–172
42. Datsyshyn OP, Kopyletst MM (2002) The development of a subsurface crack during rolling under dry friction conditions, vol 8. Mashynoznavstvo, pp 17–23 (2002)
43. Datsyshyn OP, Kopyletst MM (2003) On spalling of a rolling surface, vol 10. Mashynoznavstvo, pp 15–18 (2003)
44. Datsyshyn OP, Kopylets MM (2003) Prediction of the service life of rolling bodies according to the development of a subsurface crack. Mater Sci 39(6): 765–779
45. Datsyshyn OP, Levus AB (2003) Propagation of an edge crack under the pressure of liquid in the vicinity of the crack tip. Mater Sci 39(5):754–757
46. Datsyshyn OP, Levus AB (2000) Stress intensity factors for system of parallel surface cracks in half-plane due to hertzian load on its boundary, vol 11. Mashynoznavstvo, pp 9–15
47. Datsyshyn OP, Marchenko HP (2002) Calculation of the rolling surface durability at the stage of shear mode growth of edge cracks. In: Troshchenko VT (ed) "Trybofatyka" ("Tribofatigue"), Proceedings of the 4th international symposium on tribofatigue (ISTF 4), vol 1. Ternopil, pp 420–425
48. Datsyshyn OP, Marchenko HP (1995) Crack propagation path and residual durability of solids under rolling contact. In: Aliabadi MH, Alessandri C (eds) Contact mechanics II. Computational techniques. Computational Mechanics Publications Southampton, Boston, pp 377–383
49. Datsyshyn OP, Marchenko HP (2012) Evaluation of the influence of residual stresses on the surface fracture of railroad rails, vol 6. Zaliznychnyy Transport Ukrainy, pp 38–41
50. Datsyshyn OP, Marchenko HP (2003) Estimation of mode II surface crack growth period under rolling contact, vol 7. Mashynoznavstvo, pp 17–23
51. Datsyshyn OP, Marchenko HP (2008) Stressed state of a half plane with shallow edge crack under Hertzian loading (a survey). Mater Sci 44(1):22–34
52. Datsyshyn OP, Marchenko HP, Glazov AY (2019) On the special angle of surface cracks propagation in the railway rail heads. Eng Fract Mech 206:452–462

53. Datsyshyn OP, Marchenko HP, Hlazov AY, Levus AB (2015) Influence of compressive stresses on the propagation of surface shear cracks in railroad rails. Mater Sci 51(2):235–243
54. Datsyshyn OP, Marchenko HP, Kravchuk OA (2016) Sear surface cracks and longitudinal residual stresses in railway rail head, vol 3–4. Zaliznychnyy Transport Ukrainy, pp 53–59
55. Datsyshyn OP, Marchenko HP, Levus AB (2004) On the methodology of durability calculation at near-threshold parts of fatigue cracks growth. In: Panasyuk VV (ed) Mekhanika ruynuvannya materialiv ta mitsnist' konstruktsiy (Fracture mechanics of materials and strength of structures). Lviv, pp 107–112
56. Datsyshyn OP, Panasyuk VV (1996) Durability and fracture calculational model of solids under their contact interaction. In: Petit J (ed) ECF-11, vol II. Mechanism and mechanics of damage and failure. EMAS LTD, Warley, pp 1381–1386
57. Datsyshyn OP, Panasyuk VV (2001) Pitting of the rolling bodies contact surface. Wear 251:1347–1355
58. Datsyshyn OP, Panasyuk VV, Glazov AY (2011) Modelling of fatigue contact damages formation in rolling bodies and assesment of their durability. Wear 271(1–2):186–194
59. Datsyshyn OP, Panasyuk VV, Glazov AY (2009) The model of fatigue contact damages formation in rolling bodies and estimation of their durability. In: Proceedings of the 8th international conference on contact mechanics and wear of rail/wheel systems, vol 1. AB Editore, Firenze, pp 35–43
60. Datsyshyn OP, Panasyuk VV, Glazov AY (2016) The model of the residual lifetime estimation of trybojoint elements by formation criteria of the typical contact fatigue damages. Int J Fatigue 83(2):300–312
61. Datsyshyn OP, Panasyuk VV, Pryshlyak RE, Terlets'kyi AB (2001) Paths of edge cracks in rolling bodies under the conditions of boundary lubrication. Mater Sci 37(3), 363–373
62. Datsyshyn OP, Shchur RB (1998) Development of edge cracks under conditions of fretting fatigue. Probl Tribol 2: 7–16
63. Datsyshyn OP, Tkachov VI, Hlazov AY, Khrunyk RA (2006) Prediction of the contact durability of back-up rolls of forge-rolling mills in the process of development of pitting. Mater Sci 42(6):823–836
64. Donzella G, Faccoli M, Ghidini A, Mazzu A, Roberti R (2005) The competitive role of wear and RCF in rail steel. Eng Fract Mech 72:287–308
65. Dubourg MC, Villechaise B (1992) Stress intensity factors in a bent crack: a model. Eur J Mech A/Solids 11(2): 169–179
66. Ekberg A (2009) Fatigue of railway wheels, In: Lewis R, Olofsson U (eds) Wheel-rail interface handbook. Woodhead Publishing, pp 211–244
67. Ekberg A, Åkesson B, Kabo E (2014) Wheel/rail rolling contact fatigue—Probe, predict, prevent. Wear 314(1–2):2–12
68. Ekberg A, Kabo E (2005) Fatigue of railway wheels and rail under rolling contact and thermal loading—an overview. Wear 258:1288–1300
69. Ekberg A, Kabo E, Åkesson B (2002) An engineering model for prediction of rolling contact fatigue of railway wheels. Fatigue Fract Eng Mater Struct 25:899–909
70. Fajdiga G, Flašker J, Glodež S, Hellen TK (2003) Numerical modeling of micropitting of gear teeth flanks. Fatigue Fract Eng Mater Struct 26(12):1135–1143
71. Fehlbeck DK, Orowan EO (1955) Energy criteria of fracture. Weld J Res Suppl 34:157–160
72. Flašker J, Fajdiga G, Glodež S, Hellen TK (2001) Numerical simulation of surface pitting due to contact loading. Int J Fatigue 23:599–605
73. Fleming JR, Suh NP (1977) Mechanics of crack propagation in delamination wear. Wear 44(1):39–56
74. Fletcher DI, Beynon JH (1999) A simple method of stress intensity factors calculation for inclined surface-breaking crack with crack face friction under contact loading. Proc Inst Mech Engrs Part J J Eng Tribol 213: 481–486
75. Fletcher DI, Franklin FJ, Kapoor A (2003) Image analysis to reveal crack development using a computer simulation of wear and rolling contact fatigue. Fatigue Fract Eng Mater Struct 25(10):957–967

76. Fletcher DI, Franklin FJ, Kapoor A (2009) Rail surface fatigue and wear. In: Lewis R, Olofsson U (eds) Wheel-rail interface handbook. Woodhead Publishing, pp 280–310
77. Fletcher DI, Smith L, Kapoor A (2009) Rail rolling contact fatigue dependence on friction, predicted using fracture mechanics with a three-dimensional boundary element model. Eng Fract Mech 76:2612–2625
78. Fridman YB (1974) Mekhanicheskiye svoistva materialov (Mechanical properties of materials), vol 1, vol 2. Mashinostroenie, Moscow
79. Frolish MF, Fletcher DI, Beynon JH (2002) A quantitative model for predicting the morphology of surface initiated rolling contact fatigue cracks in back-up roll steels. Fatigue Fract Eng Mater Struct 25:1073–1086
80. Glodež S, Flasker J, Ren Z (1997) A new method for the numerical determination of pitting resistance of gear teeth flanks. Fatigue Fract Eng Mater Struct 20(1):71–83
81. Goshima T (2003) Thermomechanical effects on crack propagation in rolling contact fatigue failure. J Therm Stress 26:615–639
82. Griffith AA (1921) The phenomena of rupture and flow in solids. Phil Trans R Soc Lond 1(ser A), 221: 163–198
83. Griffith AA (1924) The theory of rupture. In: Proceedings of 1st international congress. Delft, pp 53–63
84. Grilitskyi ND, Kit GS (1978) On the stressed state in the vicinity of a crack with partial contact of the faces. Matematicheskiye Metody i Fiziko-Mekhanicheskiye Polya, vol 8, pp 35–39
85. Grylits'kyi DV, Lutsyshyn RM (1975) Napruzhennya v plastynkakh z kolovoyu liniyeyu rozmezhuvannya granychnykh umov (Stresses in plates with a circular line of separation of the boundary conditions). Vyshcha Shkola, Lviv
86. Guagliano M, Vergani L (2005) Experimental and numerical analysis of sub-surface cracks in railway wheels. Eng Fract Mech 72(2):255–269
87. Hattori T, Kien VT, Yamashita M (2011) Fretting fatigue life estimations based on fretting mechanisms. Tribol Int 44:1389–1393
88. Hearly AD, Johnson KL (1985) Mode II stress intensity factors for a crack parallel to the surface of an elastic half-space subjected to a moving point load. J Mech Phys Solids 33(1):61–81
89. Hebda M, Chichinadze AV (eds) (1989) Spravochnik po tribotekhnike (A handbook of tribology). Mashinostroenie–VKL, Moscow–Warsaw
90. Hills DA, Ashelby DW (1980) On the application of fracture mechanics to wear. Wear 54:321–330
91. Hills DA, Nowell D (1994) Mechanics of fretting fatigue. Kluwer Academic Publishers, Dordrecht
92. Hojjati-Talemi R, Wahab MA, Pauw JD, Baets PD (2014) Prediction of fretting fatigue crack initiation and propagation lifetime for cylindrical contact configuration. Tribol Int 76:73–91
93. Holmberg K (2001) Tribology in reliability engineering. In: Franek F, Bartz WJ, Pauschitz A (eds) Trybology 2001: scientific achievements, industrial applications, future challenges, Plenary and session key papers from the 2-nd world tribology congress (WTC–2001), Vienna. pp 13–19
94. Ichimaru K, Nakajima A, Hirano F (1981) Effect of asperity interaction on pitting in rollers and gears. J Mech Design 103:482–491
95. Ioakimidis NI (1997) Conditions for contact/lack of contact along a loaded simple straight crack in plane isotropic elasticity. Eng Fract Mech 56(5):675–689
96. Ioakimidis NI, Theocaris PS (1979) A system of curvilinear cracks in an isotropic elastic half-plane. Int J Fract 15(4):299–309
97. Irwin GR (1957) Analysis of stress and strain near and of a crack traversing a plate. J Appl Mech 24(3):361–364
98. Ishida M, Akama M, Kashiwaya K, Kapoor A (2003) The current status of theory and practice on rail integrity in Japanese railway—rolling contact fatigue and corrugations. Fatigue Fract Eng Mater Struct 26(10):909–919
99. Ivanova VS, Terent'ev VF (1975) Priroda ustalosti metallov (Nature of the fatigue of metals), Metallurgiya, Moscow

100. Ivanyts'kyi YL, Shtayura S (2004) Methodical guidelines. Determination of the characteristics of crack-growth resistance of materials under the conditions of complex stressed state (normal opening + transverse shear and normal fracture + longitudinal shear. In: Panasyuk VV (ed) Mekhanika ruynuvannya materialiv ta mitsnist' konstruktsiy (Fracture mechanics of materials and strength of structures). Lviv, pp 723–732
101. Johnson KL (1985) Contact mechanics. Cambridge University Press, Cambridge
102. Kaneta M, Matsuda K, Murakami Y, Nishikawa H (1998) A possible mechanism for rail dark spot defects. Trans ASME J Tribol 120: 304–309
103. Kaneta M, Murakami Y (1987) Effects of oil hydraulic pressure on surface crack growth in rolling/sliding contact. Tribol Int 20(4):210–217
104. Kaneta M, Murakami Y (1991) Propagation of semi-elliptical surface crack in lubricated rolling/sliding elliptical contact. J Trib ASME 113:270–275
105. Kaneta M, Yatsuzuka H, Murakami Y (1985) Mechanism of crack growth in lubricated rolling/sliding contact. ASLE Trans 28(3):407–414
106. Kapadia BM, Marsden KW (1997) Spalling behaviour of back-up roll materials. In: 39th MSWP conference, pp 1–38
107. Kashimura H, Tsushima N (1984) Improvement of rolling contact fatigue life of bearing steels. SAE technical paper series 84123, pp 1–9
108. Keer LM, Bryant MD (1983) A pitting model for rolling contact fatigue. Trans ASME J Lubric Technol 105(2):198–205
109. Keer LM, Bryant MD, Haritos GK (1982) Subsurface and surface cracking due to hertzian contact. Trans ASME J Lubric Technol 104(3):347–351
110. Kit GS, Krivtsun MG (1983) Ploskiye zadachi termouprugosti dlya tel s treshchinami (Plane problems of thermoelasticity for bodies with cracks). Naukova Dumka, Kiev
111. Kolesnikov YV, Morozov EM (1989) Mekhanika kontaktnogo razrusheniya (Mechanics of contact fracture). Nauka, Moscow
112. Kolosov GV (1935) Primeneniye kompleksnoy peremennoy k teorii uprugosti (Application of complex variable to the theory of elasticity). ONTI Moscow–Leningrad
113. Komvopoulos K (1996) Subsurface crack mechanisms under indentation loading. Wear 199:9–23
114. Komvopoulos K, Cho S-S (1997) Finite element analysis of subsurface crack propagation in a half-space due to a moving asperity contact. Wear 209:57–68
115. Kostetskii BI (1959) Soprotivlyeniye iznashivaniyu detaley mashin (Resistance to wear process of machine parts). Mashgiz, Moscow–Kiev, p 478
116. Kostetskii BI, Nosovskii IG, Karaulov AK, Bershadskii LI et al (1976) Poverkhnostnaya prochnost' materialov pri trenii (Surface strength of materials in friction). Tekhnika, Kiev
117. Kragel'skii IV, Dobychin MN, Kombalov VS (1977) Osnovy raschetov na treniye i iznos (Fundamentals of the numerical analyses of friction and wear). Mashinostroenie, Moscow
118. Kudish II (1989) Numerical analysis of wear and fatigue crumbling in rolling bearings: a survey [in Russian], Series X—"Bearing industry". Central research institute of information and techniko-economical investigations of the automotive industry, Moscow
119. Kudish II, Burris KW (2000) Modern state of experimentation and modeling in contact fatigue phenomenon: Part I—Contact fatigue. Normal and tangential contact and residual stresses. Nonmetallic inclusions and lubricant contamination. Crack initiation and crack propagation. Surface and subsurface cracks. Tribol Trans 43(2): 187–196
120. Kuz'menko AG (1997) Wear as a process of initiation and propagation of cracks. Probl Tribol 1, 46–64
121. Lansler E, Kabo E (2005) Subsurface crack face displacements in railway wheels. Wear 258:1038–1047
122. Lenkovs'kyi TM (2014) Determination of the characteristics of cyclic crack resistance of steels under transverse shear (a survey). Mater Sci 50(3): 340–349
123. Levus AB (2007) Kinetics of contact of the crack faces in rolling. In: Proceedings of the conference "Problemy koroziyno-mekhanichnogo ruynuvannya, inzheneriya poverhni, diagnostychni systemy" ("Problems of corrosion-mechanical fracture, surface engineering, and diagnostic systems"). Lviv, pp 79–82

124. Levus AB, Glazov AY, Datsyshyn OP (2008) Contact of edge crack faces under moving hertzian contact load. In: Suchasni problemy mekhaniky ta matematyky (Modern problems of mechanics and mathematics). Proceedings of the 2nd International Science Conference, vol 2. Lviv, pp 56–59
125. Li YC (1994) Analysis of fatigue phenomena in railway rails and wheels. In: Carpinteri A (ed) Handbook of fatigue crack propagation in metallic structures, vol 2. Elsevier Science Ltd, Oxford, pp 1497–1537
126. Lin'kov AM (1999) Kompleksnyy metod granichnykh integralnykh uravneniy teorii uprugosti (Complex method of boundary integral equations of the theory of elasticity). Nauka, St. Petersburg
127. Lundberg G, Palmgren A (1947) Dynamic capacity of rolling bearing. Acta Polytech Ser Mech Eng RSAEE 1(3): 50
128. Lunden R (2007) Elastoplastic modeling of subsurface crack growth in rail/wheel contact problems. Fatigue Fract Eng Mater Struct 30:905–914
129. Lunden R, Paulsson B (2009) Introduction to wheel-rail interface research. In: Lewis R, Olofsson U (eds) Wheel-rail interface handbook. Woodhead Publishing, pp 3–33
130. Mei B, Li Y, Wang C, Dou P (2002) Rolling contact fatigue crack initiation in medium carbon bainitic steel. J Tsinghua Univ Sci Tech 42(12), 1569–1575
131. Miller KJ (2001) Structural integrity—whose responsibility? The 36th John Player memorial lecture presented at an ordinary meeting of the institution of mechanical engineers. Institution of Mechanical Engineers, London
132. Morozov EM, Nikishkov GP (1980) Metod konechnykh elementov v mekhanike razrusheniya (Finite-element method in fracture mechanics). Nauka, Moscow
133. Morozov EM, Zernin MB (1999) Kontaktnyye zadachi mehaniki razrusheniya (Contact problems of fracture mechanics). Mashinostroenie, Moscow
134. Murakami Y (1986) Stress intensity factors handbook. Pergamon Press, Oxford
135. Murakami Y, Kaneta M (1989) Fracture-mechanics approach to tribology problems. In: Wei RP, Gangloff RP (eds) Twentieth Sympos. "Fracture mechanics: perspectives and directions", ASTM STP 1020. American Society for Testing and Materials, Philadelphia, pp 668–687
136. Murakami Y, Kaneta M, Yatsuzuka H (1985) Analysis of surface crack propagation in lubricated rolling contact. ASLE Trans 28(1):60–68
137. Murakami Y, Sakae C, Hamada S (1999) Mechanism of rolling contact fatigue and measurement of ΔK_{IIth} for steels. In: Beynon JH, Brown MW, Lindley et al (eds) Engineering against fatigue. A. A. Balkema Publishers, Rotterdam, pp 473–485
138. Murakami Y, Sakae C, Ichimaru K, Morita T (1997) Experimental and fracture mechanics study of the pit formation mechanism under repeated lubricated rolling/sliding contacts: effects of reversal of rotation and change of the driving roller. Trans ASME J Tribol 119:788–796
139. Muskhelishvili NI (1966) Nekotoryye osnovnyye zadachi matematicheskoy teorii uprugosti (Some basic problems of the mathematical theory of elasticity). Nauka, Moscow
140. Nayak L, Paul K (1979) Contact fatigue failure of rolls of hot strip mill. Indian J Technol 17:27–34
141. Neu RW (2011) Progress in standartization on fretting fatigue terminology and testing. Tribol Int 44:1371–1377
142. Noda N-A, Yagishita M, Kihara T (2000) Effect of crack shape, inclination angle, and friction coefficient in crack surface contact problems. Int J Fract 105:367–389
143. Ohkomori Y, Kitagawa I, Shinozuka K et al (1987) Cause and prevention of spalling of back-up rolls for hot strip mill. Tetsu- to Hagane 73:691–697
144. Ohkomori Y, Sakae C, Murakami Y (2000) Mode II crack growth analysis of spalling behaviour for strip mill back-up roll. In: 42nd MSWP conference. Proceedings of the ISS, 2000, vol XXXVIII, pp 723–729
145. Opanasovich VK, Kundrat NM (1980) Elastic equilibrium of a plate, the edges of which are partially in contact. Sov Mater Sci 15(6):611–614

146. O'Regan SD, Hahn GT, Rubin CA (1985) The driving force mode II crack growth under rolling contact. Wear 101:333–346
147. Orringer O, Morris JM, Steele RK (1984) Applied research on rail fatigue and fracture in the United States. Theor Appl Fract Mech 1:23–49
148. Ostash OP (2015) Struktura materialiv i vtomna dovgovichnist elementiv konstrukciy (Structure of materials and fatigue durability of structural elements). In: Panasyuk VV (ed) Fracture mechanics and strength of materials, vol 15. SPOLOM, Lviv
149. Panasyuk VV (1991) Mekhanika kvazikhrupkogo razrusheniya materialov (Mechanics of brittle fracture of materials). Naukova Dumka, Kiev
150. Panasyuk VV, Andreikiv AE, Parton VZ (1988) Osnovy mekhaniki razrusheniya materialov (Fundamentals of the fracture mechanics of materials). In: Fracture mechanics and strength of materials: a handbook, vol 1. Naukova Dumka, Kiev
151. Panasyuk VV, Datsyshin AP (1974) The equilibrium limit of a half plane with an arbitrarily oriented crack at its boundary. Sov Mater Sci 7(6):751–752
152. Panasyuk VV, Datsyshyn OP (2014) Material damages and life time of solids under a cyclic contact. Procedia Mater Sci 3:1250–1256
153. Panasyuk VV, Datsyshyn OP, Glazov AY (2007) Prediction of the contact durability of rails by pitting development, vol 3. Mashynoznavstvo, pp 3–10
154. Panasyuk VV, Datsyshyn OP, Levus AB (2002) Evolution of a system of edge cracks in the region of rolling bodies cyclic contact. In: Neimitz A et al (eds) ECF-14, Fracture mechanics, V. I/III, Beyond 2000. EMAS Publishing, Sheffield, pp 609–616
155. Panasyuk VV, Datsyshyn OP, Marchenko HP (1996) Contact problem for a half plane with cracks subjected to the action of a rigid punch on its boundary. Mater Sci 31(6):667–678
156. Panasyuk VV, Datsyshyn OP, Marchenko HP (2001) Crack growth in rolling bodies under the conditions of dry friction and wetting. Mater Sci 37(1):1–11
157. Panasyuk VV, Datsyshyn OP, Marchenko HP (2000) Stress state of a half-plane with cracks under rigid punch action. Int J Fract 101(4):347–364
158. Panasyuk VV, Datsyshyn OP, Marchenko HP (1995) To crack propagation theory under rolling contact. Eng Fract Mech 52(1):179–191
159. Panasyuk VV, Datsyshyn OP, Shchur RB (2000) Residual durability of solids contacting under conditions of fretting fatigue. Mater Sci 36(2):153–169
160. Panasyuk VV, Savruk MP, Datsyshin AP (1977) Application of singular integral equations to the solution of two-dimensional problems in crack theory. Sov Mater Sci 12(3):245–259
161. Panasyuk VV, Savruk MP, Datsyshyn AP (1976) Raspredeleniye napryazheniy okolo treshchin v plastinakh i obolochkakh (Distribution of stresses near cracks in plates and shells). Naukova Dumka, Kiev
162. Panasyuk VV, Savruk MP, Datsyshyn AP (1977) A general method of solution of two-dimensional problems in the theory of cracks. Eng Fract Mech 9(2):481–497
163. Paris P, Erdogan F (1963) A critical analysis of crack propagation laws. Trans ASME J Basic Eng 85(4):528–534
164. Paris PC, Gomes MP, Anderson WE (1961) A rational analytic theory of fatigue. Trend Eng 13:54–61
165. Paris PC, Sih GC (1965) Stress analysis of cracks. In: Fracture toughness testing and its applications. ASTM–NASA, Philadelphia, pp 30–83
166. Peng JF, Zhu MH, Cai ZB, Liu JH et al (2014) On the damage mechanisms of bending fretting fatigue. Tribol Int 76:133–141
167. Pinegin SV (1969) Kontaktnaya prochnost' i soprotivleniye kacheniyu (Contact strength and rolling resistance). Mashinostroenie, Moscow
168. Pisarenko GS, Lebedev AA (1976) Deformirovaniye i prochnost' materialov pri slozhnom napryazhennom sostoyanii (Deformation and strength of materials in the complex stressed state). Naukova Dumka, Kiev
169. Polukhin VP, Nikolaev VA, Tylkin MA et al (eds) (1976) Nadezhnost' i dolgovechnost' valkov kholodnoy prokatki (Reliability and durability of cold-rolled rolls). Metallurgiya, Moscow

170. RD 50-345-82 (1983) Metodologicheskiye ukazaniya. Raschety i ispytaniya na prochnost'. Metody mekhanicheskikh ispytaniy metalov. Opredeleniye kharasteristik treshchinostoykosti (vyazkosti razrusheniya) pri tsiklicheskom nagruzhenii (Methodical recommendations. Strength analyses and tests. Methods for mechanical testing of metals. Determination of the crack-growth resistance characteristics (fracture toughness) under cyclic loading). Izdatel'stvo Standartov, Moscow
171. Ritchie RO (1984) Threshold for fatigue crack propagation: questions and anomalies. In: Advances in fracture research. Proceedings of 6th international conference on fracture. Pergamon press, Oxford, pp 235–260
172. Ringsberg JW (2005) Shear mode growth of short surface-breaking RCF cracks. Wear 258:955–963
173. Ringsberg JW, Bergkvist A (2003) On propagation of short rolling contact fatigue cracks. Fatigue Fract Eng Mater Struct 26(10):969–983
174. Romaniv ON, Yarema SY, Nikiforchin GN, Makhutov NA, Stadnik MM (1990) Ustalost' i ciclicheskaya treshchinostoykost' konstrukcionnyh materialov (Fatigue and cyclic crack resistance of structural materials). In: Panasyuk VV (ed) Fracture mechanics and strength of materials: a handbook, vol 4. Naukova Dumka, Kiev
175. Rooke DP, Cartwright DJ (1976) The compendium of stress intensity factors. Her Majesty's Stationery Office, London
176. Rooke DP, Jones DA (1979) Stress intensity factors in fretting fatigue. J Strain Anal 14(1):1–6
177. Rosenfield AR (1980) Fracture mechanics approach to wear. Wear 61:125–132
178. Salehizaden H, Saka N (1992) Crack propagation in rolling line contacts. J Tribol 114:690–697
179. Sato M, Anderson PM, Rigney DA (1993) Rolling-sliding behavior of rail steels. Wear 162–164:159–172
180. Savruk MP (1988) Koefficienty intensivnosti napryazheniy v telakh s treshchinami (Stress intensity factors in bodies with cracks). In: Panasyuk VV (ed) Fracture mechanics and strength of materials: a handbook, vol 2. Naukova Dumka, Kiev
181. Savruk MP (1981) Dvumernyye zadachi uprugosti dlya tel s treshchinsmi (Two-dimensional problems of elasticity for bodies with cracks). Naukova Dumka, Kiev
182. Savruk MP, Datsyshin AP (1974) Interaction between a system of cracks and the boundaries of an elastic body. Sov Appl Mech 10(7):755–761
183. Savruk MP, Datsyshin AP (1972) Limiting equilibrium state of a body weakened by a system of randomly oriented cracks. Termomekhanicheskiye metody razrusheniya gornykh porod (Thermomechanical methods of fracture of rocks). Part 2. Naukova Dumka, Kiev, pp 97–102
184. Savruk MP, Prokopchuk IV, Osiv PN (1989) Chislennyy analiz v ploskikh zadachah teorii treshchin (Numerical analysis in plane problems of the theory of cracks). Naukova Dumka, Kiev
185. Savruk MP, Tomczyk A (2010) Pressure with friction of a perfectly rigid die upon an elastic half space with cracks. Mater Sci 46(3):283–296
186. Sheppard SD, Barber JR, Comninou M (1985) Short subsurface cracks under conditions of slip and stick caused by a moving compressive load. J Appl Mech 52:811–817
187. Sheppard SD, Barber JR, Comninou M (1987) Subsurface cracks under conditions of slip, stick, and separation caused by a moving compressive load. Trans ASME J Appl Mech 54(2):393–398
188. Shiratori M, Miyoshi T, Matsushita H (1986) Vycheslitel'naya mekhanika razrusheniya (Computational fracture mechanics). Mir, Moscow
189. Shur E (1971) Povrezhdeniye rel'sov (Damage of rails). Transport, Moscow
190. Sih GC (1974) Handbook of stress intensity factors, vol 1. Lehigh University Press, Bethlehem, p 420 (1973, vol 2, p 406)
191. Smith RA (2003) The wheel-rail interface–some recent accidents. Fatigue Fract Eng Mater Struct 26(10):901–907
192. Sosnovskii LA (2002) Fundamentals of the mechanics of wear-fatigue damage and fracture. In: Troshchenko VT (ed) "Trybofatyka" ("Tribofatigue"), Proceedings of the 4th international symposium on tribofatigue (ISTF 4), vol 1, pp 9–22, Ternopil

193. Sosnovskii LA, Makhutov NA, Shurinov VA (1992) Contact-mechanical fatigue: basic regularities, vol 11. Zavodskaya Laboratoriya, pp 44–61
194. Sosnovskii LA, Makhutov NA, Shurinov VA (1992) Fretting fatigue: basic regularities, vol 8. Zavodskaya Laboratoriya, pp 45–62
195. Sosnovskii LA, Makhutov NA, Shurinov VA (1992) Friction-mechanical fatigue: basic regularities, vol 9. Zavodskaya Laboratoriya, pp 46–63
196. Stadnyk MM, Didukh IV (2010) Evaluation of the durability of pipe elements in the stage of growth of a semielliptic fatigue surface crack. Visnyk Lvivs'kogo Natsional'nogo Universytetu im. I.Franka, Ser. Mekhanika-Matematyka, vol 73, pp 23–29
197. Stock R, Stanlake L, Hardwick C, Yu M et al (2016) Material concepts for top of rail friction management—classification, characterization and application. Wear 366–367:225–232
198. Su X, Clayton P (1996) Surface-initiated rolling contact fatigue of pearlitic and low carbon bainitic steels. Wear 197:137–144
199. Suh NP (1977) An overview of the delamination theory of wear. Wear 44:1–16
200. Tada H, Paris PC, Irwin GR (1985) The stress analysis of cracks: handbook. Del Research Corporation, St. Louis
201. Tait WH (1990) Roll shop. Part 1: The nature and causes of in service defects, Rolls for the metalworking industries. ISS, pp 35–149
202. Tallian TE (1988) Influence of the properties of materials and operating conditions on the durability of rolling bearings. Part 1. Description of a model and calculation of the basic durability. Part 2. Correction coefficients. Problemy Treniya i Smazki, vol 4, pp 1–13
203. Taylor D (1985) A compendium of fatigue thresholds and growth rates. Engineering Materials Advisory Services Ltd, Warley
204. Theocaris PS, Panagiotopoulos PD, Bisbos C (1993) Unilateral contact, friction and related interactions in cracks. The direct boundary integral method. Int J Solids Struct 30(11): 1545–1561
205. Timoshenko SP (1953) History of strength of materials. McGraw-Hill, New York
206. Toth L, Yarema SY (2006) Formation of the science of fatigue of metals. Part 1: 1825–1870. Mat Sci 42(5): 673–680
207. Troshchenko VT (1981) Deformirovaniye i razrusheniye metallov pri mnogotsiklovom nagruzhenii (Deformation and fracture of metals under high-cycle loading). Naukova Dumka, Kiev
208. Troshchenko VT, Tsybanev GV, Khotsyanovskii AO (1988) Determination of the durability of steels under fretting fatigue. Probl Prochn 6:3–8
209. Tsamasphyros G, Theocaris PS (1983) Integral equation solution for half-planes bonded together or in contact and contacting internal cracks or holes. Ing Arch 53(4):225–241
210. Tyfour WR, Beynon JH (1995) The steady state wear behaviour of pearlitic rail steel under dry rolling-sliding contact conditions. Wear 180:79–89
211. Tyfour WR, Beynon JH, Kapoor A (1996) Deterioration of rolling contact fatigue life of pearlitic rail steel due to dry-wet rolling-sliding line contact. Wear 197:255–265
212. Waterhause RB (1972) Fretting corrosion. Pergamon Press, Oxford-New York
213. Way S (1935) Pitting due to rolling contact. J Appl Mech Trans ASME 2: A49–A58
214. Wu XR, Carlsson AJ (1991) Weight functions and stress intensity factor solutions. Pergamon Press, Oxford
215. Xu X, Cho D-H, Chang Y-S et al (2011) Evaluation of slant crack propagation under RCF in railway rail. J Mech Sci Technol 25(5):1215–1220
216. Yarema SY (1987) Fundamentals and certain problems of fatigue fracture mechanics. Sov Mater Sci 23(5): 454–464
217. Yarema SY (1975) Stages of fatigue fracture and their consequences. Sov Mater Sci 9(6): 681–686
218. Yarema SY (1978) Assessment of metal and alloy resistance to cracking. Studies of fatigue crack growth and kinetic fatigue fracture diagrams. Sov Mater Sci 13(4): 351–368
219. Yarema SY, Mikitishin SI (1976) Analytical description of the fatigue-failure diagrams of materials. Sov Mater Sci 11(6): 660–666

220. Zafošnik B, Ren Z, Flašker J, Mishuris G (2005) Modelling of surface crack growth under lubricated rolling–sliding contact loading. Int J Fract 134:127–149
221. Zang WL, Gudmundson P (1991) Frictional contact problems of kinked cracks modeled by a boundary integral method. Int J Numer Meth Eng 31:427–446
222. Zazulyak VA, Darchuk AI, Legun AM, Ivanitskii YL, Darchuk AI (1986) Evaluation of the crack resistance of large rolling-mill back-up rolls in cyclic loading. Sov Mater Sci 21(4): 383–385
223. Zerbst U, Lundén R, Edel K-O, Smith RA (2009) Introduction to the damage tolerance behaviour of railway rails—a review. Eng Fract Mech 76:2563–2601
224. Zerbst U, Madler K, Hintze H (2005) Fracture mechanics in railway applications—an overview. Eng Fract Mech 72(2):163–194

Chapter 2
Calculation Model for Estimation of Cyclic Contact Lifetime of Body with Cracks

Abstract This chapter contains the key elements of the theory of fatigue fracture, namely, the main stages and fatigue characteristics of the materials; the criteria and diagrams of fatigue fracture, and the characteristics of fatigue crack growth resistance of the materials. We present a brief description of the specific features of cyclic contact interaction accompanied by the crack initiation in the nearsurface contact zone: rolling, rolling with sliding, fretting fatigue, pulsed contact, friction fatigue, etc. We also describe typical contact fatigue defects formed by cracks on the working surfaces of the tribojoint elements. In the same chapter, we propose a new computational model for the investigation of the fracture processes and evaluate the contact residual service life of the bodies (tribojoint elements) under the conditions of their cyclic interaction.

2.1 Classical and Nonclassical Approaches

To characterize the common features of the problems of formation of new hypotheses and approaches in the theory of strength and fracture of materials, we consider some initial concepts of the classical and nonclassical (new) approaches in this field of science.

Within the framework of the classical understanding of a material as a solid body, i.e., as a continuum, we assume that an element of the deformed (under the action of loads) body can be in one of the following states [83]: continuous (C) or fractured (F). In this case, the transition of the material from the C-state into the F-state, i.e., the fracture process (Fig. 2.1a), runs instantaneously if the characteristics of the stress-strain state (computed according to the accepted rheological model. e.g., the model of elastic continuum) attain certain critical values of these characteristics for the given material (e.g., the maximum tensile stresses attain their ultimate value σ_u). If the characteristics of the stress-strain state in the material do not attain their ultimate values, then the fracture (C $\rightarrow$ F transition) does not occur and the body preserves its integrity and, hence, strength.

O. Datsyshyn and V. Panasyuk, *Structural Integrity Assessment of Engineering Components Under Cyclic Contact*, Structural Integrity 9,
https://doi.org/10.1007/978-3-030-23069-2_2

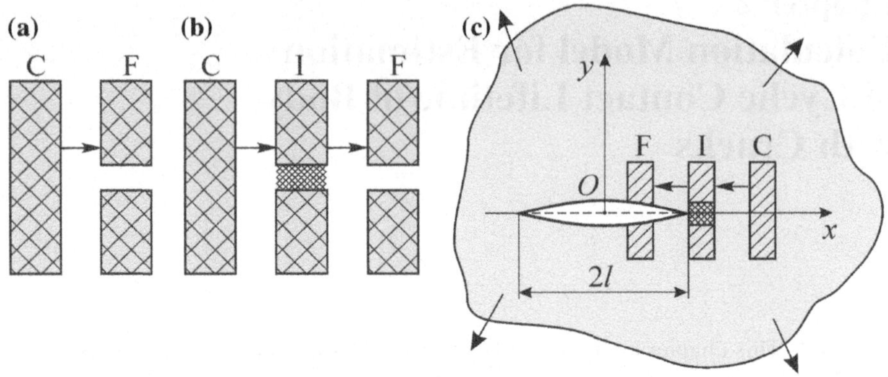

Fig. 2.1 Schemes of fracture: **a** classical scheme; **b**, **c** nonclassical schemes

The classical understanding of fracture processes resulted in the creation of appropriate experimental procedures aimed at the determination of quantitative fracture-resistance characteristics of the material, including the methods of determination of the limiting value of strength of the material σ_u, yield strength σ_T, relative elongation δ (or ε_{max}), relative reduction of area ψ of the specimen, endurance limit σ_R [40, 99], and other characteristics of specimens of the material (these are characteristics of macrovolumes, i.e., volumes much larger than the structural element of the material). The classical concept of fracture in the evaluation of strength (fracture) of structural materials is extensively used in the engineering practice and became basic for the well-known phenomenological hypotheses of strength (I, II, III, IV, etc.) [99].

In this approach to the evaluation of the strength of material, the problem is reduced to the construction (on the basis of the available theoretical and experimental data) of a function

$$F_c(J_1, J_2, J_3, C_1, C_2, C_3, \ldots) = 0, \tag{2.1}$$

where J_1, J_2, and J_3 are invariants of the stress (strain) tensor [99] (sometimes simply the principal stresses (strains) at a given point of the deformed body) and C_1, C_2, C_3,… are experimentally determined constants for the given material. This function is used as a criterion for the evaluation of the strength of material in the structure.

The classical approach of the mechanics of materials does not reveal the fracture mechanism itself. By using this approach, it is impossible to describe the fracture process and establish the factors accelerating or inhibiting this process, i.e., to formulate the principles of control over the mechanical properties of material (its fracture resistance). Moreover, within the framework of the classical approaches, it is impossible to explain the contradiction between the perfect and engineering strength of crystals; from the practical viewpoint, within the framework of these approaches, it is impossible to perform the engineering diagnostics of fracture or the absence of fracture of the material in the structure under the extreme conditions of its operation (e.g., to determine the influence of sharp stress concentrators, i.e., cracks). Finally, the analysis of experimental data shows that the scheme of instantaneous fracture of bodies over the entire cross section is in fact unrealistic.

The main idea of the nonclassical approaches (i.e., the approaches of contemporary fracture mechanics of materials and structures) can be described as follows (see Fig. 2.1b): The transition of an element of the deformed material from the C-state into the F-state is accompanied by a certain intermediate (I) state of the material. As the most important feature of the regions of deformed material characterized by the formation of I-states (process zones), we can mention the fact that, in these regions, the material is always deformed beyond the limit of elasticity and, moreover, they are characterized by the most intense processes of local plastic yield, interaction with the ambient medium, diffusion, and other phenomena responsible, as a final result, for the local fracture of the material, i.e., for the $C \to I \to F$ transition.

In other words, the nonclassical scheme of fracture takes into account the I-states of the deformed material in the evaluation of its strength and durability. Most often, these states are formed near pointed stress concentrators, i.e., cracklike defects (see Fig. 2.1c).

In the engineering practice, the following types of fracture of structural materials are encountered most frequently: plastic, brittle, fatigue, dynamic, etc.

Plastic fracture occurs after strong plastic deformation in the entire (or almost in the entire) volume of the body. As a version of plastic fracture, we can mention is the rupture of a specimen after 100% reduction of the neck area in tension, which occurs as a result of the loss of ability of the material to resist plastic deformation.

Brittle fracture occurs as a result of propagation of the main crack caused by (macroscopically insignificant) plastic deformation concentrated in the vicinity of the crack tip. In the case of perfectly brittle fracture, plastic deformation is absent and, hence, after fracture, it is possible to compose the body of the previous size from its fragments obtained in the process of fracture without gaps between them. Sometimes, in similar cases, the researchers use the term “elastic fracture”.

In the case of quasielastic (or quasibrittle) fracture, there exists a plastic zone head of the crack front and a plastically deformed (strain-hardened) material near the crack surface. The remaining (much larger) part of the volume of the body is in the elastic state. The fracture process in which the stresses acting in the net cross section are higher than the yield strength but lower than the ultimate strength of the material is also called quasibrittle fracture.

Fatigue fracture occurs under the conditions of cyclic (repeated) loading of a body as a result of accumulation of irreversible defects, which lead to the initiation and propagation of the crack. The fatigue fracture is macroscopically brittle but near the fracture surface, the material is always plastically deformed (strain-hardened).

We also distinguish high-cycle and low-cycle types of fatigue. The high-cycle fatigue (or simply fatigue) is characterized by the nominal stresses lower than the yield strength (whose repeated generation in the deformed body causes its macroscopic deformation in the elastic region) and by a large number of cycles to failure.

The low-cycle fatigue (or repeated static loading) is characterized by the nominal stresses higher than the yield strength (in each loading cycle with this level of stresses, the body suffers macroscopic plastic deformation) and by an insignificant number of cycles to failure.

The general aim of investigations in the fracture mechanics of solids (materials) is to develop the theory of the processes of deformation and fracture with regard for their structural imperfection. In this case, it is very important to establish physically justified criteria for the evaluation of strength (absence of fracture) of the structural material with regard for the presence of cracklike defects and find the ways of formation of materials and structures with high strength and durability.

As the most important concept of the fracture mechanics of solids, we can mention the fact that fracture is treated as the process of crack initiation and propagation. As compared with the classical approaches, this concept describes the mechanism of fracture and the crack is regarded as an object responsible for fracture.

Thus, the nonclassical scheme of fracture takes into account the I-states formed near the defects with sharp ends (first of all, cracklike defects) in deformed bodies, i.e., stress concentrators whose radius of rounding is comparable with the characteristic linear size of the structural elements of the material. Thus, in the evaluation of strength of a body, it is necessary to take into account its local physicomechanical properties, e.g., the ability to resist crack propagation, i.e., its crack growth resistance. The analysis of the I-states of the material within the framework of continuum mechanics requires the introduction of new (nonclassical) computational models and concepts. Since the main characteristics responsible for the behavior of the material at the crack tip are stresses, strains, and energy, all available criteria of fracture mechanics, as in the classical theories of strength, are split into energy, force, and deformation criteria.

2.2 Force Approach in Fracture Mechanics. Linear Fracture Mechanics

Despite the fact that the energy approach proposed by Griffith in [45] was historically the first criterion of fracture mechanics, the force approaches based on the concept of stress intensity factors introduced by in Irwin [50] became especially popular in the engineering practice. The description of fracture performed on this basis is quite simple and clear. The application of characteristics that can be difficultly found, such as true surface energy and the work of local plastic strains appearing in the other approaches and criteria, is also excluded. The successive realization of the force approach led to the development of a fairly rigorous complete theory of linear fracture mechanics, which served as a good basis for the analysis of brittle fracture of materials and engineering structures.

Stressed state in the vicinity of the crack contour. The starting point in the strength analyses of the structural components and structures with cracks is the investigation

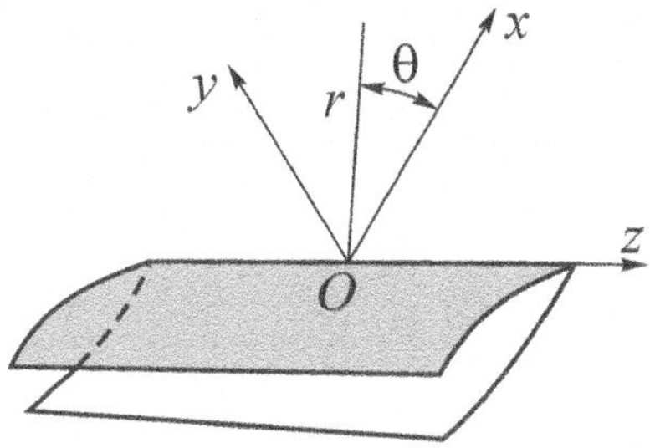

Fig. 2.2 Local coordinate systems near the crack tip

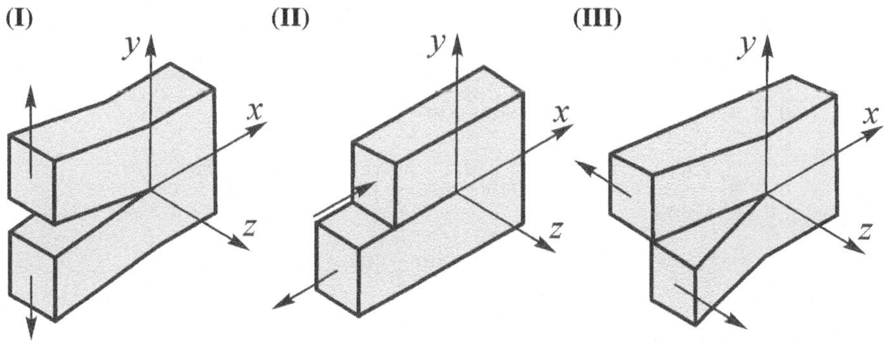

Fig. 2.3 Schemes of the principal macromechanisms of cracks propagation: (I) normal opening (mode I) mechanism; (II) transverse shear (mode II) mechanism; (III) longitudinal shear (mode III) mechanism

of the distributions of stresses and strains formed in these structures and components under the action of applied loads. In this case, it is of especial interest to analyze the zone in the immediate vicinity of the crack tip (end) because the next stage of fracture begins just in this zone and the material passes into the I-state (see Fig. 2.1). Within the framework of linear fracture mechanics based on the model of perfectly elastic body and representing the crack in the form of a cut of thickness zero whose surfaces are free of loads, the analyzed problem is reduced to a boundary-value problem of the theory of elasticity.

In the general case, the strain distribution in the vicinity of a random point O (Fig. 2.2) of the crack contour can be represented in the form of the superposition of three specific types of deformation (Fig. 2.3) corresponding to three main kinds of displacements of the crack surfaces: normal opening (mode I), transverse shear (mode II), and longitudinal shear (mode III).

The first type of deformation is connected with the normal displacements of the crack surfaces in the opposite directions (symmetrically about the planes xy and xz), the second type corresponds to the displacements in which the crack surfaces slip over each other in the direction perpendicular to the crack front (z-axis), and the third type is connected with the slip of one crack surface of over the other parallel to the crack front (z-axis).

In the realization of each indicated type of deformation of the body weakened by a crack in the vicinity of its contour, the distributions of stresses and strains can be represented in the form [50, 84]:

in the case of normal opening (I)

$$\begin{aligned}
u_x &= \frac{K_{\mathrm{I}}}{G}\sqrt{\frac{r}{2\pi}}\cos\frac{\theta}{2}\left(1-2\mu+\sin\frac{\theta}{2}\right),\\
u_y &= \frac{K_{\mathrm{I}}}{G}\sqrt{\frac{r}{2\pi}}\sin\frac{\theta}{2}\left(1-2\mu+\sin\frac{\theta}{2}\right),\\
u_z &= 0,\\
\sigma_{xx} &= \frac{K_{\mathrm{I}}}{\sqrt{2\pi r}}\cos\frac{\theta}{2}\left(1-\sin\frac{\theta}{2}\sin\frac{3\theta}{2}\right),\\
\sigma_{yy} &= \frac{K_{\mathrm{I}}}{\sqrt{2\pi r}}\cos\frac{\theta}{2}\left(1+\sin\frac{\theta}{2}\sin\frac{3\theta}{2}\right), \quad \sigma_{zz} = \frac{2\mu K_{\mathrm{I}}}{\sqrt{2\pi r}}\cos\frac{\theta}{2},\\
\tau_{xz} &= \tau_{yz} = 0, \quad \tau_{xy} = \frac{K_{\mathrm{I}}}{\sqrt{2\pi r}}\sin\frac{\theta}{2}\cos\frac{\theta}{2}\cos\frac{3\theta}{2};
\end{aligned} \tag{2.2}$$

in the case of transverse shear (II)

$$\begin{aligned}
u_x &= \frac{K_{\mathrm{II}}}{G}\sqrt{\frac{r}{2\pi}}\sin\frac{\theta}{2}\left(2-2\mu+\cos^2\frac{\theta}{2}\right),\\
u_y &= \frac{K_{\mathrm{II}}}{G}\sqrt{\frac{r}{2\pi}}\cos\frac{\theta}{2}\left(2\mu-1+\sin^2\frac{\theta}{2}\right),\\
u_z &= 0,\\
\sigma_{xx} &= -\frac{K_{\mathrm{II}}}{\sqrt{2\pi r}}\sin\frac{\theta}{2}\left(2+\cos\frac{\theta}{2}\cos\frac{3\theta}{2}\right),\\
\sigma_{yy} &= \frac{K_{\mathrm{II}}}{\sqrt{2\pi r}}\cos\frac{\theta}{2}\sin\frac{\theta}{2}\cos\frac{3\theta}{2}, \quad \sigma_{zz} = -\frac{2\mu K_{\mathrm{II}}}{\sqrt{2\pi r}}\sin\frac{\theta}{2},\\
\tau_{xz} &= \tau_{yz} = 0, \quad \tau_{xy} = \frac{K_{\mathrm{II}}}{\sqrt{2\pi r}}\cos\frac{\theta}{2}\left(1-\sin\frac{\theta}{2}\sin\frac{3\theta}{2}\right);
\end{aligned} \tag{2.3}$$

in the case of longitudinal shear (III)

$$\begin{aligned}
&u_x = 0, \quad v_y = 0, \quad u_z = \frac{K_{\mathrm{III}}}{G}\sqrt{\frac{2r}{\pi}}\sin\frac{\theta}{2},\\
&\tau_{xz} = -\frac{K_{\mathrm{III}}}{\sqrt{2\pi r}}\sin\frac{\theta}{2}, \quad \tau_{yz} = \frac{K_{\mathrm{III}}}{\sqrt{2\pi r}}\cos\frac{\theta}{2}, \quad \sigma_{xx} = \sigma_{yy} = \sigma_{zz} = \tau_{xy} = 0.
\end{aligned} \tag{2.4}$$

The stress intensity factors K_{I}, K_{II} and K_{III} serve as a measure of singularity of stresses near the crack tip, i.e., near the process zone (I-state) and play the same role for this region as the stress concentration factors in the resistance of materials to nonsharp concentrators in the body. Unlike the concentration factors, stress intensity factors are measured in kg/mm$^{3/2}$ or MPa $\sqrt{\mathrm{m}}$.

The universality of relations (2.2)–(2.4) is based on the fact that, according to these relations, the radial (along the radius r; Fig. 2.2) and angular (in the direction of changes in the angle θ; Fig. 2.2) asymptotic distributions of the stressed state do not depend on the crack length, shape of the body, and the applied loading modes. The intensity of this distribution is determined solely by the factors K_{I}, K_{II} and K_{III}, which are functions of the applied loads, geometries of the body and cracks, which are independent of the coordinates of a point near the end of the cut. If σ is a parameter that characterizes the external load upon the body and l is the characteristic linear size of the crack, then it is clear that $K_{\mathrm{I}} = K_{\mathrm{I}}(\sigma, l)$, $K_{\mathrm{II}} = K_{\mathrm{II}}(\sigma, l)$, and $K_{\mathrm{III}} = K_{\mathrm{III}}(\sigma, l)$.

Thus, the stress intensity factors are the main (basic) characteristics of the stress-strain state of a material in the vicinity of a crack and, therefore, are of very high importance in fracture mechanics. At present, the determination of these quantities for bodies with cracks of different configurations on the basis of the solution of the corresponding boundary-value problems is a large independent field of the mathematical theory of cracks (see Chap. 3).

Force criteria of brittle and quasibrittle fracture. Since for a perfectly elastic body, the stress-strain state near the tip (front) of a crack of size l caused by an external load σ is completely determined by the stress intensity factors K_{I}, K_{II} and K_{III}, these quantities are used as basic in the force criteria of the linear fracture mechanics. In this case, the criterial equation contains the parameters K_{I}, K_{II}, K_{III} and some material constants C_i, i.e., can be represented in the form

$$F\left(K_{\mathrm{I}}^{*}, K_{\mathrm{II}}^{*}, K_{\mathrm{III}}^{*}, C_i\right) = 0, \quad i = 1, 2, 3, \ldots;$$
$$K_j^{*} = K_j(\sigma_k, l), \quad j = \mathrm{I}, \mathrm{II}, \mathrm{III}, \tag{2.5}$$

where σ_k is a critical (fracture) value of the applied load. Relation (2.5) describes, on the coordinates (K_{I}, K_{II}, K_{III}), a certain ultimate surface for a given material upon attainment of which, the crack passes from the stable state into the unstable state (i.e., begins to propagate).

The presence of a zone of plastic deformation near the crack tip in metals leads to the discrepancy between the actual picture of the stress-strain state and the stress-strain state predicted by relations obtained from the solution of the problem posed for elastic bodies. However, if the zone of nonlinear properties of the material is fairly small, then the asymptotic behavior of solutions of the problems for elastic bodies can be regarded as a good approximation to the actual stress distribution. This enables us to believe that the size of the process zone, i.e., the I-state of the material, is also completely determined by the stress intensity factors. This is why, despite the fact that the actual materials do not exhibit a perfectly elastic behavior at fracture, the criteria of linear fracture mechanics remain true for these materials if the plastic yield is limited (i.e., in the case of realization of the scheme of quasibrittle fracture).

A criterion of type (2.5) was first proposed by Irwin [50] for the case of fracture of quasibrittle bodies by the normal opening mechanism. It can be formulated as follows: a crack begins to propagate in the case where the stress intensity factor attains it critical value for a given material:

in the case of plane deformation

$$K_{\mathrm{I}}^{*} = K_{\mathrm{I}C}, \tag{2.6}$$

in the plane stressed state

$$K_{\mathrm{I}}^{*} = K_{C}$$

By analogy with the Irwin criterion, the condition of start crack in the case of transverse shear ($K_{\mathrm{I}} = K_{\mathrm{III}} = 0$) is represented in the form

$$K_{\mathrm{II}}^{*} = K_{\mathrm{II}C} \tag{2.7}$$

and, for the longitudinal shear ($K_{\mathrm{I}} = K_{\mathrm{II}} = 0$), in the form

$$K_{\mathrm{III}}^{*} = K_{\mathrm{III}C}. \tag{2.8}$$

The principal difference between the quantities on the left- and right-hand sides of relations (2.6)–(2.8) is that K_{I}^{*}, K_{II}^{*} and K_{III}^{*} reflect the geometric shape of the body weakened by a crack and the conditions of its loading, whereas $K_{\mathrm{I}C}$, $K_{\mathrm{II}C}$ and $K_{\mathrm{III}C}$ are the characteristics of the material related, just as the yield and ultimate strength, a specific property of the material, i.e., the ability to resist crack propagation (i.e., the crack growth resistance). At present, the crack growth resistance of structural materials is determined by using special methods and recommendations [44, 52, 64] specifying the conditions of evaluation and application of these parameters.

For the complex loading of a body ($\sigma_1 \neq 0$, $\sigma_2 \neq 0$, $\sigma_3 \neq 0$) in which different mechanisms of fracture are realized simultaneously, the functional dependence (2.5) has a more complicated form. Thus, the criterial relation [83, 84]

$$\left(\frac{K_{\mathrm{I}}^{*}}{K_{\mathrm{I}C}}\right)^{m_1} + \left(\frac{K_{\mathrm{II}}^{*}}{K_{\mathrm{II}C}}\right)^{m_2} + \left(\frac{K_{\mathrm{III}}^{*}}{K_{\mathrm{III}C}}\right)^{m_3} = 1, \tag{2.9}$$

where m_1, m_2 and m_3 are material constants, frequently gives a good description of fracture for a broad class of structural materials under the conditions of complex loading of a body with crack, i.e., in the case where $K_{\mathrm{I}}^{*} \neq 0$, $K_{\mathrm{II}}^{*} \neq 0$, and $K_{\mathrm{III}}^{*} \neq 0$.

2.3 Fatigue Fracture. Classical Approach

The fracture processes in materials subjected to long-term loading by forces of constant or variable (as functions of time) intensities (fatigue fracture of materials) are characterized by the fact that they occur under stresses always lower than the ultimate strength of the material. According to the contemporary ideas, the fatigue fracture of materials (solids) is physically caused by the presence of structural microinhomogeneities. In the process of deformation of the body, this leads to the overloading of some zones of the material and, as a result, to the appearance of microplastic shears in its structure and the formation of various types of microdefects. An event of insignificant one-time plastic deformation does not cause noticeable changes in the structure of the material. At the same time, the long-term action of the load and multiply repeated microplastic deformations result in the accumulation of microdefects and the formation of microcracks in materials. For a certain density of these microdefects, they start to coalesce, which leads to the initiation of microcracks covering several structural elements of the material. This corresponds to the onset of the stage of propagation of macrocracks formed in the material.

The problem of fatigue resistance of the materials is one of the most urgent problems in the science of strength of deformable solids.

Main stages and characteristics of materials fatigue. Numerous experimental investigations of the behavior of microstructure of the material in the process of fatigue fracture demonstrate [51, 58, 59, 113] that microplastic shears in the nearsurface layers of the deformed body are formed in the early stages of the process and develop in a certain sequence. Depending on the character of changes caused by cyclic plastic deformation, there are three successive (partially overlapping) stages of fatigue fracture of the material [40, 99], namely: (i) the incubation stage characterized by the inhomogeneous concentration of microplastic shears, mainly in the nearsurface layers of the deformed body due to the lower yield strength of these layers of the material [101]; (ii) the stage of initiation of submicrocracks, their propagation, coalescence, and formation of microcracks, i.e., violations of the material structure on the level of its blocks, grains, and inclusions, and the formation of extrusion and intrusion steps in the surface layer [107]; (iii) the stage of crack propagation; this stage originates when one microcrack becomes predominant and turns into the macrocrack. Some researchers also distinguish the fourth stage, namely, the stage of final fracture. Since the crack growth rate in the process of final fracture is very high, the duration of this stage is quite small. In this connection, its investigation is not of high practical importance and, therefore, this process is often not considered. Sometimes, the researchers consider more detailed partitions of the process of fatigue fracture of materials but this does not introduce noticeable modifications in the main stages of fatigue fracture of materials.

In the fracture mechanics of materials, the following two periods of fatigue fracture are mainly distinguished: the period of initiation of macrocracks N_1 combining two initial stages of fatigue fracture, and the period of propagation of the macrocrack N_2, which includes both the period of crack growth to critical sizes and the stage of final fracture. Thus, the lifetime (N) of deformed bodies is determined by the following formula:

$$N = N_1 + N_2. \tag{2.10}$$

The principal difference between these periods can be described as follows. The first period is characterized by the multiple initiation of microcracks in the subsurface layers of the deformed material, determined by the structural properties of the material. The development of these microcracks is mainly controlled by tangential stresses and local properties of the structural elements. In the second period, the fracture process runs near the tip of the macrocrack under the conditions of self-similarity in the plane of action of the maximum tensile stresses. The process of crack propagation depends on the crack growth resistance of the material under the given conditions.

The duration of the period of crack initiation varies within fairly broad ranges. According to different estimates, it constitutes 20...90% of the total lifetime [109, 133]. However, these estimates cannot be regarded as well justified because the accurate definition of the time of initiation of the macrocrack has not been given up to now.

In the engineering practice, the *curves obtained as a result of testing of smooth specimens are used especially extensively as the characteristics of fatigue resistance of the materials* [122, 123]. These are the dependences of the range of acting stresses $\Delta\sigma$ or strains $\Delta\varepsilon$ (or their maximum values σ_{max} and ε_{max} in a cycle) on the number of cycles N to failure within the ranges of possible variations of the indicated quantities. The scale of N on the abscissa axis is, as a rule, logarithmic, while the scale on the ordinate axis may be either logarithmic or uniform. The stress that corresponds to the horizontal asymptote of the fatigue curve is called the fatigue limit of the material (Fig. 2.4).

The nature of the mechanisms of fatigue fracture depending on the level of acting stresses is analyzed according to the presence of inflections in the fatigue curve, the slopes of individual segments to the ON-axis, and the other specific features [51, 122]. The most characteristic segments of the curves are selected, namely, the quasistatic fracture (cyclic creep) (I), low-cycle fatigue (II, III), and high-cycle fatigue (IV,

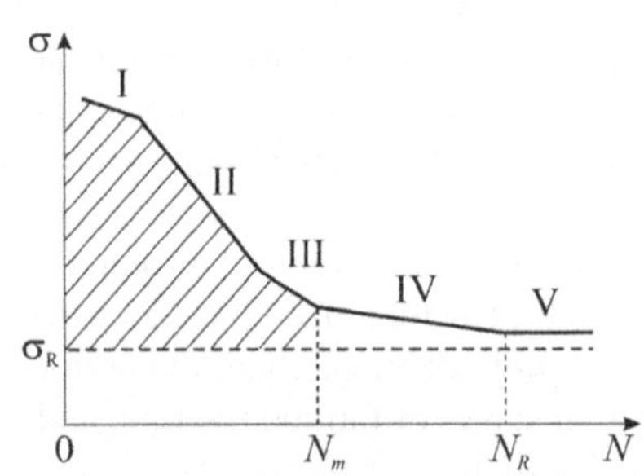

Fig. 2.4 Fatigue diagram of the metal

V). We can also observe the transition zones between these segments. For different materials and loading conditions, the duration of the period of quasistatic fracture may vary from tens to thousands of cycles. The low-cycle region ranges varies from $N \approx 5\times10^2$ to $N_m \approx 10^4$ cycles. The region of high-cycle fatigue also has two segments and corresponds to lifetimes $N > N_m$ cycles.

The presence of an almost horizontal segment V in the curve corresponds to the physical fatigue limit of the material. On a test base of 10^8 cycles, it was discovered in many pure metals with FCC, BCC, and HCP crystal lattices for specimens tested in inert media. At the same time, for numerous metals with FCC lattices, the physical fatigue limit was not found even on a test base of about 10^9–10^{11} cycles. The physical fatigue limit is usually called the fatigue limit. At the same time, in the case where it is absent, the short-time fatigue strength, i.e., the level of stresses σ_R^* on a given test base N^*, is used a characteristic of the material. If the fatigue of metals occurs in certain working media, the properties of the media introduce additional changes to the structure of the fatigue diagram of the metal.[1]

2.4 Cyclic Contact Interactions. Contact Fatigue

Among various joints of the elements of units of machines, buildings, and structures used in the engineering practice, there are two groups of joints (moving and fixed) [32]. The moving joints are, as a rule, characterized by the realization of contact interactions of rolling, rolling with slip, and sliding. The typical features of these joints are the cyclic character of contact processes, large relative displacements of contacting bodies, and low friction between them (rolling bearings, serrated joints (gearings), rolls of rolling mills, wheel-rail system, etc.). As a result of interactions of this kind, the processes of contact rolling fatigue [96, 116, 128] or friction fatigue [118] run in nearsurface zones of the bodies (elements of the joints).

In the engineering practice, fixed joints are often subjected to the in-service vibration or variable cyclic loads as a result of which, elements of the joints and their surfaces compressed by a normal force P are also subjected to the action of an additional force oscillating between the limiting values $\pm F^*$ and acting, in the general case, at an angle α to the normal of the contact surface (Fig. 2.5). In this case, the maximum and minimum values of the normal oscillating force are given by the formula $P^* = P \pm F^* \cos\alpha$, while for the maximum and minimum values of the tangential force, we can write $Q^* = \pm F^* \sin\alpha$ [54]. If $\alpha = 90°$, then we get the conditions of the process of fretting fatigue (fretting wear or fretting corrosion in the cases where the processes of wear or chemical reactions are predominant, respectively) [11, 117, 130]. For $\alpha = 0°$ ($0 \leq \alpha < 90°$), we get the conditions of contact interaction, which is called impulsive contact [11, 96, 116] when even periodic separation of the bodies

[1] V. V. Panasyuk, "H. V. Karpenko and the development of physicochemical mechanics of structural materials," in: *Lviv Scientific School on the Problems of Mechanics of Materials and Materials Science* [in Ukrainian], Spolom, Lviv (2015).

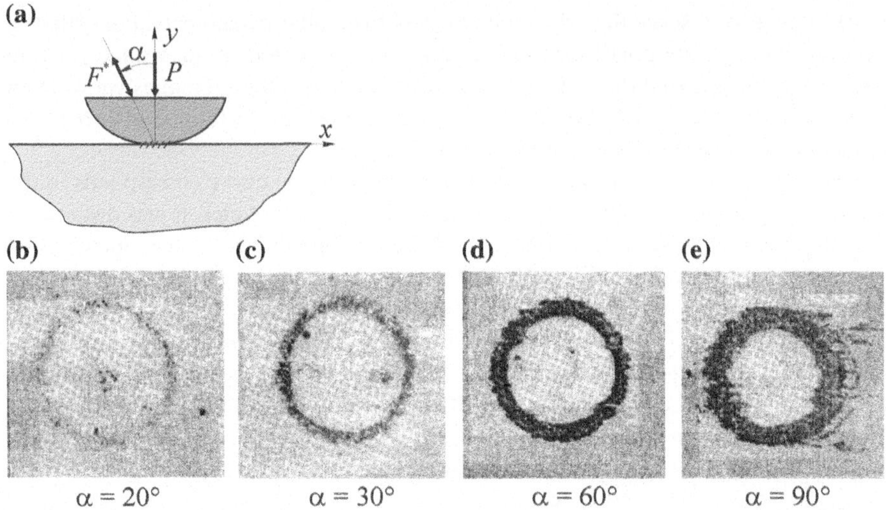

Fig. 2.5 Annular slip and fretting zones in the case of contact of a steel ball with a plane surface formed under the action of oscillating forces applied at an angle to the contact surface [54]

is possible. However, all cases are characterized by the cyclic character of contact, small relative displacements, high friction, and a typical shape of the contact zone (spot) (Fig. 2.5b–e). In the course of impulsive contact, we also observe the fatigue processes similar to the process of fretting fatigue [11, 54, 96, 116].

We now briefly consider the specific features of the process of fatigue of the material the under conditions of cyclic contact of solids (elements of tribojoints). According to the definition proposed by Pinegin [96], the phenomenon of contact fatigue of materials is understood as the process of preparation and development of fracture in the surface layers of the material subjected to long-term variable contact loads. In this case, the maximum variable loads do not cause any noticeable losses in the elastic properties of the material even in the most stressed volumes near and in the zone of force contact of the surfaces.

The stressed state of the material under contact loads is characterized by high gradients of stresses and their localization in small volumes of the material near the surface. This explains the characteristic features of the process of contact fatigue of the material. Thus, the degree and character of deformation of the material undergoes abrupt changes as the distance from the surface increases. In the upper surface layer, especially at the tops of macroasperities, we observe high levels of plastic strains. At the same time, even at depths that are only several times greater than the sizes of the contact zone, the stresses become as low as several tenths or hundredths of the limit of elasticity of the material.

Moreover, the presence of two dangerous zones, namely, the upper surface layer (in view of the stress concentration on the surfaces of irregularities) and the zone of maximal tangential stresses at a critical depth, which is lower than the sizes of the contact region. For high tangential forces in the contact region, these areas can merge into a single zone.

The phenomenon of contact fatigue fracture manifests itself in the form of fatigue cracks responsible (in the process of their development) for the separation of some volumes (particles) of the metal, which violates the process of subsequent operation of the machines. The fatigue fracture of the working surfaces differs from corrosion fracture, abrasive wear of the bodies, plastic changes in the shape (wrinkling) of the surface, and ultralocal fatigue fracture in the upper surface layer but fairly often interacts with these types of fracture and, as a result, accelerates. In the course of development of the theory of contact stresses and experimental investigations of the contact strength of materials, the ideas concerning the typical sites of initiation of the initial fatigue cracks (which is of great importance) changed many times. Depending on the location of initial fatigue cracks, the methods used for the numerical analyses of the components can be refined (with the corresponding increase in the reliability of machines). Moreover, the technology of treatment of the components can also be improved, thus guaranteeing an increase in the quality of machines and the efficiency of their manufacturing.

The analysis of the available experimental data shows that the typical sites of formation of the fatigue cracks can be determined if we consider the specific elements of tribojoints and stable values of the operating parameters. Otherwise, it is difficult to establish the required regularities because the appearance of cracks as a result of contact interaction of solids depends on a great number of factors: first of all, on the type of contact interaction, i.e., the conditions of contact loading, the shape of contacting bodies, the state of their surface layers, the presence of lubricants, residual stresses, thermal modes, defects of the material, etc.

2.5 Typical Contact Fatigue Damages

In the case of rolling of two cylinders with parallel axes over each other, the process of cracking is determined by the contact conditions [39, 96, 97, 116, 125].

During pure rolling (the friction coefficient is $0.005 \leq f \leq 0.025$), the cracks are mainly formed in the subsurface zone [32, 61, 62, 67]. In the process of growth, they appear on the surface. This process of formation of these cracks is associated with the action of maximum tangential stresses τ_{max} at a depth $h \approx (0.7\ldots0.8)a$, where a is the half length of the contact section.

During rolling with slip ($0.025 \leq f \leq 0.4$), the metal is damaged both on the surface and at a depth (Figs. 2.6, 2.7, 2.8, 2.9, 2.10 and 2.11). Although the value of τ_{max} in the subsurface layer can be higher than the stresses acting on the surface, the surface fracture is also intensified by thermal and oxidation processes, and by the influence of lubricants and other working media.

In the tests for rolling contact fatigue, two characteristic types of surface damages, namely, pitting and spalling are observed. Pitting is, in fact, crumbling of some surface areas (Fig. 2.6), which is sometimes accompanied by the cleavages of large

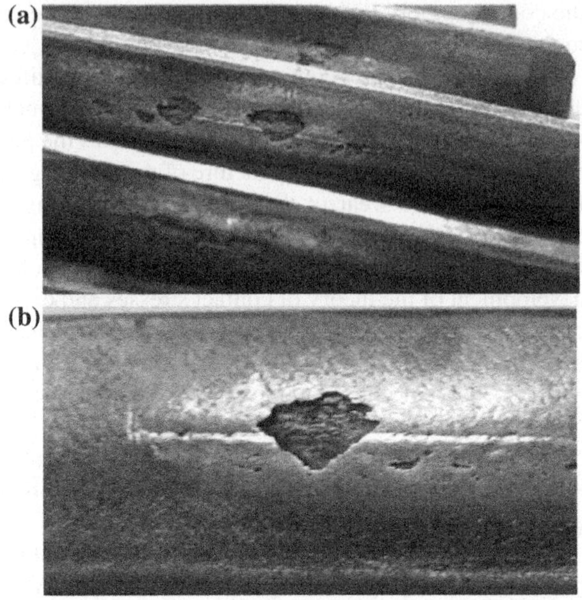

Fig. 2.6 Characteristic morphology of a gear joint teeth surfaces under contact rolling fatigue [46]: **a** individual pits on the teeth surfaces; **b** typical V-shaped pit with an opening angle of 100°

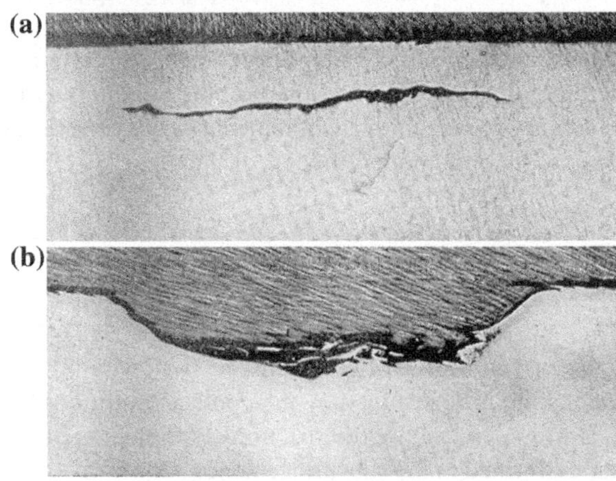

Fig. 2.7 Micrographs of surface layers cross sections of the rolling bearings elements [61]: **a** formation and growth of a subsurface crack; **b** cross sections of a pit formed after spalling; 500×

Fig. 2.8 Gap in a railroad wheel [73]

Fig. 2.9 Spalling on the running surfaces of a drill bit support (large roll and ball) [139]

fragments of the metal. The sizes of crumbling pits (and their number) increase with the number of loading cycles and the size of the contact portion. The formation of pits is possible if a system of inclined surface cracks is developed preliminarily and, moreover, lubricants, water, etc., are present in the rolling bodies contact zone (see Chap. 4). Spalling manifests itself in the form of separation of thin plates of the embrittled metal. Such damage is possible if a subsurface crack parallel to the plane of rolling at a depth comparable with the size of the contact portion is initiated. The growth of this crack results in its appearance on the surface or crossing with an edge crack (Fig. 2.7). Sometimes, the subsurface crack in the rolling body appears at a greater depth. However, for some values of operating parameters (see Chap. 4), the crack appears on the rolling surface and forms a gap (Fig. 2.8). The so-called "encircling" spallings are well known in the engineering practice (Fig. 2.9) [77, 139]. The causes and mechanisms of their formation are now extensively discussed in the literature.

In recent years, engineers and scientists studying the processes of contact rolling fatigue and the problems of rails lifetime focus their attention on two more types of contact fatigue damages: squats ("dark spots") [9, 138] (Fig. 2.10) and checks [25, 137] (Figs. 2.11, 2.12). These types of defects prove to be dangerous by the possibility to evolve into catastrophic transverse cracks.

Finally, it can be stated that a large amount of experimental data on the damages caused by contact fatigue under rolling and slipping are presented in the monographs by Kostets'kyi [60–63].

In the case of pure slipping (friction fatigue), tensile strains play a decisive role in the fracture of surface layers [118]: strain ε_x is formed behind a cylindrical indenter in the direction of its motion, while strain ε_y arises ahead the indenter in the direction perpendicular to its motion. As a result of the repeated alternating deformation of the surface layer in two mutually perpendicular directions, the initiation of fatigue cracks, namely, surface cracks (perpendicular to the direction of the indenter motion) under the influence of the amplitude values of the tensile deformations ε_x and parallel to the surface under the influence of the amplitude values of the tensile deformations ε_y is observed. The development of three-dimensional system of cracks in the result

(a)

(b)

(c)

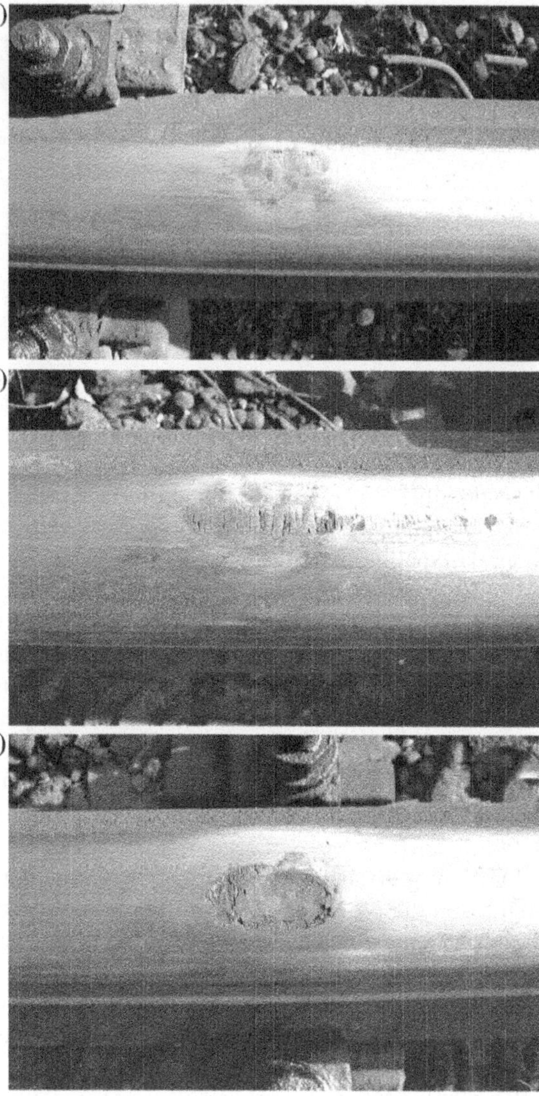

Fig. 2.10 Different stages **a–c** of the “dark spot” growth in the running surface of a rail

of friction fatigue is completed by spalling of a certain volume of the material or transversal crack (Fig. 2.13).

Several characteristic concentric zones with clear boundaries are formed on the contact surface under pulsing contact, (Figs. 2.5e, 2.14) [97]. These are:

Zone I is the central sticking zone with the maximal uniform compression. Here the initial properties of surface layers undergo insignificant changes. Under certain test conditions, even traces of mechanical treatment are reserved on the surface; in the other cases, the insignificant “peeling” of the surface is observed.

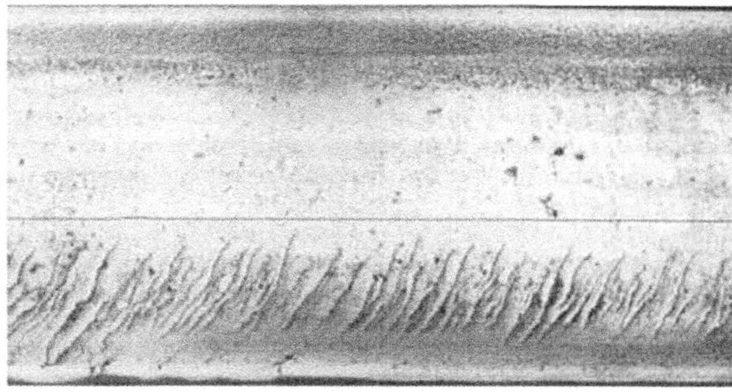

Fig. 2.11 Checks in the rail head are groups of surface cracks formed on the gauge corner of the rail with distances of 0.5–7.0 mm between them [138]

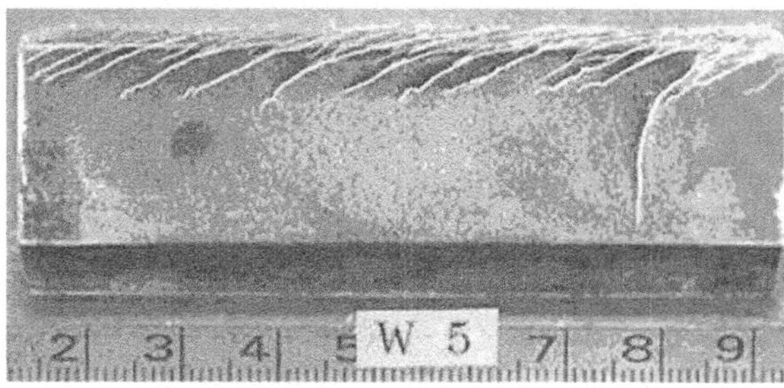

Fig. 2.12 Longitudinal section of a damage (checks) caused by contact rolling fatigue in standard carbon steel [79]

Fig. 2.13 Crack in the brake shoe of a car

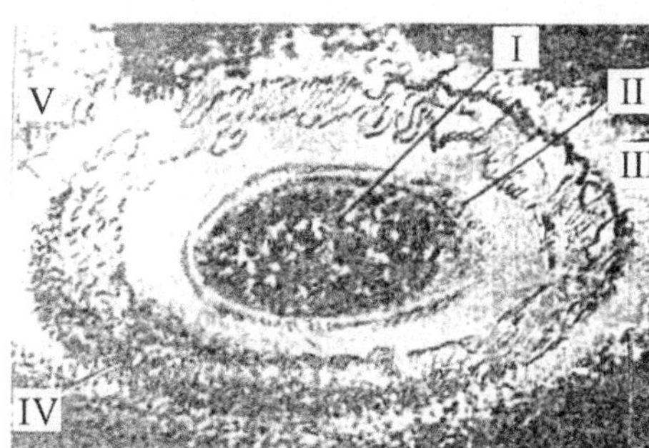

Fig. 2.14 Contact area after long-term pulsating loading [97]

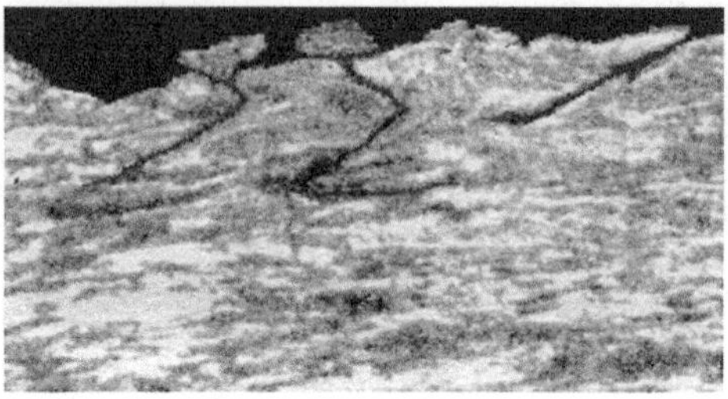

Fig. 2.15 Cracks caused by fretting fatigue on the surface of rope wire [130]

Zone II is the boundary of the sticking zone. It is characterized by the relatively deep fracture of the surface, for which corrosion-mechanical fatigue is regarded as the main cause. This zone becomes especially dangerous under slipping. Here the initial fracture has a discrete (point) character.

Zone III is a slipping zone. Depending on test conditions, different types of interaction (molecular or mechanical) and different types of surfaces fracture caused by seizure, abrasive wear, oxidation wear, and fretting corrosion may be observed.

Zone IV is the boundary of the contact region. This is the zone of maximum radial tensile stresses. In combination with compressive stresses (tangential to the contact contour), these stresses create conditions of pure shear. This promotes the initiation of fatigue cracks.

Zone V is the zone of fracture products removal by oil. This zone located beyond the boundaries of the contact zone.

Thorough investigations of materials fracture under pulsating contact loads were carried out by Pinegin et al. [98], Kennedy [57], Johnson [54], Burton and Russel [8], Tyler et al. [127] and Dawson [30], who studied the surfaces fracture under cyclic compression of steel components.

Comparing the photographs in Figs. 2.5d, e, 2.14, other numerous experimental data [3, 4, 130], and the results of their analysis, it is possible to conclude that the information presented above on the contact "spot" remains true for fretting fatigue. As the main result of this are fatigue cracks (Fig. 2.15). Moreover, on the contacting bodies surfaces, the formation of V-shaped pits and saucer-shaped cavities (see Chap. 5), strongly decreases the fatigue resistance of the components [1, 115, 130]. Note that a large amount of experimental data on fretting fatigue can be found in the monographs by Waterhause [130] and Balatskii [3, 4].

Evaluation of lifetime. During calculation of the machine parts lifetime, it is customary to use (see, e.g., [32, 96]) the endurance fatigue limit (σ_{-1})]), i.e., the stress (stresses) under which the material can withstand a large number of loading cycles without fracture. In the plots of lifetime in σ–N or log σ–log N coordinates (Fig. 2.16), this corresponds to the point of turning the $N(\sigma)$ dependence curve into a straight line parallel to the N-axis. On $N(\sigma)$ curve a point of transition from section of low-cycle fatigue *1* to section of restricted endurance limit *2* conform to σ_{1th} threshold stress of surface elastic fracture [75] or to σ^d_{yld} limit of dynamic fluidity of materials [109].

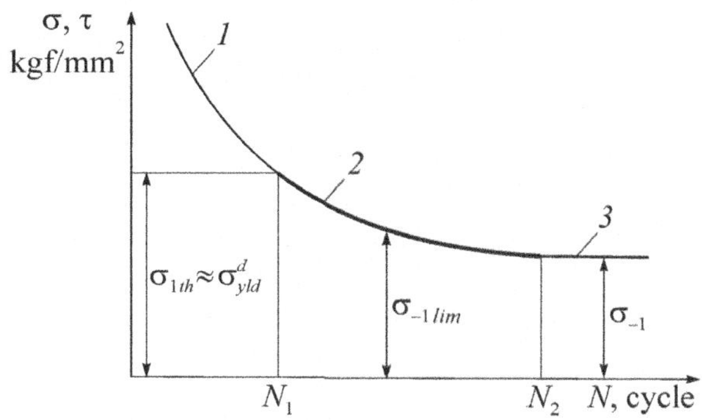

Fig. 2.16 Fatigue curve: (*1*) section of low-cycle fatigue; (*2*) section of restricted endurance limit; (*3*) section of endurance limit

For numerous materials, including certain plastics, alloys, and steels quenched for high hardness, the endurance fatigue limit does not exist, and the fatigue curve monotonically approaches the *N*-axis. Many up-to-date machines operate under severe conditions and, thus, are designed for a limited service life. Hence, the working elements of these machines should not be designed for infinitely many loading cycles. It is reasonable to make them more compact and stressed but evaluate the guaranteed number of loading cycles without fracture more accurately. That is why the values of the temporary (limited) fatigue strength, i.e., the calculated stresses $\sigma_P = \sigma_{-1\ lim}$ are used, what corresponds to one of the points on the curvilinear (or inclined in the logarithmic coordinates) segment of the fatigue curve (Fig. 2.16).

The lifetime can be also estimated using formulas of empirical type

$$N = N(\sigma_p, C_1, C_2, \ldots) \tag{2.11}$$

Formula (2.11) give an analytic description of the fatigue curve whose main argument is the calculated stress σ_p and C_1, C_2, ... are the fatigue characteristics (parameters), including $\sigma_{-1,}$ of the given material as an element of conjugation, established experimentally.

In the case of cyclic rolling contact, the calculated stress $\sigma_p = \varphi(p_0)$ is firstly established using Hertz-Dynnyk-Belyaev formula (see, e.g., [96]); here, p_0 is the maximum value of the Hertzian pressure in the contact region. Depending on the contact interaction and the type of conjugation, the role of the stress σ_p can be played by the maximum tangential, normal, or equivalent stresses at the place of the highest probability of crack initiation. As the constants C_1, C_2,..., parallel with the material fatigue limit σ_{-1}, it is also possible to use the hardness *H* of the material in the surface contact zone.

Later, the experimental curves of contact fatigue were used as a basis for the evaluation of lifetime under fatigue contact. These curves are plotted for each type of contact interaction with regard for (or depending on) main service characteristics of tribojoints and, in particular, the specific features of loading (Fig. 2.17). The experiments are usually performed for every level of loading (e.g., the value of p_0) up to the crack initiation. For lifetime estimation the fatigue limit (σ_R) corresponding

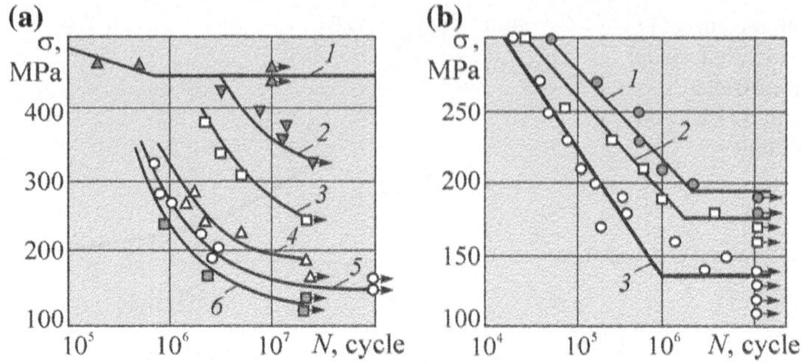

Fig. 2.17 Fretting fatigue [115] (**a**) and friction fatigue [118] (**b**) curves depending on contact load; **a** endurance of Ck35 V alloy specimens (*1*) without fretting corrosion; (*2*), (*3*), (*4*), (*5*), (*6*) specific contact load of 5, 10, 20, 50, and 100 MPa, respectively; slipping amplitude A = 0.1 mm; **b** endurance of 40X steel specimens with polymeric insert and contact pressures: (*1*) 0 MPa, (*2*) 5 MPa, (*3*) 8.5 MPa

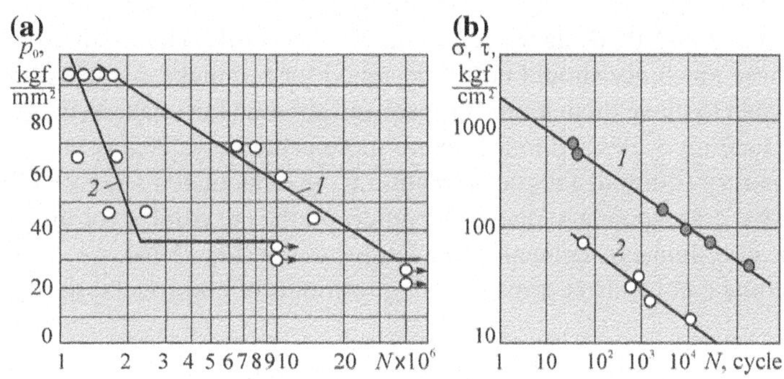

Fig. 2.18 Contact rolling fatigue curves of 9X2MΦ steel specimens [100] (**a**): (*1*) with a zone hardening break; (2) in the normalized state; comparison of the fatigue curves for rubber ($E = 32.5$ kgf/cm^2) [65] (**b**) under the conditions of: (*1*) impulsive contact; (2) friction interaction

the cycle ratio R is introduced into the calculation formula of type (2.11) as the main characteristics.

In recent years, for the contact fatigue calculation analysis, researchers try to use special curves plotted for the cases where experiments are carried out up to the appearance of pitting (Fig. 2.18a) or spalling (Fig. 2.18b) on the contact surface. Here, it is very important to note that just these defects are proposed to be used [1, 43, 48] as determining factors for the evaluation of the lifetimes of bodies under the conditions of cyclic contact.

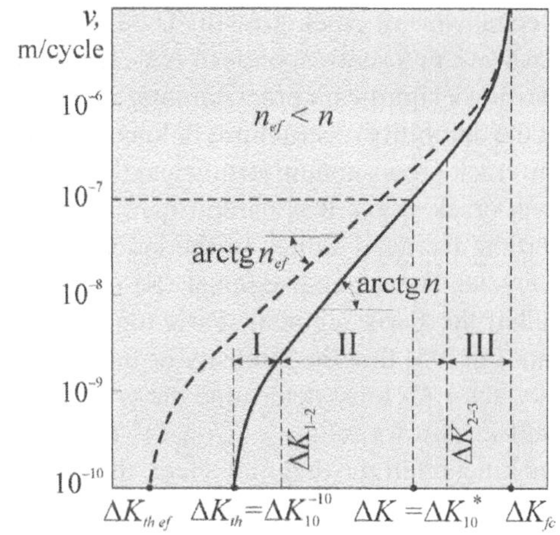

Fig. 2.19 Typical fatigue fracture diagram of the material (solid curve)

2.6 Mechanics of Materials Fatigue Fracture. Fracture Diagrams

The application of the approaches of fracture mechanics, i.e., of the nonclassical approaches to the analysis of the fatigue processes in materials in the stage of propagation of macrocracks, led to the creation of an efficient method for the quantitative description and generalization of the experimentally established regularities. This approach can be described as follows: The load applied to a body, the geometry of the body and the sizes of macrocrack used to determine the crack growth rate can be characterized by the following parameters: one of the stress intensity factors K_{I}, K_{II} and K_{III}, which describes the stress and strain fields near the crack tip for each of the main macromechanisms of crack propagation (see Fig. 2.3) under the well-known restrictions [93, 132]. This is why it is generally accepted to represent the experimental data on the development of fatigue macrocracks in materials in the form of fatigue fracture diagrams (FFD) of the material, which are also called the diagrams of cyclic crack growth resistance and show the dependence of the crack growth rate $v = \Delta l/\Delta N$ (increment Δl of crack length for the corresponding number of cycles ΔN) on the range (difference between the maximum and minimum values) of the SIF in a cycle $\Delta K = K_{\max} - K_{\min}$ or on its maximum value $K_{\max}$ [109, 134]. The complete FFD on the lg v, lg ΔK (or lg $K_{\max}$) logarithmic coordinates are S-shaped curves whose length along the abscissa axis is limited by two vertical asymptotes $\Delta K = \Delta K_{th}$ (or $K_{\max} = K_{th}$) on the left and $\Delta K = \Delta K_{fc}$ (or $K_{\max} = K_{fc}$) on the right (Fig. 2.19), where ΔK_{th} (K_{th}) is the threshold value of the range (or the maximum value) of the SIF below which the crack does not grow; ΔK_{fc} (K_{fc}) is the critical value of the range of the SIF (or $K_{\max}$) above which the crack loses its stability and begins to propagate spontaneously.

In the kinetic FFD, there are three important sections, which, as follows from microfractographic and microstructural investigations, correspond to different mechanisms of crack growth. The first curvilinear segment (I) of low crack growth rates ($v < v_{1\text{–}2}$) corresponds to ΔK close to the threshold values ΔK_{th} (see Fig. 2.19). Despite a significant practical interest, this segment is poorly studied, although 90% of the durability of structures belongs to this range of crack growth rates. In this stage, the crack grows nonuniformly and the jumps of increase of the crack length alternate with crack stops. It is natural that, in the first segment of the diagram, we present certain averaged values of the crack growth rate from the range $v = 10^{-11}$–10^{-9} m/cycle. The second segment (II) of the FFD ($v_{1\text{–}2} \le v \le v_{2\text{–}3}$) is rectilinear; it is called the Paris segment. Paris together with his colleagues showed (for the first time) in [95] that the intensity of the stress field near the crack tip characterized by the factor K_I must determine the crack growth rate and then proposed (in [94]) the well-known formula $v = C(\Delta K)^n$ for the description of the rectilinear segment of the kinetic fatigue diagram. Here, the constant n is the slope of this straight line and the constant C specifies its position along the ordinate axis for $\Delta K = 1$. Later [131], it was proposed to write this equation in a more convenient form

$$v = C(\Delta K)^n = 10^{-7}\left(\Delta K / \Delta K^*\right)^n, \quad \Delta K_{1\text{–}2} < \Delta K < \Delta K_{2\text{–}3}, \tag{2.12}$$

where ΔK^* is the value of ΔK for $v = 10^{-7}$ m/cycle (see Fig. 2.19) and has the dimensions of the SIF and a clearer physical meaning unlike the constant C whose dimensions depend on n. The length of the second segment of the diagram determined by the values of the SIF $\Delta K_{1\text{–}2}$ and $\Delta K_{2\text{–}3}$, and depends on the material and testing conditions. In most cases, $v_{1\text{–}2}$ lies within the range 10^{-9}–10^{-8} m/cycle and $v_{2\text{–}3}$ lies within the range 5×10^{-7}–5×10^{-6} m/cycle. The third segment (III) of the FFD is the segment of rapid crack growth ($v > v_{2\text{–}3}$); it is curvilinear, and its slope increases as ΔK approaches the critical value of ΔK_{fc}. The value of ΔK_{fc}, obtained under cyclic loading can be lower or higher that the values of K_{IC}, K_{IIC} and K_{IIIC}, determined under static loading. In this segment, the typical process of crack propagation is jumpwise [109].

For the analytic description of complete FFD (within the range from ΔK_{th} to ΔK_{fc}), numerous dependences were proposed in [84, 109]. Thus, in particular, the following fairly general dependence was proposed by Romaniv et al. in [109]:

$$v = AK^s_{\mathrm{Imax}}\left(K^{m_1}_{\mathrm{Imax}} - K^{m_2}_{th}\right)^{m_3} \Big/ \left(K^{m_4}_{fc} - K^{m_4}_{\mathrm{Imax}}\right)^{m_5}, \tag{2.13}$$

where A, s and m_i ($i = 1, 2, 3, 4, 5$) are constants for a given material and testing conditions. A simpler formula

$$v = v_0\left(\frac{\Delta K - \Delta K_{th}}{\Delta K_{fc} - \Delta K}\right)^q, \quad \Delta K_{th} \le \Delta K \le K_{fc}, \tag{2.14}$$

used later (in Chaps. 4 and 5) was proposed by Yarema and Mikitishin [135].

The electron-fractographic analysis of fatigue fractures revealed [82, 103] the presence of various micromechanisms of crack growth depending on the level of loading. In these fracture surfaces, we see various microscopic features, such as banded structure, striation, coalescence of pores, and rupture of the bridges between them (formation of pits), spalling inside and between the grains, etc. The mechanism of striation is mainly typical of the second segment of the diagram [82, 108]. For low values of ΔK (the first segment of the kinetic diagram), the banded structure and spalling areas along slip planes (ductile cleavage) are typical microscopic topographies of the fracture surfaces. In this case, we sometimes observe the process of growth crack by the cleavage mechanism. However, cleavage inside grains and between them and the pit mechanism are more typical of the third segment of the FFD, where the size of the plastic zone near the crack tip significantly exceeds the structural parameter of the material (e.g., the grain size) and the micromechanisms typical of the fracture under loading are realized. In the first and third segments of the diagram, the growth rate of fatigue macrocrack strongly depends on the microstructure of the material, the parameters of loading cycles, and the ambient medium [103, 108].

In the evaluation of the lifetime of structures, the kinetic fatigue diagrams play the same role as tensile diagrams in the evaluation of static strength. On the basis of the FFD, we establish the characteristics of cyclic crack growth resistance of materials according to which it is possible to choose the materials for various structures and determine the influence of the operating conditions on the serviceability of materials, etc. Under certain assumptions, on the basis of these diagrams, it is possible to find the lifetime of structures weakened by cracks, i.e., the number of cycles to failure as a result of integration of the equation of crack growth rate

$$N_2 = \int_{l_0}^{l_c} dl / v[\Delta K(l)], \qquad (2.15)$$

where l_0 and l_c are the initial and critical crack lengths, respectively.

The development of fatigue fracture is a complex process depending on numerous factors and, therefore, it is important to determine parameters used for its description. The role of these parameters is, to a large extent, played by the characteristics of cyclic crack growth resistance of the materials. The standards required for the determination of the corresponding characteristics were created in [102]. It is known that, as the characteristics of cyclic crack growth resistance of materials in the stage of crack propagation, it is customary to use the threshold value ΔK_{th}, the critical value ΔK_{fc}, and the parameters C and n or ΔK^* and n. The values of $\Delta K_{1\text{-}2}$, $\Delta K_{2\text{-}3}$ and ΔK, at a certain crack growth rate serve as auxiliary characteristics. In particular, in most cases, the value of ΔK for $v = 10^{-10}$ m/cycle (ΔK_{10}^{-10}) is chosen in practice as the threshold value ΔK_{th}. At present, there is no generally accepted method for the evaluation of the critical value ΔK_{fc} (cyclic crack growth resistance of the final fracture). It was proposed [109] to assume that the time of onset of jumpwise crack growth corresponds to the beginning of unstable development of fatigue fracture. However, this phenomenon is not typical of many materials. This is why the quantity

ΔK_{fc} is specified as the value of ΔK for sufficiently high values of fatigue crack growth rates, e.g., $v = 10^{-5}$ m/cycle (ΔK_{10}^{-5}).

It has been recently established that crack closure is one of the most important phenomena of the subthreshold growth of fatigue cracks [105]. This phenomenon can be described as follows: a fatigue crack may closed due to the joining of its faces at a certain distance from the tip in the tensile part of a loading cycle. In this case, the stressed state near the crack tip is determined not by the range of the SIF (ΔK) but by a certain "efficient" range $\Delta K_{ef} = K_{max} - K_0$, where K_0 is the minimum SIF for which the crack opens near its tip. According to Elber [35] who studied this phenomenon for the first time, it is caused by plastic yield of the material near the crack tip. However, as follows from the results of performed investigations (see [109]), the crack closure can be caused by several different processes: the action of residual compressive stresses caused by variable plastic strains near the crack tip, the roughness of the fracture surface, the formation of a layer of oxides in the crack mouth in the process of fretting, the phase transformations in the process zone, etc. The effect of crack closure can be evaluated [106] by the coefficient k_3 determined from the relation

$$k_3 = (K_{\mathrm{Imax}} - K_0)/(K_{\mathrm{Imax}} - K_{\mathrm{Imin}}) = \Delta K_{ef}/\Delta K.$$

In numerous cases, the diagrams plotted on the basis of effective ranges $\Delta K_{ef} = k_3 \Delta K$ (see the dashed curve in Fig. 2.19) within the boundaries of the first and, partially, the second segments appear to be less variable than the conventional diagrams (v, ΔK). Sometimes they remain absolutely invariant under changes in the shapes and sizes of the specimens, asymmetry of the loading cycle, structure of the material, action of the medium, etc.

Note that, at present, the methods aimed at testing materials for cyclic crack growth resistance in the presence normal opening (mode I) cracks (see Fig. 2.3) are well developed and brought to the level of standards [2, 102]. Moreover, there are numerous efficient procedures of testing for the cyclic crack growth resistance in the case of longitudinal shear (mode III) cracks (see, e.g., [53]). However, there are only separate procedures [6, 77] that can be used for transversal shear (mode II) cracks.

2.6.1 Calculation Model for Estimation of Movable and Immovable Joints Contact Lifetime by Criteria of Contact Fatigue Damages Formation

In the 1970–90s, the first theoretical models of the mechanics of fatigue contact fracture, namely, the computational models for the evaluation of the lifetime of elements of moving and immobile (fixed) joints, were developed. For the case of contact rolling interaction, various models were proposed by Keer and Bryant [56]; Hearly and Johnson [47]; Kaneta and Murakami [55]; Sheppard et al. [114], and Bower

[7]; Glodež et al. [42]. For the case of fretting-fatigue interaction, the computational models were proposed by Rooke and Jones [110]; Dekhovich and Makhutov [31]; Dubourg and Villechaise [33]; Hills and Nowell [49]; Faanes [36]; Fernando et al. [38], and Troshchenko and Tsybanev [124]. Somewhat later, it was shown that the proposed model inaccurately take into account the action of operating factors in tribojoints (residual stresses, lubrication, effect of crack closure, friction between the crack faces, temperature, etc.). For this reason, new models of contact rolling interaction were formulated by T.L. Anderson, J.H. Beynon, S. Bogdanski, A. Ekberg, M. Endo, G. Faidiga, J. Flašker, D.I. Fletcher, M.F. Frolish, S. Glodež, S. Hamada, T. Hellen, K. Ichimaru, E. Kabo, M. Kaneta, I.I. Kudish, T. Morita, Y. Murakami, M. Olzhak, J.W. Rinsberg, C. Sakae, J. Stupnicki, W. Tyfour and other researchers [5, 34, 37, 41, 69, 76–78, 104, 126]. For the contact fretting fatigue interaction, new models were proposed by M. Ciavarella, G. Demelio, D. Dini, J. Dominguez, M.-C. Dubourg, S. Faanes, T.N. Farris, M. Garcia, D.A. Hills, M. Kuno, V. Lamacq, A. Navarro, D. Nowell, M.P. Sholwinski, C. Vallelanio, R.B. Waterhause and other researchers [10, 36, 71, 72, 80, 81, 119, 129].

The proposed models are based on the concepts of fatigue fracture mechanics, the solutions of contact problems of the theory of elasticity for bodies with cracks, and the analysis of the main operating parameters of various types of contact interaction in tribojoints. In the major part of these models, the path of crack propagation and durability are predicted in a simplified way according to the growth of a rectilinear shear (edge or subsurface) macrocrack in the contact zone and later, when the crack penetrates into the zone of tension, according to the direction (angle) of its initial deviation in an elastic half-plane. In a specific case discussed in [37, 42], the authors evaluate the durability of fear teeth flanks according to the development of micropitting by using a curvilinear path of propagation of an edge crack constructed by them. Actually, under the conditions of fatigue contact, cracks grow curvilinearly in the contact zone and realize various types of fracture of the products. Crumbling often occurs as a result of the evolution of a system of cracks.

In the literature, one can also find the models based on the probabilistic approach developed by Tallian [120, 121], Kudish [66, 68, 70] and Dang and Maitournam [12]. However, these models prove to be inefficient for the evaluation of the lifetime of tribojoints. Note that the Dang Van model based on the use of the elements of multiaxial fatigue has found fairly extensive applications.

In what follows, we describe a computational model for the investigation of the propagation of macrocracks and evaluation of the residual lifetime of solids in the case of their contact interaction. The fracture criteria for the material in the process of growth of fatigue cracks and the algorithms of step-by-step construction of the paths of crack propagation, developed with the use of singular integral equations of two-dimensional contact problems of the theory of elasticity for bodies with curvilinear cracks serve as a basis of the model [15, 16, 25, 26, 88, 91, 92, 111]. The model takes into account the redistribution of stresses caused by crack propagation and the variations of loading in a contact cycle. Elements of the model were partially published earlier, in particular, for the general case of cyclic contact in [18–20, 27,

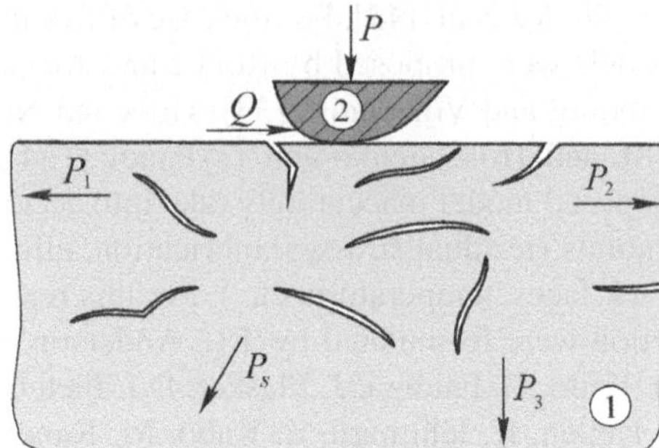

Fig. 2.20 General scheme of the model

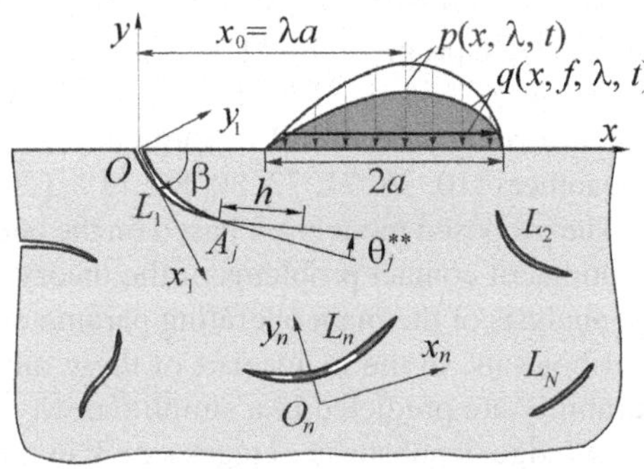

Fig. 2.21 Computational scheme of the model

86, 87], for the rolling fatigue in [13, 21, 24, 26, 29, 89] and for the fretting fatigue in [16, 21, 22, 90].

Main statements of the model [20]. Consider a problem of cyclic contact interaction of two bodies one of which is damaged by cracks. We model this body by an elastic half-plane weakened by a system of cracks (Fig. 2.20). The second body (counterbody) is modeled either as a rigid punch (Fig. 2.20) or as a set of normal $p(x, \lambda, t)$ and tangential $q(x, f, \lambda, t)$ forces distributed over the boundary of the half-plane (Fig. 2.21). These forces depend on the shape and sizes of the counterbody, mechanical characteristic of the materials and contact surfaces, and the features of contact interaction in a contact cycle (as functions of time t). As the most extensively used force schemes of model loading $p(x)$, we can mention a concentrated force, constant pressure, elliptic (Hertzian) distribution of forces, etc. The friction forces are taken into account as the tangential forces $q(x, f, \lambda, t)$. The simplest distribution of tangential forces is represented by their relationship with normal forces according to the Amonton-Coulomb law $q(x) = fp(x)$ under the conditions of complete slip between the bodies, where f is the friction coefficient for these bodies.

As already indicated (Sect. 2.3), in the fatigue fracture mechanics of materials, the lifetime N of a damaged body is evaluated by analyzing two components: N_i, i.e., the period (number of loading cycles) prior to the initiation of macrocrack l_0, and N_g, i.e., the period of growth of the macrocrack from the initial length up to the critical (admissible) length l_c. The period N_g is called the residual lifetime. We now consider this period.

By analyzing the experimental data [6, 41, 74, 77], we conclude that, in the contact zone, the (edge and subsurface) macrocracks first rectilinearly develop by the mode II mechanism and then curvilinearly propagated by the mode I mechanism. Thus, we can represent the residual lifetime N_g of cyclically contacting bodies with cracks in the form

$$N_g = N_{g\tau} + N_{g\sigma}, \tag{2.16}$$

where

$$N_{g\tau} = \int_{l_{0\tau}}^{l_{c\tau}} v^{-1}\left(\Delta K_\tau(l),\ C_{1\tau},\ \ldots,\ C_{m\tau}\right) dl,$$
$$N_{g\sigma} = \int_{l_{0\sigma}}^{l_{c\sigma}} v^{-1}\left(\Delta K_\sigma(l),\ C_{1\sigma},\ \ldots,\ C_{m\sigma}\right) dl. \tag{2.17}$$

Here, $N_{g\tau}$ and $N_{g\sigma}$ are the lifetimes in the stage of macrocrack propagation by the mode II and mode I mechanisms, respectively; $l_{0\tau}$, $l_{0\sigma}$, $l_{c\tau}$ and $l_{c\sigma}$ are the initial and admissible (critical) lengths of the macrocrack in the stages of shear and normal opening, respectively, $v = dl/dN$ is the crack growth rate, l is the length of propagating crack. The parameter of the stress-strain state $K(l, \lambda, t, \theta)$ responsible for fracture at the crack tip is chosen with regard for the probable fracture mechanism. We choose the dependences $v(\Delta K)$ according to the experimental data in the form of fatigue fracture diagrams (FFD) of the material. On the basis of these diagrams, we also determine the constants $C_1, \ldots, C_m$ characterizing the cyclic crack growth resistance of the material.

Within the framework of linear fracture mechanics, we determine the parameter $K(l, \lambda, t, \theta)$ via the stress intensity factors (SIF) K_{I} and K_{II} from the relations of the corresponding criterion of local fracture:

$$K = K\left(l,\ \lambda,\ t,\ \theta^*\right) = \tilde{K}\left[K_{\mathrm{I}}(l,\ \lambda,\ t),\ K_{\mathrm{II}}(l,\ \lambda,\ t)\,,\theta^*(l,\ \lambda,\ t)\right];$$
$$\theta^* = F(l,\ \lambda,\ t) = \tilde{F}[K_{\mathrm{I}}(l,\ \lambda,\ t),\ K_{\mathrm{II}}(l,\ \lambda,\ t)]; \tag{2.18}$$

here, θ is a polar angle measured from the tangent to the crack at its tip (see, e.g., Figure 2.2) and θ^* is the angle at which the parameter K attains its extreme value (maximum in modulus) for fixed l, λ and t. Since the stresses and, hence, the quantities $K(l, \lambda, t, \theta^*)$ and $\theta^*(l, \lambda, t)$ vary in a contact cycle, it is possible to assume that the crack grows in a cycle only at the time when the parameter $K(l, \lambda, t, \theta^*)$ attains its extreme value both in the angle θ and in the arguments t and λ (i.e., for the values, $t = t^*$ and $\lambda = \lambda^*$). Then the direction of growth of the crack at the point A (Fig. 2.21) is determined by the angle

$$\theta^{**} = \theta^{*}(l, \lambda, t^{*})$$

In this case, the range of the parameter ΔK in a contact cycle must be larger than the range of the threshold of fatigue growth of the crack in the material ΔK_{th}, i.e., the following conditions must be satisfied:

$$\max\left|K(l, \lambda, t, \theta^{*})\right| = \left|K(l, \lambda^{*}, t^{*}, \theta^{**})\right|; \tag{2.19}$$

$$\Delta K = \max K(l, \lambda, t, \theta^{**}) - \min K(l, \lambda, t, \theta^{**}) \geq \Delta K_{th}. \tag{2.20}$$

In the stage of propagation of the macrocrack by the mechanism of shear, the maximum shear stresses are assumed to be responsible for fracture at the crack tip. Thus, relations (2.18) for the corresponding criterion take the form [55]:

$$K = K_{\mathrm{II}\theta}(l, \lambda, t, \theta^{*}) = \frac{1}{2}\cos\frac{\theta^{*}}{2}\left[K_{\mathrm{I}}(l, \lambda, t)\sin\theta^{*} + K_{\mathrm{II}}(l, \lambda, t)(3\cos\theta^{*} - 1)\right];$$
$$2K_{\mathrm{II}}(l, \lambda, t)\mathrm{tg}^{3}\frac{\theta^{*}}{2} - 2K_{\mathrm{I}}(l, \lambda, t)\,\mathrm{tg}^{2}\frac{\theta^{*}}{2} - 7K_{\mathrm{II}}(l, \lambda, t)\,\mathrm{tg}\frac{\theta^{*}}{2} + K_{\mathrm{I}}(l, \lambda, t) = 0. \tag{2.21}$$

The conditions of crack growth (2.19) and (2.20) are specified as follows:

$$\max\left|K_{\mathrm{II}\theta}(l, \lambda, t, \theta^{*})\right| = \left|K_{\mathrm{II}\theta}(l, \lambda^{*}, t^{*}, \theta^{**})\right|; \tag{2.22}$$

$$\max K_{\mathrm{II}\theta}(l, \lambda, t, \theta^{**}) - \min K_{\mathrm{II}\theta}(l, \lambda, t, \theta^{**}) \geq K_{\mathrm{II}th}. \tag{2.23}$$

Here, $K_{\mathrm{II}th}$ is the threshold of fatigue growth of the macrocrack under the conditions of transverse shear. Condition (2.23) is also the condition of transition from the stage of macrocrack initiation to the stage of its growth by the shear mechanism and can be used for the evaluation of the initial (nucleus) crack length $l_{0\tau}$.

In the second stage of macrocrack propagation by the mode I mechanism, the maximum tensile hoop stresses are responsible for fracture. In this case, we describe the parameter $K(l, \lambda, t, \theta^{*})$ by using the relation of generalized criterion of normal opening (σ_θ-criterion) [84]:

$$K = K_{\mathrm{I}\theta}(l, \lambda, t, \theta^{*}) = \cos^{3}\frac{\theta^{*}}{2}\left[K_{\mathrm{I}}(l, \lambda, t) - 3K_{\mathrm{II}}(l, \lambda, t)\,\mathrm{tg}\frac{\theta^{*}}{2}\right];$$
$$\theta^{*} = 2\mathrm{arctg}\frac{K_{\mathrm{I}}(l, \lambda, t) - \sqrt{K_{\mathrm{I}}^{2}(l, \lambda, t) + 8K_{\mathrm{II}}^{2}(l, \lambda, t)}}{4K_{\mathrm{II}}(l, \lambda, t)}. \tag{2.24}$$

Hence, the conditions of crack growth take the form

$$\max K_{\mathrm{I}\theta}(l, \lambda, t, \theta^{*}) = K_{\mathrm{I}\theta}(l, \lambda^{*}, t^{*}, \theta^{**}); \tag{2.25}$$

$$\max K_{\mathrm{I}\theta}(l,\ \lambda,\ t,\ \theta^{**}) - \min K_{\mathrm{I}\theta}(l,\ \lambda,\ t,\ \theta^{**}) \geq K_{\mathrm{I}th}; \tag{2.26}$$

where $K_{\mathrm{I}th}$ is the threshold of fatigue growth of the macrocrack by the normal opening. Condition (2.26) can be also treated as the condition of transition from the stage of macrocrack propagation by the mechanism of shear to the stage of growth by the normal opening mechanism and used to determine the critical length of the shear macrocrack $l_{c\tau}$. In general, the shear crack can also grow if condition (2.26) is already satisfied; in this case, relation (2.26) turns into the branching condition. Here, we assume that $l_{0\sigma} = l_{c\tau}$. In order to determine the critical length $l_{c\sigma}$, we use the following conditions:

$$\Delta K_{\mathrm{I}\theta} = \Delta K_{\mathrm{I}fc}\ \text{(a)}\ \text{ or }\ \Delta K_{\mathrm{I}\theta} = \Delta K_{\mathrm{I},2\text{-}3}\ \text{(b)}, \tag{2.27}$$

where $\Delta K_{\mathrm{I}fc}$ is the critical range of the SIF K_{I} upon attainment of which the crack begins to grow spontaneously.

The transient conditions (2.23) and (2.26), which can be used to find the crack lengths $l_{0\tau}$, $l_{c\tau}$ and $l_{0\sigma}$ depending on the structure of the formulas that describe the diagram of fatigue crack growth rate in the material in relations (2.17), were applied in our calculations in a different form, namely,

$$\Delta K_{\mathrm{II}\theta}(l_{0\tau}) = \Delta K_{\mathrm{II},v_{th}}\ \text{(a)}\ \text{ or }\ \Delta K_{\mathrm{II}\theta}(l_{0\tau}) = \Delta K_{\mathrm{II},1-2}\ \text{(b)}, \tag{2.28}$$

and

$$\Delta K_{\mathrm{I}\theta}(l_{0\sigma}) = \Delta K_{\mathrm{I},v_{th}}\ \text{(a)}\ \text{ or }\ \Delta K_{\mathrm{I}\theta}(l_{0\sigma}) = \Delta K_{\mathrm{I},1-2}\ \text{(b)}, \tag{2.29}$$

where $\Delta K_{v_{th}} = \Delta K_{10^{-10}}$ is the range of the conventional SIF threshold [109] corresponding to a fatigue crack growth rate $v_{th} = 10^{-10}$ m/cycle and $\Delta K_{1\text{–}2}$ and $\Delta K_{2\text{–}3}$ are the SIF ranges corresponding to the lower and upper bounds of the Paris rectilinear segment in the fatigue fracture diagram of the material.

Algorithm for the construction of the propagation paths of fatigue cracks in the zone of cyclic contact interaction of solids. The methods of step-by step evaluation of the quasistatic paths of crack propagation in the two-dimensional case were proposed by Savruk and Osiv [112] and Yarema, and Zboromyrskii [136]. Later, these methods were developed in [25, 88] and generalized for the case of problems of cyclic contact interaction of solids [13, 24, 28, 90].

The analysis of the crack path (trajectory) by the step-by-step method is performed by using the solutions of the corresponding singular integral equations. We determine the geometry of propagating cracks with regard for the redistribution of stresses at the crack tips in the course of crack propagation and control its growth rate by the range of the parameter (stress intensity factor of mixed type) in a loading cycle, which is connected with the SIF (K_{I} and K_{II}) in a certain way depending on the mechanism (criterion) of local fracture at the crack tip (see relation (2.18)). For simplicity, we

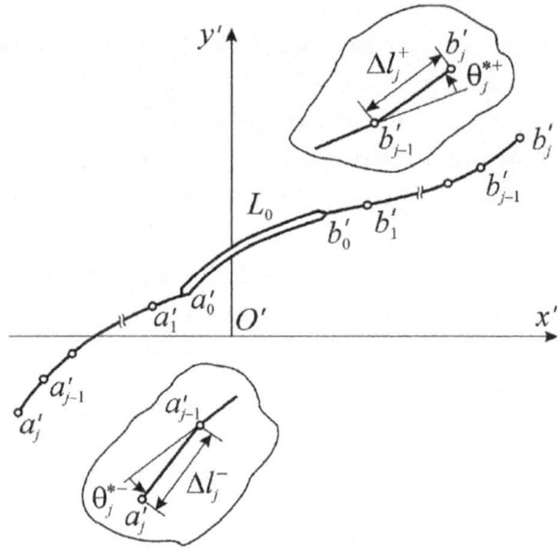

Fig. 2.22 Scheme of the algorithm for the construction of the fatigue cracks propagation paths

assume that the crack propagates in the absence of contact of its faces and branching at the tips.

Assume that a plate weakened by a curvilinear crack is subjected to the action of cyclic loads and that the contour of the initial crack referred to a local coordinate system *x'O'y'* have its origin at a point a_0' and the end at a point b_0' (Fig. 2.22).

The numerical analysis of the path of fatigue crack is performed step by step. Suppose that the crack is subjected to the action of external loading and begins to grow from both tips in directions specified by certain angles θ_1^{*+} and θ_1^{*-}. These angles are determined via the SIF $K_{\mathrm{I}}^{\pm}$ and $K_{\mathrm{II}}^{\pm}$ on the basis of the solution of the singular integral equation of the plane problem of the theory of elasticity for the analyzed domain (the superscripts "+" and "–" correspond to the right and left tips of the crack, respectively). Note that the relation

$$f\left(K_{\mathrm{I}},\ K_{\mathrm{II}},\ \theta^*\right) = 0 \tag{2.30}$$

specifying the direction of crack growth in each stage via the SIF at the crack tip, is established from the chosen criterion of local fracture (see, e.g., relations (2.21) and (2.24)). If we now assume now that, in the process of growth, the crack got small increments of length Δl_1^+ and Δl_1^- from the right and left tips, respectively, then the relationship between them is established by the formula for the crack growth rate

$$\frac{\Delta l}{\Delta N} \approx v(\Delta K,\ C_1,\ \ldots,\ C_m), \tag{2.31}$$

where Δl is the increment of crack length for ΔN loading cycles; ΔK is the range of the parameter K regarded as constant for ΔN loading cycles, and $C_1, \ldots, C_m$ are the characteristics of crack growth resistance of the material.

As a result of the solution of the singular integral equation of the problem for a plate with newly formed crack, we find the SIF $K_{\mathrm{I}}^{\pm}$ and $K_{\mathrm{II}}^{\pm}$ in the second stage of construction of the path. By using these SIF, we determine new values of the angles of deviation of the crack ($\theta_2^{*\pm}$), the ranges of the parameters K ($K_2^{\pm}$), and current increments of the path ($\Delta l_2^{\pm}$). Consecutively repeating the outlined procedure, we eventually arrive at a crack path in the form of a broken line (Fig. 2.22). In this case, to establish the relationship between the increments of the right (+) and left (−) branches of the path in the k-th stage, by using formula (2.31), we obtain

$$\frac{\Delta l_k^+}{\Delta l_k^-} = \frac{v_k^+}{v_k^-}, \qquad k = 1, \ldots, j. \tag{2.32}$$

It should be emphasized that, for the sake of simplicity, in the process of construction of the path, the increment at one of the tips can be regarded as constant and independent of the velocity of the motion of the tip. Then setting a constant increment ($\Delta l_k^+ = H = \text{const}$, $k = 1, \ldots, j$), e.g., at the right tip, we can find the increment of the path at the left tip in the next stage according to relation (2.31) by the following formula: $\Delta l_k^- = \frac{v(\Delta K^-,\ C_1,\ \ldots,\ C_m)}{v(\Delta K^+,\ C_1,\ \ldots,\ C_m)} H, \qquad k = 1, \ldots, j.$

In the presence of kinks on the path, the procedure of construction of the solution by the method of singular integral equations becomes much more complicated. In this case, in order to use the algorithm efficiently, it is reasonable to approximate segments of the path in the vicinity of the angular points by smooth curves. We now consider the case of construction, e.g., of the right branch of the path or of the path of edge crack in more detail.

In the next stage of construction of the path, we determine the abscissa of the right crack tip after its increment by the formula [13, 14]

$$x_j' = x_0' + \sum_{k=1}^{j} h_k, \tag{2.33}$$

where $x_0' = \mathrm{Re}\, b_0'$ is the abscissa of the right tip of the initial crack, $h_k = H \cos \gamma_k$ is the increment of the abscissa in each stage of construction of the path, and γ_k is the angle between the $O'x'$-axis and the direction of crack increment (Fig. 2.23).

In the first stage, the equation of the contour coincides with the equation of the initial crack. We approximate the next increments of the path at the right tip by the third-degree polynomials [112]

$$y_j'(x') = a_j(x' - q_j)^3 + b_j(x' - q_j)^2 + c_j(x' - q_j) + d_j, \qquad q_j \le x' \le x_j', \tag{2.34}$$

where

$$q_j = x_{j-1}' - \delta_j \qquad (\delta_j \ge 0) \tag{2.35}$$

Fig. 2.23 Scheme of the path construction in j-th stage

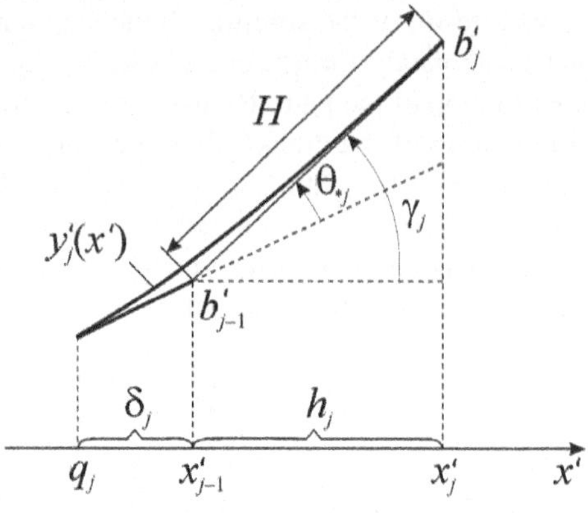

was is a parameter introduced with to smooth the path on the segment $\left[q_j,\ x'_j\right]$ and

$$\delta_j = \begin{cases} h_j/2, & \theta_j^{*+} \neq 0; \\ 0, & \theta_j^{*+} = 0. \end{cases} \tag{2.36}$$

Assume that

$$y' = \Gamma_j\left(x' - x'_{j-1}\right) + Y_{j-1} \qquad \left(\Gamma_j = \operatorname{tg}\gamma_j,\quad Y_{j-1} = y'_{j-1}\left(x'_{j-1}\right)\right) \tag{2.37}$$

is the equation of a straight line passing through the tip $(x_{j-1},\ Y_{j-1})$ in the direction of crack propagation, which is determined by the angle θ^*. Then the unknown coefficients of Eq. (2.34) can be found from the conditions that curve (2.34) is smoothly connected with the curve $y' = y'_{j-1}(x')$ and the straight line (2.37) at the points $x' = q_j$ and $x' = x'_j$. Hence, we get

$$a_j = \frac{(c_j + \Gamma_j)(h_j + \delta_j) + 2(d_j - Y_{j-1} - \Gamma_j h_j)}{(h_j + \delta_j)^3};$$
$$b_j = \frac{3(Y_{j-1} - d_j + \Gamma_j h_j) - (\Gamma_j + 2c_j)(h_j + \delta_j)}{(h_j + \delta_j)^2};$$
$$c_j = \left.\frac{dy'_{j-1}(x')}{dx'}\right|_{x'=q_j}; \qquad d_j = y'_{j-1}(q_j). \tag{2.38}$$

Similar results can be obtained for the left branch of the path.

Indeed, in the next stage of construction of the path, after the crack increment, we determine the abscissa of the left crack tip by the formula

$$\tilde{x}'_j = \tilde{x}'_0 - \sum_{k=1}^{j} \tilde{h}_k, \tag{2.39}$$

where $\tilde{x}'_0 = \operatorname{Re} a'_0$, $\tilde{h}_k = \Delta l_k^- \cos \gamma_k$, Δl_k^- is an increment determined by equality (2.32), and γ_k is the angle between the $O'x'$-axis and the direction of the next increment of the crack.

The approximating polynomials for the left crack tip in the next stage are given by the formula

$$\tilde{y}'_j(x') = \tilde{a}_j(x' - q_j)^3 + \tilde{b}_j(x' - q_j)^2 + \tilde{c}_j(x' - q_j) + \tilde{d}_j, \quad \tilde{x}'_j \le x' \le q_j \tag{2.40}$$

Here,

$$\tilde{a}_j = \frac{\left(\tilde{c}_j + \tilde{\Gamma}_j\right)\left(\tilde{h}_j + \tilde{\delta}_j\right) + 2\left(\tilde{d}_j - \tilde{Y}_{j-1} - \tilde{\Gamma}_j \tilde{h}_j\right)}{\left(\tilde{h}_j + \tilde{\delta}_j\right)^3};$$

$$\tilde{b}_j = \frac{3\left(\tilde{Y}_{j-1} - \tilde{d}_j + \tilde{\Gamma}_j \tilde{h}_j\right) - \left(\tilde{\Gamma}_j + 2\tilde{c}_j\right)\left(\tilde{h}_j + \tilde{\delta}_j\right)}{\left(\tilde{h}_j + \tilde{\delta}_j\right)^2};$$

$$\tilde{c}_j = \left.\frac{d\tilde{y}'_{j-1}(x')}{dx'}\right|_{x' = \tilde{q}_j}; \qquad \tilde{d}_j = \tilde{y}'_{j-1}(\tilde{q}_j). \tag{2.41}$$

$$\tilde{q}_j = \tilde{x}'_j + \tilde{\delta}_j \qquad (\delta_j \ge 0); \tag{2.42}$$

$$\tilde{\delta}_j = \begin{cases} \tilde{h}_j / 2, & \theta_j^{*-} \ne 0; \\ 0, & \theta_j^{*-} = 0. \end{cases} \tag{2.43}$$

and the quantities Γ_j, $\tilde{Y}_{j-1}$ determine the equation of the straight line

$$y' = \tilde{\Gamma}_j(x' - \tilde{x}'_{j-1}) + \tilde{Y}_{j-1} \qquad \left(\tilde{\Gamma}_j = \operatorname{tg} \tilde{\gamma}_j, \quad \tilde{Y}_{j-1} = \tilde{y}'_{j-1}(\tilde{x}'_{j-1})\right), \tag{2.44}$$

passing through the left crack tip $(\tilde{x}_{j-1}, \tilde{Y}_{j-1})$ in the direction of its propagation at an angle θ_j^{*-}.

Thus, the coefficients of Eq. (2.34) and Eq. (2.40) can be found from relations (2.35), (2.38) and (2.41), (2.42) if the geometry of the crack in the previous segments $\left[q_{j-1}, x'_{j-1}\right]$ and $\left[\tilde{x}'_{j-1}, \tilde{q}_{j-1}\right]$ is known. In this case, the equation of the crack in the next stage of construction of the path on the segment can be written in the form

$$y'(x') = \begin{cases} \tilde{y}'_{j-1}(x'), \ \tilde{x}'_{j-1} \le x' \le \tilde{q}_{j-1}; \\ \tilde{y}'_k(x'), \quad \tilde{q}_{k+1} \le x' \le \tilde{q}_k, \quad k = 1, \ldots, j-2; \\ y'_0(x'), \quad \tilde{q}_1 \le x' \le q_1; \\ y'_k(x'), \quad q_k \le x' \le q_{k+1}, \quad k = 1, \ldots, j-2; \\ y'_{j-1}(x'), \ q_{j-1} \le x' \le x'_{j-1}; \end{cases} \tag{2.45}$$

where the function $y'_0(x')$ describes the contour of the initial crack on the segment $[\mathrm{Re}\, a'_0, \ \mathrm{Re}\, b'_0]$.

For the edge crack, the methods used for the construction of its path are simplified because it is not necessary to use formula (2.32). If, e.g., the crack approaches the edge of the plate by its left end a'_0, then the increments of the path occur only for the crack right tip with constant steps H, and we can use relations (2.33)–(2.38) for our calculations. This simplified version of the algorithm is also used in the case where one of the crack tips becomes fixed due to the attainment of the threshold value of the range of the parameter K at this tip.

Note that the determination of the crack path is completely terminated if the range of the parameter K becomes equal to its threshold value at both tips of the growing crack or if it attains its critical value ΔK_{fc} at one tip.

References

1. Arzamasov BN, Sidorin II, Kosolapov GF et al (1986) Materialovedeniye (Materials Science). Mashinostroenie, Moscow
2. ASTM Designation E647-00: Standard test method for measurement of fatigue crack growth rates (2000)
3. Balatskii LT (1972) Ustalost' valov v soedineniyakh (Fatigue of Shafts in Joints). Tekhnika, Kiev
4. Balatskii LT (1982) Prochnost' pressovykh soedineniy (Strength of Pressed Joints). Tekhnika, Kiev
5. Bogdanski S, Olzak M, Stupnicki J (1998) Numerical modeling of a 3D rail RCF "squat"-type crack under operating load. Fatigue Fract Eng Mater Struct 21:923–935
6. Bold PE, Brown MW, Allen RJ (1991) Shear mode crack growth and rolling contact fatigue. Wear 144:307–317
7. Bower AF (1988) The influence of crack face friction and trapped fluid on surface initiated rolling contact fatigue cracks. J Tribol Trans ASME 110(4):704–711
8. Burton RA, Russel JA (1966) Lubricant effects on fatigue in a stationary concentrated contact under vibratory looding. Tr ASME J Basic Eng 88(3):573–580
9. Cannon DF, Edel KO, Grassie SL, Sawley K (2003) Rail defects: an overview. Fatigue Fract Eng Mater Struct 26(10):865–886
10. Ciavarella M, Demelio G (2001) A review of analytical aspects of fretting fatigue, with extension to damage parameters, and application to dovetail joints. Int J Solids Struct 38(10–13):1791–1811
11. Collins JA (1981) Failure of materials in mechanical design. Wiley, New York
12. Van Dang K, Maitournam MH (2003) Rolling contact in railways: modeling, simulation and damage prediction. Fatigue Fract Eng Mater Struct 26(10):939–948
13. Datsishin OP, Marchenko HP, Panasyuk VV (1993) Theory of crack growth in rolling contact. Mater Sci 29(4):373–383

14. Datsyshin AP, Marchenko HP (1992) Application of singular integral equations to the calculation of trajectories of crack propagation in bounded plates. In: Proceedings of All-Union Conference "Integralnyye uravneniya i krayevyye zadachi matematicheskoy fiziki" ("Integral Equations and Boundary-Value Problems of Mathematical Physics"). Vladivostok, pp 89–100
15. Datsyshin AP, Marchenko HP (1985) Interaction of curvilinear cracks with the boundary of an elastic half plane. Sov Mater Sci 20(5):466–473
16. Datsyshin AP, Savruk MP (1974) Integral equations of the plane problem of crack theory. J Appl Math Mech 38(4):677–686
17. Datsyshyn OP (1999) Durability and fracture calculate model for structural materials under fretting fatigue. Naukovyy Visnyk Ukrayinskogo Derzhavnogo Lisotekhnichnogo Universytetu 9:139–149
18. Datsyshyn OP (2000) Fatigue fracture of calculation model of solids under their contact interaction. Naukovi Notatky Luts'kogo Derzhavnogo Universytetu, Issue 7:74–78
19. Datsyshyn OP (1996) Fracture and wear processes simulating under cyclic contact of solid bodies. In: Petit J (ed) ECF-11, vol II. Mechanism and mechanics of damage and failure. EMAS LTD, Warley, pp 1411–1416
20. Datsyshyn OP (2005) Service life and fracture of solid bodies under the conditions of cyclic contact interaction. Mater Sci 41(6):709–733
21. Datsyshyn OP, Glazov AYu (2004) Durability prognosis of rolling bodies by pitting development. In: Panasyuk VV (ed) Mekhanika ruynuvannya materialiv ta mitsnist' konstruktsiy (Fracture Mechanics of Materials and Strength of Structures), Lviv, pp 243–248
22. Datsyshyn OP, Kadyra VM (2006) A fracture mechanics approach to prediction of pitting under fretting fatigue conditions. Int J Fatigue 28(4):375–385
23. Datsyshyn OP, Kalakhan OS, Kadyra VM, Shchur RB (2004) Pitting formation under the conditions of fretting fatigue. Mater Sci 40(2):159–172
24. Datsyshyn OP, Kopylets' MM (2003) Prediction of the service life of rolling bodies according to the development of a subsurface crack. Mater Sci 39(6):765–779
25. Datsyshyn OP, Levus AB (2000) Stress intensity factors for system of parallel surface cracks in half-plane due to Hertzian load on its boundary. Mashynoznavstvo 11:9–15
26. Datsyshyn OP, Marchenko HP (2003) Estimation of mode II surface crack growth period under rolling contact, Mashynoznavstvo Mashynoznavstvo 7:17–23
27. Datsyshyn OP, Marchenko HP, Hlazov AYu, Levus AB (2004) One approach to the evaluation of durability of solid bodies. Mater Sci 40(4):484–490
28. Datsyshyn OP, Marchenko HP, Levus AB (2001) Edge cracks growth paths during rolling under dry friction. Mashynoznavstvo 4–5, 38–41
29. Datsyshyn OP, Panasyuk VV (2001) Pitting of the rolling bodies contact surface. Wear 251:1347–1355
30. Dawson PH (1967) Contact fatigue in soft steel with random loading. J Mech Engn Sci 9(1):79 80
31. Dekhovich LA, Makhutov NA (1981) Use of failure mechanics for evaluating fretting fatigue strength. Sov Mater Sci 17(3):280–283
32. Dobrovol's'kii VO (1951) Detali mashin (Machine Parts). Gostekhizdat Ukrainy, Kiev
33. Dubourg MC, Villechaise B (1992) Stress intensity factors in a bent crack: a model. Eur J Mech A/Solids 11(2):169–179
34. Ekberg A, Kabo E, Åkesson B (2002) An engineering model for prediction of rolling contact fatigue of railway wheels. Fatigue Fract Eng Mater Struct 25:899–909
35. Elber W (1970) Fatigue crack closure under cyclic tension. Eng Fract Mech 2(1):37–45
36. Faanes S (1995) Inclined cracks in fretting fatigue. Eng Fract Mech 52(1):1–82
37. Fajdiga G, Flašker J, Glodež S, Hellen TK (2003) Numerical modeling of micropitting of gear teeth flanks. Fatigue Fract Eng Mater Struct 26(12):1135–1143
38. Fernando US, Brown MW, Miller KJ (1996) Linear elastic fracture mechanics interpretation of crack growth behaviour in fretting fatigue. In: Advances in Fracture Resistance in Materials (ICF–8 Proceeding), Vol II. Tat McGraw-Hill Publishing, New Delhi, pp 207–215
39. Firsov VT (1990) Vestn. Mashinostroyeniya 10:38–40

40. Fridman YaB (1974) Mekhanicheskiye svoistva materialov (Mechanical Properties of Materials), Vol 1, Vol 2. Mashinostroenie, Moscow
41. Frolish MF, Fletcher DI, Beynon JH (2002) A quantitative model for predicting the morphology of surface initiated rolling contact fatigue cracks in back-up roll steels. Fatigue Fract Eng Mater Struct 25:1073–1086
42. Glodež S, Flasker J, Ren Z (1997) A new method for the numerical determination of pitting resistance of gear teeth flanks. Fatigue Fract Eng Mater Struct 20(1):71–83
43. GOST 25.501–78 (1979) Raschety i ispytaniya na prochnost' v mashinostroyenii. Metody ispytaniy na kontaktnuyu ustalost' (Strength Analyses and Tests in Mechanical Engineering. Contact Fatigue Testing Methods). Izd. Standartov, Moscow
44. GOST 25.506–85 (1985) Raschety i ispytaniya na prochnost'. Metody mekhanicheskih ispytaniy metalov. Opredeleniye kharakteristik treshchinostoykosti (vyazkosti razrusheniya) pri staticheskom nagruzhenii (Strength Analyses and Tests. Methods for Mechanical Testing of Metals. Determination of the Crack Resistance Characteristics (Fracture Toughness) Under Static Loading). Izd. Standartov, Moscow
45. Griffith AA (1924) The phenomena of rupture and flow in solids. philosophical transactions of the royal society, London, 1(ser. A. 221):163–198 (1921); Griffith AA (1924) The theory of rupture. In: Proceedings 1st international congress Delft, pp 53–63
46. Hannes D, Alfredsson B (2014) Numerical investigation of the spall opening angle of surface initiated rolling contact fatigue. Eng Fracture Mech 131:538–556
47. Hearly AD, Johnson KL (1985) Mode II stress intensity factors for a crack parallel to the surface of an elastic half-space subjected to a moving point load. J Mech Phys Solids 33(1):61–81
48. Hebda M, Chichinadze AV (eds) (1989) Spravochnik po tribotekhnike (A Handbook of Tribology). Mashinostroenie-VKL, Moscow-Warsaw
49. Hills DA, Nowell D (1994) Mechanics of fretting fatigue. Kluwer Academic Publishers, Dordrecht
50. Irwin GR (1957) Analysis of stress and strain near and of a crack traversing a plate. J Appl Mech 24(3):361–364
51. Ivanova VS, Terent'ev VF (1975) Priroda ustalosti metallov (Nature of the Fatigue of Metals), Metallurgiya, Moscow
52. Ivanyts'kyi YaL, Kun' PS (2013) Trishchynostiykist' konstruktsiynykh materialiv za skladnogo navantazhennya (Crack growth resistance of structural materials under complex loading). SPOLOM, Lviv
53. Ivanyts'kyi YaL, Shtayura S (2004) Methodical guidelines. Determination of the characteristics of crack-growth resistance of materials under the conditions of complex stressed state (normal opening + transverse shear and normal fracture + longitudinal shear. In: Panasyuk VV (ed) Mekhanika ruynuvannya materialiv ta mitsnist' konstruktsiy (Fracture Mechanics of Materials and Strength of Structures), Lviv, pp 723–732
54. Johnson KL (1985) Contact mechanics. Cambridge University Press, Cambridge
55. Kaneta M, Yatsuzuka H, Murakami Y (1985) Mechanism of crack growth in lubricated rolling/sliding contact. ASLE Trans 28(3):407–414
56. Keer LM, Bryant MD (1983) A pitting model for rolling contact fatigue. Trans ASME: J Lubric Technol 105(2):198–205
57. Kennedy NG (1956) Fatigue of curved surfaces in contact under repeated load cycles. In: International congresson fatigue of metals, London
58. Klesnil M, Lukaš P (1980) Fatigue of metallic materials. Prague: Academia, 239 p
59. Kocańda S (1976) Ustalostnoye razrusheniye metallov (Fatigue Fracture of Metals). Metallurgiya, Moscow
60. Kostetskii BI (1959) Soprotivlyeniye iznashivaniyu detaley mashin (Resistance to Wear Process of Machine Parts). Mashgiz, Moscow-Kiev, 478 p
61. Kostetskii BI (1950) Iznosostoykost' detaley mashin (Wear Resistance of Machine Parts). Mashgiz, Moscow-Kiev
62. Kostetskii BI, Nosovskii IG, Bershadskii LI, Karaulov AK (1975) Nadezhnost' i dolgovechnost' mashin (Reliability and Durability of Machines). Tekhnika, Kiev

63. Kostetskii BI, Nosovskii IG, Karaulov AK, Bershadskii LI et al (1976) Poverkhnostnaya prochnost' materialov pri trenii (Surface Strength of Materials in Friction). Tekhnika, Kiev
64. Kovchik SE, Morozov EM (1988) Characteristics of Short-Term Crack Resistance of Materials and the Methods for Their Determination. In: Panasyuk VV (ed) Mekhanika razrusheniya i prochnost' konstruktsiy (Fracture Mechanics and Strength of Materials): A Handbook (in 4 volums), Vol 3, Naukova Dumka, Kiev
65. Kragel'skii IV, Dobychin MN, Kombalov VS (1977) Osnovy raschetov na treniye i iznos (Fundamentals of the Numerical Analyses of Friction and Wear). Mashinostroenie, Moscow
66. Kudish II (2000) A new statistical model of contact fatigue. Tribol Trans 43(4):711–721
67. Kudish II (1989) Raschet iznosa i ustalostnogo vykrashivaniya v podshipnikah kacheniya: obzor (Numerical Analysis of Wear and Fatigue Crumbling in Rolling Bearings: A Survey). Series X—"Bearing Industry," Central Research Institute of Information and Techniko-Economical Investigations of the Automotive Industry, Moscow
68. Kudish II, Burris KW (2004) Modeling of surface and subsurface crack behaviour under contact load and the presence of lubricant. Int J Fracture 125:125–147
69. Kudish II, Burris KW (2000a) Modern state of experimentation and modeling in contact fatigue phenomenon: Part I—Contact fatigue. Normal and tangential contact and residual stresses. Nonmetallic inclusions and lubricant contamination. Crack initiation and crack propagation. Surface and subsurface cracks. Tribol Trans 43(2):187–196
70. Kudish II, Burris KW (2000b) Modern state of experimentation and modeling in contact fatigue phenomenon: Part II—Analysis of the existing statistical mathematical models of bearing and gear fatigue life. New statistical model of contact fatigue. Tribol Trans 43(2):293–301
71. Kuno M, Waterhouse RB, Nowell D, Hills DA (1989) Initiation and growth of fretting fatigue cracks in the partial slip regime. Fatigue Fract Eng Mater Struct 12(5):387–398
72. Lamacq V, Dubourg M-C (1999) Modeling of initial fatigue crack growth and crack branching under fretting conditions. Fatigue Fract Eng Mater Struct 22(6):535–542
73. Lunden R, Paulsson B (2009) Introduction to wheel-rail interface research. In: Lewis R, Olofsson U (eds) Wheel-rail interface handbook. Woodhead Publishing, pp 3–33
74. Miller KJ (2001) Structural integrity—whose responsibility? The 36th John Player Memorial Lecture presented at an Ordinary Meeting of the Institution of Mechanical Engineers, London: Inst. Mech. Engn
75. Morozov EM, Zernin MB (1999) Kontaktnyye zadachi mehaniki razrusheniya (Contact Problems of Fracture Mechanics). Mashinostroenie, Moscow
76. Murakami Y, Endo M (1994) Effects of defects, inclusions and ingomogeneities on fatigue strength. Int J Fatigue 16:163–182
77. Murakami Y, Sakae C, Hamada S (1999) Mechanism of rolling contact fatigue and measurement of K_{IIth} for steels. In: Beynon JH, Brown MW, Lindley et al (eds) engineering against fatigue, A.A. Balkema Publ., Rotterdam, pp 473–485
78. Murakami Y, Sakae C, Ichimaru K, Morita T (1997) Experimental and fracture mechanics study of the pit formation mechanism under repeated lubricated rolling/sliding contacts: effects of reversal of rotation and change of the driving roller. Trans ASME J Tribol 119:788–796
79. Mutton PJ, Tan M, Bartle P, Kapoor A (2009) The effect of severe head wear on rolling contact fatigue in heavy haul operations. In: Proceedings of the 8th international conference on contact mechanics and wear of rail/wheel systems, Vol 2. AB Editore, Firenze, pp 397–407
80. Navarro C, Garcia M, Dominguez J (2003) A procedure for estimating the total life in fretting fatigue. Fatigue Fract Eng Mater Struct 26:459–468
81. Nowell D, Dini D, Hills DA (2006) Recent developments in the understanding of fretting fatigue. Eng Fract Mech 73(2):207–222
82. Ostash OP, Zhmur-Klimenko VT (1987) Fatigue crack growth in metals at low temperatures (a review). Sov Mater Sci 23(2):124–135
83. Panasyuk VV (1991) Mekhanika kvazikhrupkogo razrusheniya materialov (Mechanics of Brittle Fracture of Materials). Naukova Dumka, Kiev

84. Panasyuk VV, Andreikiv AE, Parton VZ (1988) Fundamentals of the fracture mechanics of materials. In: Panasyuk VV (ed) Mekhanika razrusheniya i prochnost' konstruktsiy (Fracture Mechanics and Strength of Materials): A Handbook (in 4 volums), Vol 1, Naukova Dumka, Kiev
85. Panasyuk VV, Berezhnits'kyi LT (1964) Determination of the limit forces in tension of a plate with an arc-shaped crack. In: Karpenko GV (ed) Voprosy mekhaniki real'nogo tverdogo tela (Problems of Mechanics of Real Solid Body), Naukova Dumka, Kiev, pp 3–19
86. Panasyuk VV, Datsyshyn OP (1995) Calculational model fatigue fracture of solids under their contact interaction. In: Aliabadi MH, Alessandri C (ed), Contact Mechanics II. Computational Techniques. Computational Mechanics Publications, Southampton, Boston, pp 385–392
87. Panasyuk VV, Datsyshyn OP, Glazov AYu (2007) Prediction of the contact durability of rails by pitting development. Mashynoznavstvo 3:3–10
88. Panasyuk VV, Datsyshyn OP, Marchenko HP (2000) Stress state of a half-plane with cracks under rigid punch action. Int J Fract 101(4):347–364
89. Panasyuk VV, Datsyshyn OP, Marchenko HP (1995) To crack propagation theory under rolling contact. Eng Fract Mech 52(1):179–191
90. Panasyuk VV, Datsyshyn OP, Shchur RB (2000) Residual durability of solids contacting under conditions of fretting fatigue. Mater Sci 36(2):153–169
91. Panasyuk VV, Savruk MP, Datsyshin AP (1976) Raspredeleniye napryazheniy okolo treshchin v plastinakh i obolochkakh (Distribution of Stresses near Cracks in Plates and Shells). Naukova Dumka, Kiev
92. Panasyuk VV, Savruk MP, Datsyshyn AP (1977) A general method of solution of two-dimensional problems in the theory of cracks. Eng Fract Mech 9(2):481–497
93. Panasyuk VV, Vitvitskii PM, Yarema SYa (1973) Plastic deformations in the vicinity of cracks and fracture criteria (a survey). Problemy Prochnosti 2:3–19
94. Paris P, Erdogan F (1963) A critical analysis of crack propagation laws. Trans ASME J Basic Eng 85(4):528–534
95. Paris PC, Gomes MP, Anderson WE (1961) A rational analytic theory of fatigue. Trend Eng 13:54–61
96. Pinegin SV (1969) Kontaktnaya prochnost' i soprotivleniye kacheniyu (Contact Strength and Rolling Resistance). Mashinostroenie, Moscow
97. Pinegin SV, Gudchenko VM (1967) On the mechanism of fracture of materials under the action of cyclic contact load. In: Prochnost' metallov pri tsiklicheskikh nagruzkakh (Strength of Metals under Cyclic Loads). Nauka, Moscow, pp 133–139
98. Pinegin SV, Orlov AV, Gudchenko VM (1966) Fracture of material under the action of pulsating contact load. Mashinostroenie 1:7–9
99. Pisarenko GS, Lebedev AA (1976) Deformirovaniye i prochnost' materialov pri slozhnom napryazhennom sostoyanii (Deformation and Strength of Materials in the Complex Stressed State). Naukova Dumka, Kiev
100. Polukhin VP, Nikolaev VA, Tylkin MA et al (eds) (1976) Nadezhnost' i dolgovechnost' valkov kholodnoy prokatki (Reliability and Durability of Cold-Rolled Rolls). Metallurgiya, Moscow
101. Prokopenko AV, Torgov VN (1986) Surface properties and fatigue limit of the metal. Report I. Dependence of yield strength on layer thickness. Strength Mater 18(4):448–455
102. RD 50-345-82 (1983) Metodologicheskiye ukazaniya. Raschety i ispytaniya na prochnost'. Metody mekhanicheskikh ispytaniy metalov. Opredeleniye kharasteristik treshchinostoykosti (vyazkosti razrusheniya) pri tsiklicheskom nagruzhenii (Methodical recommendations. Strength analyses and tests. methods for mechanical testing of metals. Determination of the crack-growth resistance characteristics (fracture toughness) under cyclic loading). Izdatel'stvo Standartov, Moscow
103. Richards CE, Lindley TC (1972) The influence of stress intensity and microstructure on fatigue crack propagation in perlitic materials. Eng Fract Mech 4(4):951–978
104. Ringsberg JW (2005) Shear mode growth of short surface-breaking RCF cracks. Wear 258:955–963

105. Romaniv ON (1986) A structural concept of the fatigue limit of structural alloys. Sov Mater Sci 22(1):103–112
106. Romaniv ON (1983) Effect of crack closure and evaluation of the cyclic crack resistance of constructional alloys. Sov Mater Sci 19(3):212–225
107. Romaniv ON, Deev NA, Sorokivskii IS (1976) Some aspects of the mechanism of fatigue-crack formation in high-strength low-temper steels. Sov Mater Sci 11(1):37–43
108. Romaniv ON, Tkach AN (1987) A structural analysis of kinetic fatigue failure curves of constructional steels. Sov Mater Sci 23(5):441–453
109. Romaniv ON, Yarema SYa, Nikiforchin GN, Makhutov NA, Stadnik MM (1990) Fatigue and cyclic crack resistance of structural materials. Panasyuk VV (ed) Mekhanika razrusheniya i prochnost' konstruktsiy (Fracture Mechanics and Strength of Materials): A Handbook (in 4 volums), Vol 4, Naukova Dumka, Kiev
110. Rooke DP, Jones DA (1979) Stress intensity factors in fretting fatigue. J Strain Anal 14(1):1–6
111. Savruk MP (1981) Dvumernyye zadachi uprugosti dlya tel s treshchinsmi (Two-Dimensional Problems of Elasticity for Bodies with Cracks). Naukova Dumka, Kiev
112. Savruk MP, Osiv PN (1982) Calculation of the statistical trajectory of propagation of cracks. Problemy Prochnosti 11:19–23
113. Serensen SV (1985) Izbrannyye trudy (Selected Works): in 3 volums. Naukova Dumka, Kiev
114. Sheppard SD, Barber JR, Comninou M (1987) Subsurface cracks under conditions of slip, stick and separation caused by a moving compressive load. Trans ASME: J Appl Mech 54(2):393–398
115. Shevelya VV, Kalda GS (1998) Fretting-ustalost' metallov (Fretting Fatigue of Metals). Podillya, Khmel'nits'kii
116. Sosnovskii LA, Makhutov NA, Shurinov VA (1992a) Contact-mechanical fatigue: basic regularities. Zavodskaya Laboratoriya 11:44–61
117. Sosnovskii LA, Makhutov NA, Shurinov VA (1992b) Fretting fatigue: basic regularities, Zavodskaya Laboratoriya 8:45–62
118. Sosnovskii LA, Makhutov NA, Shurinov VA (1992) Friction-mechanical fatigue: basic regularities, Zavodskaya Laboratoriya 9:46–63 (1992)
119. Szolwinski MP, Farris TN (1996) Mechanics of fretting fatigue crack formation. Wear 198:93–107
120. Tallian TE (1982) A unified model for the prediction of the durability of rolling contact. Trans ASME: J Lubric Technol 104(3):340–346
121. Tallian TE, McCool JI (1971) An engineering model of spalling fatigue failure in rolling contact II. The surface model. Wear 17:447–461
122. Terent'ev VF, Bily M (1973) The construction of the complete fatigue curve. Strength Mater 5(2):160–165
123. Troshchenko VT, Sosnovskii LA (1987) Soprotivleniye ustaloasti metallov i splavov (Fatigue Resistance of Metals and Alloys): A Handbook, vol 1. Naukova Dumka, Kiev
124. Troshchenko VT, Tsybanev HV (2002) Fretting fatigue of metallic materials and structural components. In: Troshchenko VT (ed) "Trybofatyka" ("Tribofatigue"): Proceedings of the 4th International Symposium on Tribofatigue (ISTF 4), Vol 1, Ternopil', pp 23–28
125. Trubin GK (1962) Kontaktnaya ustalost' materialov dlya zubchactykh koles (Contact Fatigue of Materials for Gear Wheels). Mashinostroenie, Moscow
126. Tyfour WR, Beynon JH (1994) The effect of rolling direction reversal on fatigue crack morphology and propagation. Tribol Int 27(4):273–282
127. Tyler JC, Burton RA, Ku PM (1963) Contact fatigue under oscillatory normal load. ASLE Tr 6(4):255–269
128. Vakulenko IO, Anofriev VH, Hryshchenko MA, Perkov OM (2009) Defekty zaliznychnykh kolis (Defects of Railroad Wheels). Dnipropetrovs'k
129. Vallellano C, Dominguez J, Navarro A (2003) On the estimation of fatigue failure under fretting conditions using notch methodologies. Fatigue Fract Eng Mater Struct 26:469–478
130. Waterhause RB (1972) Fretting corrosion. Pergamon Press, Oxford-New York

131. Yarema SYa (1981) Correlation of the parameters of the Paris equation and the cyclic crack resistance characteristics of materials. Strength Mater 13(9):1090–1098
132. Yarema SYa (1987) Fundamentals and certain problems of fatigue fracture mechanics. Sov Mater Sci 23(5):454–464
133. Yarema SYa (1975) Stages of fatigue fracture and their consequences. Sov Mater Sci 9(6):681–686
134. Yarema SYa (1978) Assessment of metal and alloy resistance to cracking. Studies of fatigue crack growth and kinetic fatigue fracture diagrams. Sov Mater Sci 13(4):351–368
135. Yarema SYa, Mikitishin SI (1976) Analytical description of the fatigue-failure diagrams of materials. Sov Mater Sci 11(6):660–666
136. Yarema SYa, Zboromirskii AI (1985) An analytic investigation of the growth of a fatigue crack arbitrarily oriented in a biaxial nonuniform field of stresses. Sov Mater Sci 20(6):554–562
137. Zerbst U, Lundén R, Edel K-O, Smith RA (2009) Introduction to the damage tolerance behaviour of railway rails—a review. Eng Fracture Mech 76:2563–2601
138. Zerbst U, Madler K, Hintze H (2005) Fracture mechanics in railway applications—an overview. Eng Fract Mech 72(2):163–194
139. Zhidovtsev NA, Katsov KB, Karpenko GV et al (1979) Stoykost' burovykh dolot (Durability of Drill Bits). Naukova Dumka, Kiev

Chapter 3
Singular Integral Equations for Some Contact Problems of Elasticity Theory for Bodies with Cracks

Abstract This chapter is devoted to the mathematical method of modeling of the deformation and fracture process for elements of moving and fixed joints (tribojoints). In this chapter, we deduce singular integral equations of some contact problems of the elasticity theory for bodies with cracks. The Kolosov-Muskhelishvili complex potentials for the analyzed problems are represented in the form of integral representations with Cauchy-type kernels with respect to the derivatives of the discontinuities of displacements on the crack contours. In the general case, the problems are reduced to systems of singular integral equations of the first kind. We propose singular integral equations (SIE) for the elastic half-plane weakened by a system of curvilinear cracks under the action of various model contact loads applied to the boundary of the half-plane. We briefly describe the Gauss-Chebyshev method of mechanical quadratures that enables one to efficiently construct the numerical solutions of these SIE. We also deduce the relations for the stress intensity factors at the crack tips can be expressed via the solutions of SIE for inside and edge curvilinear cracks in the half-plane. In this chapter, we present both known results available from the literature and new results.

3.1 Some Input Theses and Relations for Plane Problem of Elasticity Theory

This is an auxiliary section. In what follows, one can find some data on the plane problem of the theory of elasticity required for our subsequent presentation. Thus, we present the Kolosov-Muskhelishvili relations [31, 37], which enable us to reduce the problem of determination of the stress-strain state of the elastic body to the construction of functions of complex variables. We study the method of singular integral equations (SIE) aimed at the solution of the first and second main problems of the theory of elasticity for bodies weakened by system of rectilinear [11, 44] and curvilinear [49, 53] cracks. The asymptotic representation of the stress-strain state in the vicinity of the crack tips is obtained in terms of the stress intensity factors [5, 64, 65]. We also propose a brief presentation of the Gauss-Chebyshev method of

O. Datsyshyn and V. Panasyuk, *Structural Integrity Assessment of Engineering Components Under Cyclic Contact*, Structural Integrity 9,
https://doi.org/10.1007/978-3-030-23069-2_3

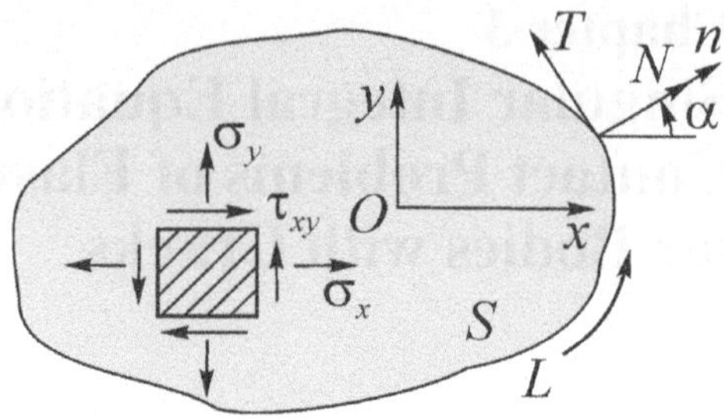

Fig. 3.1 Components of the vector of the stresses N and T on the contour L of the elastic domain S

mechanical quadratures [20, 44, 53], which is an efficient tool for the construction of numerical solutions of the SIE for numerous important problems.

Basic relations of the plane (two-dimensional) problem of the elasticity theory. The elastic equilibrium of solid bodies can be described by the equations of the plane problem of the theory of elasticity either in the case of plane deformation of the bodies with constant cross sections subjected to the action of external forces located in these cross sections or in the case of generalized plane stressed state, i.e., under the conditions of deformation of a thin plate by forces acting in its plane. For the evaluation of the stress-strain state at any point of the deformed elastic isotropic body, it is necessary to find three components of the stress tensor σ_x, σ_y, and τ_{xy} (Fig. 3.1) and two components of the vector of displacements u and v. If a Cartesian coordinate system is chosen so that the plane xOy coincides either with the cross section of a cylinder or with the middle plane of the plate, then, under the conditions of plane problem of the theory of elasticity, the indicated components, are functions of two variables (x and y).

We now introduce a stress function Airy $U(x, y)$ [37] by the formulas

$$\sigma_x = \frac{\partial^2 U}{\partial y^2}; \qquad \sigma_y = \frac{\partial^2 U}{\partial x^2}; \qquad \tau_{xy} = -\frac{\partial^2 U}{\partial x \partial y}. \tag{3.1}$$

Then the solution of the plane problem of the theory of elasticity, even in the absence of bulk forces, is reduced to the integration of a biharmonic equation

$$\frac{\partial^4 U}{\partial x^4} + 2\frac{\partial^4 U}{\partial x^2 \partial y^2} + \frac{\partial^4 U}{\partial y^4} = 0 \tag{3.2}$$

with given boundary conditions on the contour L of the domain S (Fig. 3.1).

The general solution of Eq. (3.2) can be represented in terms of two analytic functions $\varphi(z)$ and $\chi(z)$ of one complex argument $z = x + iy$ by the following Goursat formula:

$$U(x, y) = \mathrm{Re}[\bar{z}\varphi(z) + \chi(z)], \qquad \bar{z} = x - iy, \tag{3.3}$$

In [31, 37], it was shown that the components of stresses σ_x, σ_y and τ_{xy} and displacements u and v are expressed via the complex potentials $\Phi(z)$ and $\Psi(z)$ by the formulas

$$\sigma_x + \sigma_y = 2\left[\Phi(z) + \overline{\Phi(z)}\right]; \tag{3.4}$$

$$\sigma_y - \sigma_x + 2i\tau_{xy} = 2\left[\bar{z}\Phi'(z) + \Psi(z)\right]; \tag{3.5}$$

$$2G(u + iv) = æ\varphi(z) - z\overline{\Phi(z)} - \overline{\psi(z)}; \tag{3.6}$$

$$\Phi(z) = \varphi'(z); \qquad \Psi(z) = \psi'(z); \qquad \psi(z) = \chi'(z).$$

Here, u and v are the components of the vector of displacements in the directions of the x- and y-axes; $G = E/2(1 + \mu)$ is the shear modulus, E is Young's modulus; μ is Poisson's ratio; $æ = 3–4\mu$ in the case of plane deformation, and $æ = (3 - \mu)/(1 + \mu)$ for the generalized plane stressed state. The principal vector of forces acting from the right upon an arc AB located inside the plate or on its contour if we move along this arc from A to B is given by the formulas

$$X + iY = -i[\mu(z)]_A^B = -i\int_A^B \mu'(z)dz; \tag{3.7}$$

$$\mu(z) = \varphi(z) + z\overline{\varphi'(z)} + \overline{\psi(z)}, \tag{3.8}$$

where X and Y are the projections of the principal vector on the axes Ox and Oy; $[]_A^B$ is the difference between the expressions in brackets taken for the values of z at the points B and A.

For the principal moment of the analyzed forces relative to the origin, we can write

$$\begin{aligned} M &= \operatorname{Re}\left[\chi(z) - z\psi(z) - z\bar{z}\varphi'(z)\right]_A^B \\ &= \operatorname{Re}\left\{\int_A^B \overline{\mu(z)}dz - \left[\bar{z}\psi(z) + z\psi(z) + z\bar{z}\varphi'(z)\right]_A^B\right\} = \\ &= \operatorname{Re}\left\{\int_A^B \overline{\mu(z)}dz - \left[z\overline{\mu(z)}\right]_A^B\right\} = -\operatorname{Re}\int_A^B \bar{z}\mu'(z)dz. \end{aligned} \tag{3.9}$$

It is known [37] that the functions $\Phi(z)$ and $\Psi(z)$ are not invariant under the transformations of the Cartesian coordinate system. Thus, if the new coordinate system $x_1O_1y_1$ is connected with the old system xOy by the formula

$$z = z_1 e^{i\alpha} + z_1^0, \tag{3.10}$$

and the functions $\Phi_1(z_1)$ and $\Psi_1(z_1)$ play the same role in the system $x_1O_1y_1$ as the functions $\Phi(z)$ and $\Psi(z)$ in the system xOy, then we get

Fig. 3.2 The components of the stress tensor in a polar coordinate system

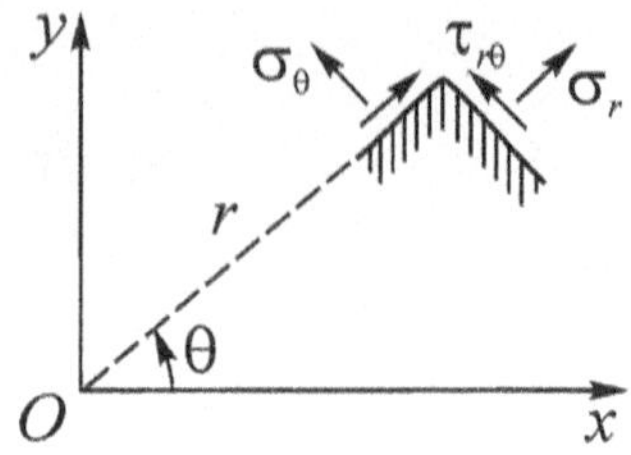

$$\begin{gathered}\Phi(z)=\Phi_1(z_1); \qquad z_1=e^{-i\alpha}\left(z-z_1^0\right);\\ \Psi(z)=e^{-2i\alpha}\left[\Psi_1(z_1)-\bar{z}_1^0 e^{i\alpha}\Phi_1'(z_1)\right];\\ \Psi(z)+\bar{z}\Phi'(z)=e^{-2i\alpha}\left[\Psi_1(z_1)+\bar{z}_1\Phi_1'(z_1)\right].\end{gathered} \tag{3.11}$$

Here, $z_1=x_1+iy_1$; $z_1^0=x_1^0+iy_1^0$; x_1^0, y_1^0 are the coordinates of the point O_1 in the old coordinate system.

The components σ_r, σ_θ and $\tau_{r\theta}$ of the stress tensor and the components of the displacements vector v_r and v_θ in a polar coordinate system r, θ (Fig. 3.2) are expressed via the corresponding components of stresses and displacements in Cartesian coordinates by the following formulas:

$$\sigma_r+\sigma_\theta=\sigma_x+\sigma_y; \qquad \sigma_\theta-\sigma_r+2i\tau_{r\theta}=e^{2i\theta}\left(\sigma_y-\sigma_x+2i\tau_{xy}\right); \tag{3.12}$$

$$v_r+iv_\theta=e^{-i\theta}(u+iv). \tag{3.13}$$

The solution of the plane problem of the theory of elasticity is reduced to finding two analytic functions $\varphi(z)$ and $\psi(z)$ in domain S taking the required boundary values of these functions on the contour L of the domain S occupied by a deformed elastic body (on the boundary of the body). For the first main problem, i.e., in the case where external stresses are given on the boundary L, the boundary condition for these functions takes the form

$$\varphi(t)+t\overline{\varphi'(t)}+\overline{\psi(t)}=i\int_0^s (X_n+iY_n)ds+C, \qquad t\in L, \tag{3.14}$$

where X_n and Y_n are the components of given forces in the zone with outer normal n, t is a variable point of the contour L, s is the corresponding arc abscissa, and C is an arbitrary complex constant.

For the second main problem, i.e., in the case where displacements are given on the contour L, we arrive at the boundary condition as a result of the limit transition from relation (3.6)

$$æ\varphi(t)-t\overline{\varphi'(t)}-\overline{\psi(t)}=2G[u(t)+iv(t)], \qquad t\in L. \tag{3.15}$$

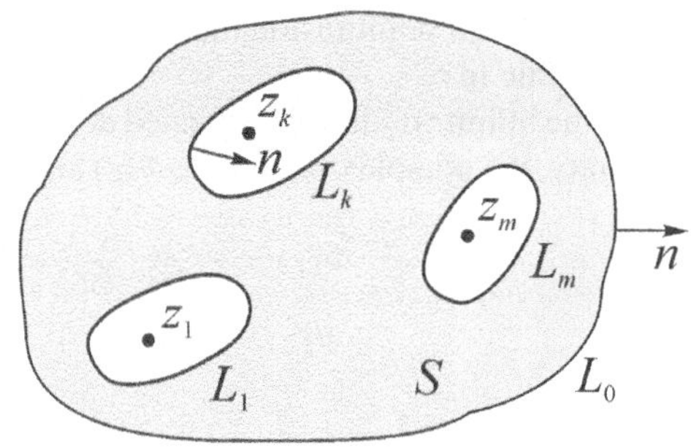

Fig. 3.3 The multiply connected domain

Here, $u(t)$ and $v(t)$ are known functions on L.

Differentiating expression (3.14) with respect to t, we obtain another form of the boundary conditions for the first main problem

$$\Phi(t) + \overline{\Phi(t)} + \frac{d\bar{t}}{dt}\left[t\overline{\Phi'(t)} + \overline{\Psi(t)}\right] = N + iT, \qquad t \in L, \tag{3.16}$$

where N and T are, respectively, the given normal and tangential components of external stresses acting upon the boundary of the body (see Fig. 3.1).

Similarly, for the second main problem, it follows from relation (3.15) that

$$æ\Phi(t) - \overline{\Phi(t)} - \frac{d\bar{t}}{dt}\left[t\overline{\Phi'(t)} + \overline{\Psi(t)}\right] = 2G\left[u'(t) + iv'(t)\right], \qquad t \in L. \tag{3.17}$$

In the boundary conditions (3.16), (3.17), the derivatives have the form

$$\frac{d\bar{t}}{dt} = \frac{d\bar{t}}{ds}\Big/\frac{dt}{ds} = -e^{-2i\alpha} \quad \text{and} \quad \frac{dt}{ds} = \frac{dx}{ds} + i\frac{dy}{ds} = -\sin\alpha + i\cos\alpha = ie^{i\alpha}, \tag{3.18}$$

where α is the angle formed by the outer normal to the contour and the Ox-axis (see Fig. 3.1).

For a multiply connected domain (Fig. 3.3), the functions $\Phi(z)$ and $\Psi(z)$ (under the conditions of single-valuedness of displacements) take the form (in any its end part)

$$\Phi(z) = -\frac{1}{2\pi(1+æ)}\sum_{k=1}^{m}\frac{X_k + iY_k}{z - z_k} + \Phi^*(z);$$

$$\Psi(z) = \frac{1}{2\pi(1+æ)}\sum_{k=1}^{m}\left[\frac{æ(X_k - iY_k)}{z - z_k} - \frac{\bar{z}_k(X_k + iY_k)}{(z - z_k)^2}\right] + \Psi^*(z). \tag{3.19}$$

Here, X_k and Y_k are the components of the principal vector of external forces applied to the closed contour L_k ($k = 1, 2, \ldots, m$), which is not self-crossing, z_k is a arbitrary fixed point inside the contour L_k, and the functions $\Phi^*(z)$ and $\Psi^*(z)$ are holomorphic in S.

For the infinite multiply connected domain, i.e., if the contour L_0 is entirely located at infinity, the complex potentials $\Phi(z)$ and $\Psi(z)$ take the form

$$\begin{aligned} \Phi(z) &= -\tfrac{X+iY}{2\pi(1+æ)z} + B + iC + \Phi_0(z); \\ \Psi(z) &= \tfrac{æ(X-iY)}{2\pi(1+æ)z} + B' + iC' + \Psi_0(z). \end{aligned} \tag{3.20}$$

Here

$$X = \sum_{k=1}^{m} X_k; \quad Y = \sum_{k=1}^{m} Y_k \tag{3.21}$$

are the components of the vector of external forces applied to the boundary of the domain S and the functions $\Phi_0(z)$ and $\Psi_0(z)$ for large $|z|$ are expanded in the series

$$\Phi_0(z) = \frac{a_1}{z^2} + \frac{a_2}{z^3} + \cdots; \quad \Psi_0(z) = \frac{b_1}{z^2} + \frac{b_2}{z^3} + \cdots, \tag{3.22}$$

where a_i and b_i are complex constants. The real constants B, B' and C' are expressed via the values of the principal stresses p and q at infinity by the formulas

$$B = \frac{p+q}{2}; \quad B' + iC' = -\frac{1}{2}(p-q)e^{-2i\alpha_1}, \tag{3.23}$$

where α_1 is the angle between the Ox-axis and the direction of action of the stress p; the constant C is caused by the rotation at infinity and does not affect the distribution of stresses.

Some statements from the theory of analytic functions. In what follows, we present results on the theory of analytic functions necessary for our subsequent presentation. The detailed information about analytic functions, Cauchy-type integrals, and singular equations can be found, e.g., in the monographs [22, 36, 37].

Cauchy-type integrals. Let L be a simple closed (or open) smooth contour in the plane of complex variable $z = x + iy$ or a collection of finitely many contours of this kind without common points and let $f(t)$ be an absolutely integrable function given on L (except, possibly, finitely many points). Then the integral

$$\mathcal{F}(z) = \frac{1}{2\pi i} \int_L \frac{f(t)dt}{t-z} \tag{3.24}$$

is an analytic function in the entire complex plane except the points of the contour L. It is customary to call this integral a Cauchy-type integral, the function $f(t)$ is called its density, and the expression $1/(t-z)$ is called the kernel. If the function $f(t)$ satisfies the Hölder condition (condition $H(\nu)$ or H) on L, i.e., if for any two points of the contour L, the inequality

$$|f(t_2) - f(t_1)| \leq A|t_2 - t_1|^{\nu}, \quad A > 0, \quad 0 < \nu \leq 1, \tag{3.25}$$

holds, then integral (3.24) has the limiting values $\mathcal{F}^+(t_0)$ and $\mathcal{F}^-(t_0)$ at all points t_0 of the contour L that do not coincide with its ends as $z \to t_0$ from the left (+) and from the right (–), respectively, relative to the chosen positive direction of traversing the contour. These limiting values also satisfy the condition $H(\nu)$ and are determined by the Sochocki-Plemelj formulas [22, 36]

$$\mathcal{F}^+(t_0) = \frac{1}{2}f(t_0) + \frac{1}{2\pi i}\int\limits_L \frac{f(t)dt}{t - t_0}; \quad \mathcal{F}^-(t_0) = -\frac{1}{2}f(t_0) + \frac{1}{2\pi i}\int\limits_L \frac{f(t)dt}{t - t_0}; \quad t_0 \in L. \tag{3.26}$$

Here, the integral on the right-hand sides is understood in the sense of the Cauchy principal value, i.e.,

$$\int\limits_L \frac{f(t)dt}{t - t_0} = \lim_{\varepsilon \to 0} \int\limits_{L - L_\varepsilon} \frac{f(t)dt}{t - t_0}, \quad t_0 \in L, \tag{3.27}$$

where L_ε is a part of the curve L located inside the circle $|z - t_0| < \varepsilon$. It is customary to say that integrals (3.27) are singular integrals. Relations (3.26) remain true in the case where L is a arbitrary piecewise-smooth line under the condition that the point t_0 differs from the nodes (including the ends) and $f(t)$ satisfies the condition H in the vicinity of t_0.

Consider an integral

$$\Phi(z) = \frac{1}{2\pi i}\int\limits_L \frac{\varphi(t)\overline{dt}}{t - z}, \tag{3.28}$$

where $\varphi(t) \in H$, i.e., satisfies condition (3.25). In view of the relation

$$d\bar{t}/dt = e^{-2i\theta(t)}, \quad t \in L, \tag{3.29}$$

where $\theta(t)$ is the angle formed by the positive tangent to the contour L at the point t and the Ox-axis, under the assumption that L is a Lyapunov curve, i.e., the angle $\theta(t)$ satisfies the condition H, for the limiting values of integral (3.28), we obtain the following expression:

$$\Phi^{\pm}(t_0) = \pm\frac{1}{2}\varphi(t_0)\frac{d\bar{t}_0}{dt_0} + \frac{1}{2\pi i}\int\limits_L \frac{\varphi(t)\overline{dt}}{t - t_0}, \quad t_0 \in L. \tag{3.30}$$

Relation (3.30) follows from relations (3.26) if the function $\varphi(t)\overline{dt}/dt$ satisfies the condition H. Note that the direction of traversing the contour is assumed to be the positive direction of the tangent.

We now find the limiting values of the integral

$$F(z,\bar{z}) = \frac{1}{2\pi i}\int\limits_L \frac{(\bar{t}-\bar{z})f(t)}{(t-z)^2}dt, \tag{3.31}$$

where L is a closed or open Lyapunov contour and $f(t)$ satisfies the condition $H(v)$. As above (see also [53], p. 12), we conclude that the limiting values of integral (3.31) at the points t_0 of the contour L that do not coincide with the nodes are given by the formulas

$$F^{\pm}(t_0,\bar{t}_0) = \pm\frac{1}{2}f(t_0)\frac{d\bar{t}_0}{dt_0} + \frac{1}{2\pi i}\int\limits_L \frac{(\bar{t}-\bar{t}_0)f(t)}{t-t_0}dt, \tag{3.32}$$

where the derivative on the right-hand side of the equality is given by formula (3.29).

Inversion formulas for the Cauchy-type integral. Let L be a collection of finitely many closed smooth contours without common points and let the positive direction be chosen so that, in the case of motion along L, the domain S remains to the left (see Fig. 3.3). Consider an integral equation

$$\frac{1}{\pi}\int\limits_L \frac{\varphi(t)dt}{t-\tau} = f(\tau), \tag{3.33}$$

where $f(t)$ is a function from the class H given on L and $\varphi(t)$ is an unknown function, which also belongs to the class H. The unique solution of this equation is given by the following formula [36]:

$$\varphi(\tau) = -\frac{1}{\pi}\int\limits_L \frac{f(t)dt}{t-\tau} = f(\tau), \quad \tau\in L. \tag{3.34}$$

Assume that the contour L in Eq. (3.33) consists of a collection of m smooth disjoint open arcs L_k ($k = 1, \ldots, m$) whose ends are a_k and b_k (positive traversing of the contour is performed from a_k to b_k). Also let the function $f(t)$ belong to the class H and let the unknown function $\varphi(t)$ belong to the class H^*, i.e., let $\varphi(t)$ satisfy the condition H in each closed sections of the contour that does not contain nodes and can be represented, near each node (c), in the form

$$\varphi(t) = \frac{\varphi_*(t)}{|t-c|^{\alpha}}, \qquad 0\le\alpha = \text{const} < 1,$$

where $\varphi_*(t)$ belongs to the class H in the vicinity of c. Then Eq. (3.33) has solutions of different classes (bounded or unbounded near the ends a_k and b_k). In particular, the solution of this equation which is not bounded at all ends of the contours L_k has the form [22, 36]

$$\varphi(\tau) = -\frac{1}{\pi R(\tau)} \int_L \frac{R(t) f(t)}{t-\tau} dt + \frac{P_{m-1}(\tau)}{R(\tau)}, \tag{3.35}$$

where $P_{m-1}(\tau)$ is an arbitrary polynomial of degree no greater than $m - 1$ and $R(t)$ is the limit value of the canonical function

$$R(z) = \prod_{k=1}^{m} \sqrt{(z-a_k)(z-b_k)} \tag{3.36}$$

on approaching the contour from the left, i.e.,

$$R(t) = R^+(t) = -R^-(t).$$

We now write Eq. (3.35) for the case where the contour L is a segment of the real axis $-l \le t \le l$,

$$\varphi(\tau) = \frac{1}{\pi\sqrt{l^2-\tau^2}} \left[-\int_{-l}^{l} \frac{\sqrt{l^2-\tau^2} f(t)}{t-\tau} dt + C \right]. \tag{3.37}$$

Here, the constant C is determined from the additional condition imposed on the solution $\varphi(t)$. The value of the integral of $\varphi(t)$ along L is, as a rule, known. Thus, it follows from solution (3.37) that

$$C = \int_{-l}^{l} \varphi(t) dt. \tag{3.38}$$

The constant C can be chosen to guarantee the boundedness of the solution at one end of the segment $|t| \le l$. Thus, the solution bounded at the point $\tau = l$ and unbounded for $\tau = -l$, has the form

$$\varphi(\tau) = -\frac{1}{\pi}\sqrt{\frac{l-\tau}{l+\tau}} \int_{-l}^{l} \sqrt{\frac{l+t}{l-t}} \frac{f(t)dt}{t-\tau}. \tag{3.39}$$

A large number of works (see the bibliography in the monograph [36]) are devoted to the solution of Eq. (3.33), which plays an important role in various problems of mathematical physics.

3.2 Stress-Strain State of Elastic Plate With Smooth Curvilinear Cut

Integral representations of the complex stress potentials [53]. Consider the main boundary problems of the two-dimensional theory of elasticity for an infinite isotropic plane weakened by a smooth curvilinear cut L with origin at a point a and end at a point b.

First, we obtain the solution of the auxiliary problem for the case where the jumps of stresses and the derivatives of displacements are given on the cut

$$(N+iT)^{+}-(N+iT)^{-}=2q(t), \quad t\in L; \tag{3.40}$$

$$\frac{d}{dt}\left[(u+iv)^{+}-(u+iv)^{-}\right]=\frac{i(1+æ)}{2G}g'(t), \quad t\in L. \tag{3.41}$$

In this case, at the ends of the cut, the jump of displacements is absent

$$g(a)-g(b)=0. \tag{3.42}$$

Moreover, the stresses and rotations are absent at infinity.

Assume that the functions $q(t)$ and $g'(t)$ belong to the class H^{*}. By using relations (3.16) and (3.17) and representing the left-hand sides of equalities (3.40) and (3.41) via the complex potentials $\Phi(z)$ and $\Psi(z)$, after simple transformations, we obtain

$$\Phi^{+}(t)-\Phi^{-}(t)=i\left[g'(t)-2iq(t)/(1+æ)\right]=iQ(t), \quad t\in L; \tag{3.43}$$

$$\left[\bar{t}\Phi'(t)+\Psi(t)\right]^{+}-\left[\bar{t}\Phi'(t)+\Psi(t)\right]^{-}=i\left[\overline{Q(t)}-2i\overline{q(t)}-Q(t)\right]\frac{d\bar{t}}{dt}, \quad t\in L. \tag{3.44}$$

Equality (3.43) represents a problem of conjugation for the piecewise-holomorphic function $\Phi(z)$. The solution of this problem vanishing at infinity is given by the following Cauchy-type integral [37]:

$$\Phi(z)=\frac{1}{2\pi}\int_{L}\frac{Q(t)dt}{t-z}. \tag{3.45}$$

We seek the potential $\Psi(z)$ in the form

$$\Psi(z)=-\frac{1}{2\pi}\int_{L}\frac{\bar{t}Q(t)dt}{(t-z)^{2}}+\Omega(z), \tag{3.46}$$

where the unknown function $\Omega(z)$ is also piecewise holomorphic in the entire plane, including the infinitely distant point. By using relations (3.32) and (3.46), and equality (3.44), we get the following problem of conjugation for the function $\Omega(z)$:

$$\Omega^+(t) - \Omega^-(t) = i\left[\overline{Q(t)} - 2i\overline{q(t)}\right]\frac{d\bar{t}}{dt}, \qquad t \in L. \tag{3.47}$$

The solution of this problem has the form

$$\Omega(z) = \frac{1}{2\pi}\int_L \frac{\overline{Q(t)} - 2i\overline{q(t)}}{t - z}\overline{dt}. \tag{3.48}$$

In view of (3.46), we get

$$\Psi(z) = \frac{1}{2\pi}\int_L \left[\frac{\overline{Q(t)} - 2i\overline{q(t)}}{t - z}\overline{dt} - \frac{\bar{t}Q(t)dt}{(t - z)^2}\right]. \tag{3.49}$$

Thus, functions (3.45) and (3.49) give the solution of the posed auxiliary problem (3.40), (3.41) in the general case of a nonself-balanced load $q(t)$. These solutions can also be regarded as integral representations of the complex stress potentials $\Phi(z)$ and $\Psi(z)$ for an infinite plane containing a cut made along the contour L.

Integral equations of the main boundary problems. By using the representations of complex potentials (3.45) and (3.49), we can consider various boundary problems for the infinite plane weakened by a cut L. Assume that nonself-balanced forces (the first main problem)

$$N^{\pm} + iT^{\pm} = p(t) \pm q(t), \qquad t \in L \tag{3.50}$$

or the derivatives of displacements (the second main problem)

$$2G\frac{d}{dt}\left(u^{\pm} + iv^{\pm}\right) = f'(t) \mp \frac{1 + æ}{2i}g'(t), \qquad t \in L, \tag{3.51}$$

are given on the faces of the cut. In this case, the stresses and rotations are absent at infinity. In the first main problem, we also assume that the cut (crack) faces are not in contact. The functions $p(t)$ and $q(t)$ in (3.50) and $f'(t)$ and $g'(t)$ in (3.51) belong to the class H.

For these problems, we seek the complex potentials $\Phi(z)$ and $\Psi(z)$ in the form (3.45) and (3.49) under the assumption that the function $g'(t)$ is unknown in the first main problem and the function $q(t)$ is unknown in the second main problem. Satisfying the boundary condition (3.50) with the help of relations (3.16), (3.26), (3.30), and (3.32), we get the following singular integral equation for the function $g'(t)$ [49]:

$$\frac{1}{2\pi}\int_L \left[\frac{2Q(t) + 2iq(t)}{t - t'}dt + k_1(t, t')(Q(t) + 2iq(t))dt + k_2(t, t')\overline{Q(t)}\,\overline{dt}\right] = p(t'), \quad t' \in L, \tag{3.52}$$

where the regular kernels $k_1(t, t')$ and $k_2(t, t')$ have the form

$$k_1(t, t') = \frac{d}{dt'} \ln \frac{t - t'}{\bar{t} - \bar{t}'}; \quad k_2(t, t') = -\frac{d}{dt'} \frac{t - t'}{\bar{t} - \bar{t}'}. \tag{3.53}$$

The solution of Eq. (3.52) must satisfy the condition

$$\int_L g'(t)dt = 0, \tag{3.54}$$

which guarantees the single-valuedness of displacements in traversing the contour L. This condition also follows from (3.42).

In the case where the cut (crack) faces are subjected to the action of self-balanced loads, in relation (3.50), we get $q(t) = 0$ and $Q(t) = g'(t)$ [see (3.43)]. Then the singular equation (3.52) can be simplified as follows:

$$\frac{1}{2\pi} \int_L \left[K(t, t') g'(t)dt + L(t, t') \overline{g'(t)dt}\right] = p(t'), \quad t' \in L, \tag{3.55}$$

where the kernels $K(t, t')$ and $L(t, t')$ are given by the formulas

$$\begin{gathered} K(t, t') = \frac{1}{t - t'} + \frac{1}{2} \frac{d}{dt'} \left(\ln \frac{t - t'}{\bar{t} - \bar{t}'} \right) = \frac{1}{2} \left[\frac{1}{t - t'} + \frac{1}{\bar{t} - \bar{t}'} \frac{\overline{dt'}}{dt'} \right]; \\ L(t, t') = -\frac{1}{2} \frac{d}{dt'} \left(\frac{t - t'}{\bar{t} - \bar{t}'} \right) = \frac{1}{2} \left[\frac{1}{\bar{t} - \bar{t}'} - \frac{t - t'}{(\bar{t} - \bar{t}')^2} \frac{\overline{dt'}}{dt'} \right]. \end{gathered} \tag{3.56}$$

Moreover, condition (3.54) must be satisfied.

Similarly, by using relations (3.17), (3.26), (3.30), and (3.32) and satisfying the boundary conditions (3.51), we write the singular integral equation of the second main problem

$$\begin{aligned} \frac{1}{2\pi} \int_L \Bigg[& \frac{(æ - 1)Q(t) - 2iq(t)}{t - t'} dt + k_1(t, t')(Q(t) + 2iq(t))dt \\ & - k_2(t, t') \overline{Q(t)\, dt} \Bigg] \\ = {} & f'(t'), \quad t' \in L, \end{aligned} \tag{3.57}$$

where the kernels $k_1(t, t')$ and $k_2(t, t')$ are the same as in Eq. (3.52).

Assume that, as in the solution of the second main problem for the plane with cuts [37], the principal vector of forces applied to the cut L (with projections X and Y onto the Ox- and Oy-axes, respectively). Then the unknown function in Eq. (3.57) satisfies the additional condition

$$\int_L q(t)dt = \frac{i}{2}(X + iY), \tag{3.58}$$

which follows from equality (3.7). Indeed, integrating equality (3.7) along the faces of the cut L and taking into account the fact that

$$[\mu'(t)]^+ - [\mu'(t)]^- = 2q(t), \quad t \in L \tag{3.59}$$

we arrive at relation (3.58).

The singular integral equations (3.52) and (3.57) belong to the type of equations studied in detail in [36]. In the class of functions unbounded near the ends a and b, i.e., in the class h_0 with the index $\kappa = 1$ ([36], p. 256), Eqs. (3.52) and (3.57) are always solvable. Under conditions (3.54) and (3.58), the solution of these equations is unique. Note that Eqs. (3.52) and (3.57) can also be represented in the form

$$\Phi(t) + \overline{\Phi(t')} + \frac{d\bar{t}'}{dt'}\left[t'\overline{\Phi'(t')} + \overline{\Psi(t')}\right] = p(t'), \quad t' \in L, \tag{3.60}$$

$$æ\Phi(t) - \overline{\Phi(t')} - \frac{d\bar{t}'}{dt'}\left[t'\overline{\Phi'(t')} + \overline{\Psi(t')}\right] = f'(t'), \quad t' \in L, \tag{3.61}$$

where $\Phi(t')$ and $\Psi(t')$ are direct values of the potentials $\Phi(z)$ and $\Psi(z)$, i.e., these values are obtained by the direct substitution of $z = t'$ in the expressions for $\Phi(z)$ and $\Psi(z)$. Relations (3.60) and (3.61) enable us to easily construct singular integral equations of the first and second main problems for domains with cuts if the integral representations of the complex potentials $\Phi(z)$ and $\Psi(z)$ in terms of the jumps of stresses and displacements on these cuts are known.

3.3 Stress Intensity Factors

Distributions of stresses and displacements in the vicinity of tips of the cracks. Consider the case of curvilinear crack (cut) whose faces are subjected to the action of the external load (3.50). The solution of this problem is given by potentials (3.45) and (3.49) in which the function $q(t)$ satisfies the condition H and the function $g'(t)$, i.e., the solution of Eq. (3.52), belongs to the class H^*. Thus, near the tips of the crack L, the principal term of the asymptotic expansions of stresses and displacements depends only on the function $g'(t)$. It follows from the theory of singular integral equations [36] [see also solution (3.35)] that the function $g'(t)$ near the beginning of the crack $a = l^-$ and its end $b = l^+$ can be represented in the form

$$g'(t) = \frac{g_*(t)}{\sqrt{t - l^\pm}}, \tag{3.62}$$

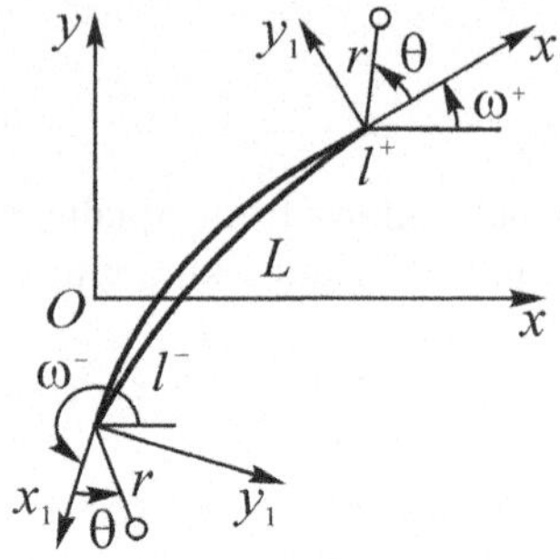

Fig. 3.4 Curvilinear crack in elastic plane

where $g_*(t)$ is a function that belongs to the class H on L in the vicinity of the ends $l^\pm$ and $\sqrt{t-l^\pm}$ is any of these branches, which continuously varies on L.

By using the well-known results on the behavior of the Cauchy-type integral near the ends of the line of integration [36] and the value of the integral

$$\int_L \frac{(\bar t - \bar z)dt}{\sqrt{t-l^\pm}(t-z)^2} = \int_L \frac{(\bar t - \bar l^\pm)}{(t-l^\pm)} \frac{\sqrt{t-l^\pm}dt}{(t-z)^2} + (\bar l^\pm - \bar z)\times$$
$$\times \int_L \frac{dt}{\sqrt{t-l^\pm}(t-z)^2} = \frac{1}{2}\int_L \frac{(\bar t - \bar l^\pm)}{(t-l^\pm)} \frac{dt}{\sqrt{t-l^\pm}(t-z)} +$$
$$+(\bar l^\pm - \bar z)\int_L \frac{dt}{\sqrt{t-l^\pm}(t-z)^2} + O(r^0) = \frac{\pi i}{2\sqrt{z-l^\pm}}\left(e^{-2i\omega^\pm} + \frac{\bar z - \bar l^\pm}{z-l^\pm}\right) + O(r^0) \tag{3.63}$$

in the vicinity of $z = l^\pm$, we can represent the functions $\Phi(z)$ and $\Psi(z) + \bar z\Phi'(z)$ at points quite close to $l^\pm$ but are not located on L in the form

$$\Phi(z) = \frac{ig_*(l^\pm)}{2\sqrt{z-l^\pm}} + O(r^0);$$
$$\Psi(z) + \bar z\Phi'(z) = -\frac{ig_*(l^\pm)}{4\sqrt{z-l^\pm}}\left(e^{-2i\omega^\pm} + \frac{\bar z - \bar l^\pm}{z-l^\pm}\right) - \overline{\frac{ig_*(l^\pm)}{2\sqrt{z-l^\pm}}}e^{-i\omega^\pm} + O(r^0), \tag{3.64}$$

where $\omega^+ = \theta(l^+)$, $\omega^- = \theta(l^-) + \pi$, $\theta(l^\pm)$ is the angle between the direction of the positive tangent to L at the ends $l^\pm$ and the Ox-axis (Fig. 3.4); $\sqrt{z-l^\pm}$ is a branch holomorphic near $l^\pm$ in the plane weakened by a crack made along L and taking the value $\sqrt{t-t^\pm}$ on the left side of L; $r = |z-l^\pm|$, and $O(r^n)$ means that the ratio $O(r^n)/r^n$ is bounded as $z \to l^\pm$.

To determine the distribution of stresses and displacements in a small vicinity of the crack ends, we pass to a new (polar) coordinate system with origin at the crack tip $z = l^+$ (or $z = l^-$) (Fig. 3.4), i.e., we set

$$z = l^{\pm} + z_1 e^{i\omega^{\pm}}; \qquad z_1 = re^{i\theta}, \qquad -\pi < \theta < \pi. \tag{3.65}$$

By using the formulas of transformation of the complex potentials in passing to the new coordinate system (3.11), in view of (3.64), we get

$$\begin{aligned} \Phi_1(z_1) &= \frac{K_{\mathrm{I}}^{\pm} - iK_{\mathrm{II}}^{\pm}}{2\sqrt{2\pi z_1}} + O(r^0); \\ \Psi_1(z_1) + \bar{z}_1 \Phi_1'(z_1) &= \frac{iK_{\mathrm{II}}^{\pm}}{\sqrt{2\pi z_1}} + \frac{K_{\mathrm{I}}^{\pm} - iK_{\mathrm{II}}^{\pm}}{4\sqrt{2\pi z_1}} \left(1 - \frac{\bar{z}_1}{z_1}\right) + O(r^0), \end{aligned} \tag{3.66}$$

where

$$K_{\mathrm{I}}^{\pm} - iK_{\mathrm{II}}^{\pm} = \mp \lim_{t \to l^{\pm}} \left[\sqrt{2\pi |t - l^{\pm}|}\, g'(t)\right]. \tag{3.67}$$

Here and in what follows, the real quantities $K_{\mathrm{I}}^{\pm}$ and $K_{\mathrm{II}}^{\pm}$ with lower signs correspond to the beginning of the crack ($z = a = l^-$), while the real quantities with the upper sign correspond to its end ($z = b = l^+$) (Fig. 3.4).

On the basis of relations (3.4)–(3.6) and (3.66), we determine the asymptotic distributions of stresses and displacements in the vicinities of the tips of curvilinear cracks [as a two-dimensional analog of relations (2.2)–(2.4)]:

$$\begin{aligned} \begin{pmatrix} \sigma_y \\ \sigma_x \\ \tau_{xy} \end{pmatrix} &= \frac{K_{\mathrm{I}}^{\pm}}{4\sqrt{2\pi r}} \begin{bmatrix} 5\cos\dfrac{\theta}{2} - \cos\dfrac{5\theta}{2} \\ 3\cos\dfrac{\theta}{2} + \cos\dfrac{5\theta}{2} \\ -\sin\dfrac{\theta}{2} + \sin\dfrac{5\theta}{2} \end{bmatrix} + \frac{K_{\mathrm{II}}^{\pm}}{4\sqrt{2\pi r}} \begin{bmatrix} -\sin\dfrac{\theta}{2} + \sin\dfrac{5\theta}{2} \\ -7\sin\dfrac{\theta}{2} - \sin\dfrac{5\theta}{2} \\ 3\cos\dfrac{\theta}{2} + \cos\dfrac{5\theta}{2} \end{bmatrix} \\ &+ O(r^0);\; 4G\begin{pmatrix} u \\ v \end{pmatrix} = K_{\mathrm{I}}^{\pm}\sqrt{\frac{r}{2\pi}} \begin{pmatrix} (2æ - 1)\cos\dfrac{\theta}{2} - \cos\dfrac{3\theta}{2} \\ (2æ + 1)\sin\dfrac{\theta}{2} - \sin\dfrac{3\theta}{2} \end{pmatrix} \\ &+ K_{\mathrm{II}}^{\pm}\sqrt{\frac{r}{2\pi}} \begin{pmatrix} (2æ + 3)\sin\dfrac{\theta}{2} + \sin\dfrac{3\theta}{2} \\ -(2æ - 3)\cos\dfrac{\theta}{2} - \cos\dfrac{3\theta}{2} \end{pmatrix} + O(r). \end{aligned} \tag{3.68}$$

Here, σ_x, σ_y, τ_{xy} and u, v are the components of stresses and displacements in the local coordinate system $x_1 y_1$ with origin at the crack tip (see Fig. 3.4). The quantities $K_{\mathrm{I}}^{\pm}$ and $K_{\mathrm{II}}^{\pm}$ are called the stress intensity factors for the distributions of stresses symmetric or antisymmetric about the crack line, respectively (in what follows, we sometimes omit the superscripts "+" and "–" and understand the quantities K_{I} and K_{II} as the stress intensity factors either at the beginning of the crack or at its end. They are functions of the load and the parameters characterizing the configuration of the body and the shape of the crack and can be found from the solution of the problem of the theory of elasticity.

The stress intensity factors play an extremely important role in the mechanics of brittle fracture [5, 27, 39, 46, 57]. The fact that the distributions of stresses and displacements near the crack tip always have the same functional dependence in the polar coordinates (r, θ) with origin at the crack tip for any configuration of the body, shape of the crack, and external load is quite important. Thus, the stress intensity factors can be regarded as the parameters reflecting the redistribution of stresses in the body as a result of crack initiation. Therefore, the distribution of stresses in a small vicinity of the crack tip can be found if we determine the values of the stress intensity factors K_{I} and K_{II}.

The independence of the asymptotic distributions (3.68) of stresses and displacements over r and θ on the crack shape the follows from the representation of solution of the singular integral equation (3.52) near the crack tip in the form (3.62). Thus, relations (3.68) are true in the general case of configuration of the body and the crack shape. In [5], Cherepanov rigorously proved the general character of distribution (3.68).

Relations (3.68) can be also obtained from the asymptotic distributions of the exact solutions of various partial problems in a small vicinity of the crack tip. The asymptotic relations (3.68) in [64, 65] were established just by using this method.

In a similar way, we can deduce the asymptotic relations for the stress distributions near the ends of a curvilinear cut with displacements specified on its faces. In this case, the function $q(t)$ satisfies representation (3.62) and the function $g'(t)$ is bounded near the ends of the cut. These distributions are not presented because, in what follows, we mainly consider the cases where external loads are applied on the faces of the cuts.

Stress intensity factors for rectilinear and curvilinear cracks.

Rectilinear cracks [44]. Consider the problem of determination of the stress-strain state of an infinite plane containing a rectilinear crack (cut) $|x| \le l$ along the Ox-axis whose faces are loaded by nonself-balanced forces (3.50) and the stresses are absent at infinity. For this problem, equality (3.52) yields the following integral equation:

$$\frac{1}{\pi}\int_{-l}^{l}\frac{Q(t)+iq(t)}{t-x}dt = p(x), \qquad |x| < l. \tag{3.69}$$

Under condition (3.54), in view of (3.37) and (3.38), this equation has the following solution:

$$g'(x) = -i\frac{æ-1}{æ+1}q(x) - \frac{1}{\pi\sqrt{l^2-x^2}}\left[\int_{-l}^{l}\frac{\sqrt{l^2-t^2}\,p(t)dt}{t-x} - i\frac{æ-1}{æ+1}\int_{-l}^{l}q(t)dt\right]. \tag{3.70}$$

Substituting solution (3.70) in relations (3.45) and (3.49), we easily obtain the expressions for the complex potentials $\Phi(z)$ and $\Psi(z)$, which coincide with the expressions presented in [37]. On the basis of relations (3.67) and (3.70), we determine the stress intensity factors

$$K_{\mathrm{I}}^{\pm} - iK_{\mathrm{II}}^{\pm} = -\frac{1}{\sqrt{\pi l}}\left[\int_{-l}^{l}\sqrt{\frac{l \pm t}{l \mp t}}\,p(t)dt \pm i\frac{æ - 1}{æ + 1}\int_{-l}^{l} q(t)dt\right]. \tag{3.71}$$

Here, the point $x = -\,l$ is the beginning of the crack and the point $x = +\,l$ is its end. Formula (3.71) was deduced for the first time (in a different way) in [58].

We now present the stress intensity factors for some special cases of loading. Assume that the crack faces are loaded by self-balanced ($q(t) = 0$) constant normal σ and tangential τ forces ($p(t) = -\sigma - i\tau = \text{const}$). Thus, it follows from relation (3.71) that

$$K_{\mathrm{I}}^{\pm} - iK_{\mathrm{II}}^{\pm} = (\sigma - i\tau)\sqrt{\pi l}, \tag{3.72}$$

whence, in particular, by the superposition method, we can get the values of the stress intensity factors for the cases where an infinite plane weakened by a load-free crack is subjected to tension by external forces p and q applied at infinity and acting in mutually perpendicular directions [see relations (3.20) and (3.23)]. In this case,

$$K_{\mathrm{I}}^{\pm} - iK_{\mathrm{II}}^{\pm} = \frac{1}{2}\left[p + q - (p - q)e^{2i\alpha_1}\right]\sqrt{\pi l}. \tag{3.73}$$

Assume that normal P and shear Q forces of the same magnitude but with opposite directions are applied at points $x = \xi$ $(|\xi| < l)$ on the upper and lower crack faces, i.e.,

$$p(x) = -(P - iQ)\,\delta(x - \xi); \quad q(x) = 0, \tag{3.74}$$

where $\delta(x - \xi)$ is the delta-function [62]. Substituting (3.74) in relation (3.71), we find

$$K_{\mathrm{I}}^{\pm} - iK_{\mathrm{II}}^{\pm} = \frac{P - iQ}{\sqrt{\pi l}}\sqrt{\frac{l \pm \xi}{l \mp \xi}}. \tag{3.75}$$

Note that the delta-function $\delta(x)$ is equal to zero for $x \neq 0$ and to infinity for $x = 0$. Hence, in this case,

$$\int_{-\infty}^{\infty}\delta(x)dx = 1; \quad \int_{a}^{b} f(\xi)\delta(\xi - x)d\xi = \begin{cases} 0 & \text{for } \; x < a \; \text{ and } \; x > b; \\ f(x) & \text{for } \; a < x < b; \\ f(x)/2 & \text{for } \; x = a \; \text{ or } \; x = b. \end{cases}$$

The obtained exact solutions can be used to construct approximate solutions for the curvilinear crack slightly different from a rectilinear (or arcwise) curve by the perturbation method (see [50, 53]).

Curvilinear crack [50, 53]. Consider the integral equation (3.55). Assume that the shape of a smooth curvilinear crack L is determined by the following parametric equation:

$$x = x(\xi), \quad y = y(\xi), \quad |\xi| \le 1. \tag{3.76}$$

In view of the fact that, for smooth contours, the functions $x(\xi)$ and $y(\xi)$ have continuous derivatives $x'(\xi)$ and $y'(\xi)$, which are not simultaneously, equal to zero (see [36], p. 13), we can perform the following change of variables in Eq. (3.55) [22, 50]:

$$t = l\omega(\xi) = l[x(\xi) + iy(\xi)], \quad |\xi| \le 1, \quad t' = l\omega(\eta), \quad |\eta| \le 1. \tag{3.77}$$

As a result, Eq. (3.55) and condition (3.54), take the following canonical dimensionless form [49, 53]:

$$\int_{-1}^{1} \left[M(\xi,\eta)\,\varphi(\xi) + N(\xi,\eta)\,\overline{\varphi(\xi)}\right]d\xi = \pi p(\eta), \quad |\eta| < 1. \tag{3.78}$$

$$\int_{-1}^{1} \varphi(\xi)d\xi = 0, \tag{3.79}$$

where

$$M(\xi,\ \eta) = lK(l\omega(\xi), l\omega(\eta)) = \frac{1}{2}\left[\frac{1}{\omega(\xi) - \omega(\eta)} + \frac{1}{\overline{\omega(\xi)} - \overline{\omega(\eta)}}\frac{\overline{\omega'(\eta)}}{\omega'(\eta)}\right];$$
$$N(\xi,\ \eta) = lL(l\omega(\xi),\ l\omega(\eta)) = \frac{1}{2}\left[\frac{1}{\overline{\omega(\xi)} - \overline{\omega(\eta)}} - \frac{\omega(\xi) - \omega(\eta)}{\left(\overline{\omega(\xi)} - \overline{\omega(\eta)}\right)^2}\frac{\overline{\omega'(\eta)}}{\omega'(\eta)}\right];$$
$$\varphi(\xi) = g'(l\omega(\xi))\,\omega'(\xi)/\,p; \quad p(\eta) = p(l\omega(\eta))\,/\,p. \tag{3.80}$$

The stress intensity factors are given by relation (3.67). By the change of variables (3.77), this relation can be represented in the form

$$K_{\mathrm{I}}^{\pm} - iK_{\mathrm{II}}^{\pm} = \mp p\sqrt{\pi l}\lim_{\eta\to\pm1}\left[\sqrt{(1-\eta^2)|\omega'(\eta)|}\frac{\varphi(\eta)}{\omega'(\eta)}\right], \tag{3.81}$$

where the lower signs correspond to the left crack tip (beginning), and the upper signs correspond to the right crack tip (end).

Note, two parameters p and l are distinguish here for convenience; these parameters have the dimension of stress and length respectively and will acquire concrete values (content), depending on the mode of loading and the shape of the region in the tasks discussed below.

3.4 System of Curvilinear Cracks in Elastic Plane. Singular Integral Equations

Integral equations of the main boundary problems [53]. Assume that the isotropic elastic plane referred to a Cartesian coordinate system xOy contains N smooth curvilinear cracks L_n ($n = 1, 2, \ldots, N$) without common points (Fig. 3.5). Since relations (3.26), (3.30), (3.38), and (3.42) are true not only for a single but also for a collection of smooth contours, it is easy to see that the integral representations for the complex potentials (3.45) and (3.49) and the integral equations (3.52) and (3.57) for the first and second main problems remain true for a system of smooth curvilinear cracks L_n ($n = 1, 2, \ldots, N$) if L is regarded as a collection of the contours L_n, i.e., $L = L_1 + L_2 + \ldots + L_N$. The additional conditions (3.54) and (3.58) should be written for each contour L_n separately. Combining these conditions, we get

$$\int_{L_n} Q(t)dt = -\frac{2i}{\text{æ}+1}\int_{L_n} q(t)dt = \frac{X_n + iY_n}{\text{æ}+1}, \quad n = 1, 2, \ldots, N, \tag{3.82}$$

where X_n and Y_n are the projections of the principal vector of forces applied to the crack L_n on the Ox- and Oy-axes, respectively.

Note that the integral equation (3.69) is also true for any system of cracks along the same straight line. It follows from the form of this equation and formula (3.35) that problems of this kind are reduced to the evaluation of the corresponding quadratures.

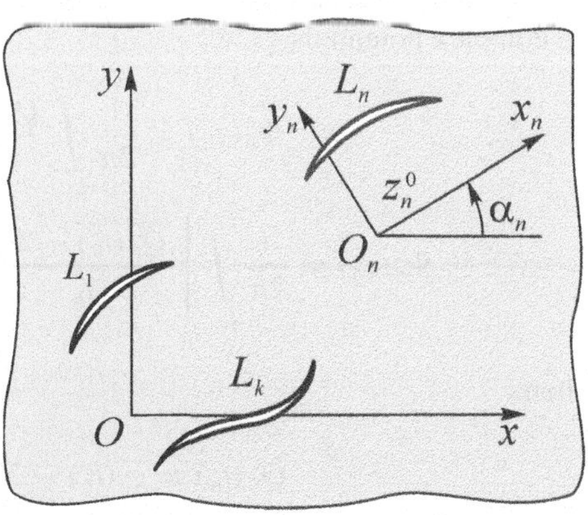

Fig. 3.5 System of curvilinear cracks in elastic plane

In this case, the complex integral equation (3.69) splits into two identical real equations corresponding to symmetric and antisymmetric stress distributions.

As in the case of a single contour, the integral equations (3.52) and (3.57) for the system of contours L_n ($n = 1, 2, \ldots, N$) always possess solutions in the class of functions unbounded at all ends of the contours L_n and this solution is unique in the case where N conditions (3.82) are satisfied.

In what follows, we represent the integral expressions for the complex potentials $\Phi(z)$ and $\Psi(z)$ and the integral equations of the main boundary problems for the infinite plane weakened by a system of curvilinear cracks in a somewhat different form. To do this we refer each contour L_n to a local coordinate system $x_nO_ny_n$ (Fig. 3.5). In the principal Cartesian coordinate system xOy, the points O_n are determined by the complex coordinates $z_n^0 = x_n^0 + iy_n^0$, and the O_nx_n-axes make angles α_n with the Ox-axis. Then the relationship between the coordinates of points of the plane in the local and principal coordinates is determined by the following relations:

$$z = z_n e^{i\alpha_n} + z_n^0; \qquad z = x + iy; \qquad z_n = x_n + iy_n. \tag{3.83}$$

Assume that the jumps of stresses and the derivatives of displacements

$$(N_n + iT_n)^+ - (N_n + iT_n)^- = 2q_n(t_n), \qquad t_n \in L_n; \tag{3.84}$$

$$\frac{d}{dt_n}\left[(u_n + iv_n)^+ - (u_n + iv_n)^-\right] = \frac{i(1 + æ)}{2G} g_n'(t_n), \qquad t_n \in L_n \tag{3.85}$$

are specified on the crack L_n referred to the coordinate system $x_nO_ny_n$. Moreover, the jump of displacements at the ends of the crack and the stresses and rotations at infinity are set equal to zero (here and in what follows, we use the same notation as in Sect. 3.3; the subscript n indicates that the corresponding quantity is presented in the local coordinates and corresponds to the contour L_n; in particular, t_n stands for the complex coordinate of a point of the contour L_n in the local coordinate system $x_nO_ny_n$).

According to relations (3.45) and (3.49), we have the following expressions for the complex potentials:

$$\Phi_n(z_n) = \frac{1}{2\pi} \int_{L_n} \frac{Q_n(t_n)dt_n}{t_n - z_n};$$

$$\Psi_n(z_n) = \frac{1}{2\pi} \int_{L_n} \left[\frac{\overline{Q_n(t_n)} - 2i\overline{q_n(t_n)}}{t_n - z_n} \overline{dt_n} - \frac{\overline{t_n} Q_n(t_n)dt_n}{(t_n - z_n)^2} \right], \tag{3.86}$$

where

$$Q_n(t_n) = g_n'(t_n) - 2iq_n(t_n) / (æ + 1). \tag{3.87}$$

In view of the linearity of the problem, the functions [49]

$$\Phi(z) = \frac{1}{2\pi}\sum_{k=1}^{N}\int_{L_k}\frac{Q_k(t_k)e^{i\alpha_k}}{T_k - z}dt_k, \qquad T_k = t_k e^{i\alpha_k} + z_k^0;$$

$$\Psi(z) = \frac{1}{2\pi}\sum_{k=1}^{N}\int_{L_k}\left[\frac{\overline{Q_k(t_k)} - 2i\overline{q_k(t_k)}}{T_k - z}e^{-i\alpha_k}\overline{dt_k} - \frac{\bar{T}_k Q_k(t_k)}{(T_k - z)^2}e^{i\alpha_k}dt_k\right], \tag{3.88}$$

obtained as a result of superposition of the complex potentials (3.86) for isolated cracks describe the stressed state of the elastic plane caused by the discontinuities of displacements $g_k(t_k)$ and stresses $q_k(t_k)$ on the N contours L_k ($k = 1, 2, \ldots, N$). Indeed, the functions $\Phi_n(z_n)$ and $\Psi_n(z_n)$ (3.86) are single-valued everywhere except the contour L_n, where they have discontinuities of displacements $g_n(t_n)$ and stresses $q_n(t_n)$. This is why the fields of stresses and displacements caused by these functions are continuous on the contours L_k ($k = 1, 2, \ldots, N$; $k \neq n$). Thus, the sums of these functions, i.e., the complex potentials (3.88), specify the solution of the problem in the case where the discontinuities of displacements $g_k(t_k)$ and stresses $q_k(t_k)$ are given on the N contours L_k ($k = 1, 2, \ldots, N$).

By using representation (3.88), we construct the integral equations of the problem for the case where, the stresses are given on some contours but displacements are specified on the other contours. Let the crack faces L_n ($n = 1, 2, \ldots, m$) be loaded by nonself-balanced forces

$$N_n^{\pm}(t_n) + iT_n^{\pm}(t_n) = p_n(t_n) \pm q_n(t_n), \qquad t_n \in L_n; \quad n = 1, 2, \ldots, m \tag{3.89}$$

and let the derivatives of displacements be specified on the other cracks L_n ($n = m + 1, \ldots, N$):

$$2G\frac{d}{dt_n}\left(u_n^{\pm} + iv_n^{\pm}\right) = f_n'(t_n) \mp \frac{æ + 1}{2i}g_n'(t_n), \qquad t_n \in L_n; \quad n = m + 1, \ldots, N. \tag{3.90}$$

Moreover, at infinity, the stresses and rotations are absent.

Satisfying the boundary conditions (3.89) and (3.90) with the help of potentials (3.88) on each crack, for the N unknown functions $g'_n(t_n)$ ($n = 1, 2, \ldots, m$) and $q_n(t_n)$ ($n = m + 1, \ldots, N$), we get the following system of N singular integral equations [52]:

$$\frac{1}{\pi}\sum_{k=1}^{N}\left[K_{nk}\left(t_k, t_n'\right)Q_k(t_k)dt_k + L_{nk}\left(t_k, t_n'\right)\overline{Q_k(t_k)}\,\overline{dt}_k + M_{nk}\left(t_k, t_n'\right)q_k(t_k)dt_k\right] =$$

$$= \begin{cases} p_n\left(t_n'\right), & n = 1, 2, \ldots, m; \\ -f_n'\left(t_n'\right), & n = m + 1, \ldots, N, \qquad t_n' \in L_n. \end{cases} \tag{3.91}$$

Here,

$$K_{nk}(t_k, t'_n) = \frac{e^{i\alpha_k}}{2}\left(\frac{1}{T_k - T'_n} + \frac{e^{-2i\alpha_n}}{\bar{T}_k - \bar{T}'_n}\frac{\overline{dt'_n}}{dt'_n}\right), \quad n = 1, 2, \ldots, m;$$

$$K_{nk}(t_k, t'_n) = \frac{e^{i\alpha_k}}{2}\left(-\frac{æ}{T_k - T'_n} + \frac{e^{-2i\alpha_n}}{\bar{T}_k - \bar{T}'_n}\frac{\overline{dt'_n}}{dt'_n}\right), \quad n = m + 1, \ldots, N;$$

$$L_{nk}(t_k, t'_n) = \frac{e^{-i\alpha_k}}{2}\left(\frac{1}{\bar{T}_k - \bar{T}'_n} - \frac{T_k - T'_n}{(\bar{T}_k - \bar{T}'_n)^2}\frac{\overline{dt'_n}}{dt'_n}e^{-2i\alpha_n}\right), \quad n = 1, 2, \ldots, N;$$

$$M_{nk}(t_k, t'_n) = \frac{ie^{i(\alpha_k - 2\alpha_n)}}{\bar{T}_k - \bar{T}'_n}\frac{\overline{dt'_n}}{dt'_n}; \qquad T'_n = t'_n e^{i\alpha_n} + z_n^0, \qquad n = 1, 2, \ldots, N. \quad (3.92)$$

The system of Eqs. (3.91) can be represented in the compact form as follows:

$$\Phi(T'_n) + \overline{\Phi(T'_n)} + e^{-2i\alpha_n}\frac{\overline{dt'_n}}{dt'_n}\left[T'_n\overline{\Phi'(T'_n)} + \overline{\Psi(T'_n)}\right] = p_n(t'_n), \quad t'_n \in L_n;$$
$$n = 1, 2, \ldots, m; \quad (3.93)$$

$$æ\Phi(T'_n) - \overline{\Phi(T'_n)} - e^{-2i\alpha_n}\frac{\overline{dt'_n}}{dt'_n}\left[T'_n\overline{\Phi'(T'_n)} + \overline{\Psi(T'_n)}\right] = f'_n(t'_n), \quad t'_n \in L_n;$$
$$n = m + 1, \ldots, N, \quad (3.94)$$

where $\Phi_n(T'_n)$ and $\Psi_n(T'_n)$ are the direct values of the complex potentials $\Phi(z)$ and $\Psi(z)$, i.e., the values obtained by the direct substitution of $z = T'_n$ in the expressions for $\Phi(z)$ and $\Psi(z)$.

The solution of system (3.91) must satisfy the conditions

$$\int_{L_n} Q_n(t_n)dt_n = -\frac{2i}{æ + 1}\int_{L_n} q_n(t_n)dt_n = \frac{X_n + iY_n}{æ + 1}e^{-i\alpha_n}, \quad n = 1, 2, \ldots, N \quad (3.95)$$

guaranteeing the single-valuedness of displacements in traversing the contours L_n.

The integral equations (3.91) remain true in the case where the displacements and the principal vector of total forces acting upon all indicated contours are given on the contours L_n ($n = m + 1, \ldots, N$). However, in this case, the additional conditions (3.95) for $n = m + 1, \ldots, N$ should be replaced by different conditions [35, 37]. Equations (3.91) belong to the already analyzed type of singular integral equations (but represented in a different form). Hence, if conditions (3.95) are satisfied, then these equations always possess a unique solution in the class of functions unbounded at all endpoints of the contours L_k.

System of rectilinear cracks arbitrarily located in the elastic plane [43–45]. Assume that the elastic isotropic plane referred to a Cartesian coordinate system xOy contains N rectilinear cuts (cracks) of length $2l_k$ ($k = 1, 2, \ldots, N$). The centers of the cracks O_k are specified by the coordinates $z_k^0 = x_k^0 + iy_k^0$. The origins of the local

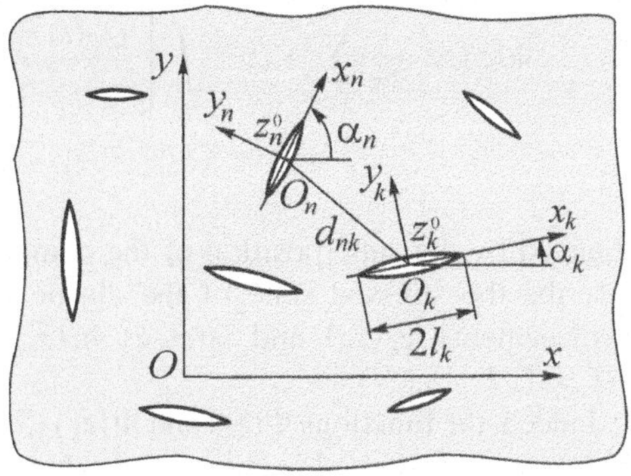

Fig. 3.6 System of rectilinear cracks arbitrarily located in the elastic plane

coordinate systems $x_kO_ky_k$ are located at the points O_k. The O_kx_k-axes coincide with the crack lines and form angles α_k with the Ox-axis (Fig. 3.6). The crack faces are loaded by forces

$$\sigma_k^{\pm} - i\tau_k^{\pm} = p_k(x_k) \pm q_k(x_k), \quad |x_k| < l_k, \quad k = 1, 2, \ldots, N, \tag{3.96}$$

and at infinity the stresses are absent. It is necessary to determine the stress-strain state of this domain. Here, σ_k and τ_k are the normal and tangential components of stresses in the local coordinate system $x_kO_ky_k$.

First, we consider the problem of determination of stresses in an unbounded elastic plane containing a single crack $|x_k| \leq l_k, \ y_k = 0$. Assume that discontinuities of displacements $g_k(x_k)$ and stresses $q_k(x_k)$ are specified on this crack. For the indicated problem, by virtue of formula (3.86), the complex potentials $\Phi(z_k)$ and $\Psi(z_k)$ in the coordinate system $x_kO_ky_k$ take the form

$$\Phi(z_k) = \frac{1}{2\pi}\int_{-l_k}^{l_k} \frac{Q_k(t)dt}{t - z_k};$$

$$\Psi(z_k) = \frac{1}{2\pi}\int_{-l_k}^{l_k} \left[\frac{\overline{Q_k(t)} - 2i\overline{q_k(t)}}{t - z_k} - \frac{tQ_k(t)}{(t - z_k)^2}\right] dt, \tag{3.97}$$

where

$$z_k = x_k + iy_k; \quad Q_k(x) = g_k'(x) - i\frac{2q_k(x)}{æ + 1}. \tag{3.98}$$

In view of the linearity of the problem, the functions

$$\Phi(z) = \frac{1}{2\pi}\sum_{k=1}^{N}\int_{-l_k}^{l_k} \frac{Q_k(t)dt}{t - z_k};$$

$$\Psi(z) = \frac{1}{2\pi}\sum_{k=1}^{N} e^{-2i\alpha_k} \int_{-l_k}^{l_k} \left[\frac{\overline{Q_k(t)} - 2i\overline{q_k(t)}}{t - z_k} - \frac{\bar{T}_k e^{i\alpha_k}}{(t - z_k)^2} Q_k(t) \right] dt;$$
$$T_k = te^{i\alpha_k} + z_k^0; \quad z_k = e^{-i\alpha_k}\left(z - z_k^0\right), \tag{3.99}$$

obtained by the superposition of the complex potentials (3.91) for isolated cracks describe the stressed state of the elastic plane caused by the discontinuities of displacements $g_k(x_k)$ and stresses $q_k(x_k)$ on N segments $|x_k| \le l_k, \ y_k = 0\,(k = 1, 2, \ldots, N)$.

Indeed, the functions $\Phi(z_k)$ and $\Psi(z_k)$ (3.97) are single-valued everywhere except the segment $|x_k| \le l_k, \ y_k = 0,$, where they have discontinuities of displacements $g_k(x_k)$ and stresses $q_k(x_k)$. This is why the fields of stresses and displacements caused by these functions are continuous on the segments $|x_n| \le l_n, \ y_n = 0\,(n = 1, 2, \ldots, N, \ n \ne k)$. Thus, the sums of these functions, i.e., the complex potentials (3.99), give the solution of the problem with discontinuities of displacements $g_k(x_k)$ and stresses $q_k(x_k)$ given on the N segments $|x_k| \le l_k, \ y_k = 0\,(k = 1, 2, \ldots, N)$.

We now pass to the solution of the problem of stress-strain state of an infinite plane with loading (3.96) given on N cracks $|x_k| \le l_k, \ y_k = 0\,(k = 1, 2, \ldots, N)$ in the absence of stresses at infinity. We seek the complex potentials for this problem in the form (3.99) under the assumption that, in this case, the discontinuities of displacements $g_k(x_k)$ are unknown functions. Since, in relation (3.99), the discontinuities of stresses $q_k(x_k)$ have already been taken into account, it remains to satisfy the following conditions on the cracks:

$$\sigma_n^+ + \sigma_n^- - i\left(\tau_n^+ + \tau_n^-\right) = 2p_n(x_n), \quad |x_n| < l_n, \quad n = 1, 2, \ldots, N. \tag{3.100}$$

By using the formulas of passing to the new coordinate system (3.11), we write the complex potentials $\Phi_n(z_n)$ and $\Psi_n(z_n)$ of the analyzed problem in the coordinate system $x_nO_ny_n$ as follows:

$$\Phi_n(z_n) = \frac{1}{2\pi}\sum_{k=1}^{N}\int_{-l_k}^{l_k} \frac{Q_k(t)dt}{t - z_k};$$

$$\Psi_n(z_n) = \frac{1}{2\pi}\sum_{k=1}^{N} e^{2i\alpha_{nk}} \int_{-l_k}^{l_k} \left[\frac{\overline{Q_k(t)} - 2i\overline{q_k(t)}}{t - z_k} - \frac{(\bar{T}_k - \bar{z}_n^0)e^{i\alpha_k}}{(t - z_k)^2} Q_k(t) \right] dt, \tag{3.101}$$

where $z_k = e^{-i\alpha_k}\left(z_n e^{i\alpha_n} + z_n^0 - z_k^0\right)$ and $\alpha_{nk} = \alpha_n - \alpha_k$.

We now find the stresses on the axes x_n by using relations (3.4) and (3.5) and substitute them in the boundary conditions (3.100). As a result, we arrive at the system of N singular integral equations of the analyzed problem:

$$\sum_{k=1}^{N}\int_{-l_k}^{l_k}\left[Q_k(t)K_{nk}(t,x)+\overline{Q_k(t)}L_{nk}(t,x)+\frac{iq_k(t)}{\bar{T}_k-\bar{X}_n}e^{i(\alpha_k-2\alpha_n)}\right]dt=\pi p_n(x),$$
$$|x|<l_n,\qquad n=1,\,2,\,\ldots,\,N. \tag{3.102}$$

Here,

$$K_{nk}(t,x)=\frac{e^{i\alpha_k}}{2}\left(\frac{1}{T_k-X_n}+\frac{e^{-2i\alpha_n}}{\bar{T}_k-\bar{X}_n}\right);$$
$$L_{nk}(t,x)=\frac{e^{-i\alpha_k}}{2}\left(\frac{1}{\bar{T}_k-\bar{X}_n}-\frac{T_k-X_n}{\left(\bar{T}_k-\bar{X}_n\right)^2}e^{-2i\alpha_n}\right); \tag{3.103}$$
$$X_n=xe^{i\alpha_n}+z_n^0.$$

For the sake of convenience, here and in what follows, we omit the subscript in x_n.

Kernels (3.103) of the obtained system of equations are regular except the case $n=k$ when $K_{nk}(t,x)$ turns into a singular Cauchy kernel and $L_{nk}(t,x)=0$. In this case, system (3.102) can be represented in the form

$$\int_{-l_n}^{l_n}\frac{g_n'(t)dt}{t-x}+\sum_{k\neq n}\int_{-l_k}^{l_k}\left[g_k'(t)K_{nk}(t,x)+\overline{g_k'(t)}L_{nk}(t,x)\right]dt=$$
$$=\pi p_n(x)+\frac{2i}{æ+1}\sum_{k=1}^{N}\int_{-l_k}^{l_k}\left\{q_k(t)\left[K_{nk}(t,x)-\frac{æ+1}{2(\bar{T}_k-\bar{X}_n)}e^{i(\alpha_k-2\alpha_n)}\right]\right.$$
$$\left.-\overline{q_k(t)}L_{nk}(t,x)\right\}dt,\ |x|<l_n,\qquad n=1,\,2,\,\ldots,\,N, \tag{3.104}$$

where $\sum_{k\neq n}$ denotes the sum $\sum_{k=1}^{N}$ and, for $k=n$, the corresponding term is equal to zero.

If a self-balanced load is applied to the crack faces, i.e., the boundary conditions (3.96) on cracks take the form

$$\sigma_k^+ + i\tau_k^+ = \sigma_k^- + i\tau_k^- = p_k(x_k),\qquad |x_k|<l_k,\qquad k=1,\,2,\,\ldots,\,N, \tag{3.105}$$

then the system of integral equations is simplified, namely,

$$\int_{-l_n}^{l_n}\frac{g_n'(t)dt}{t-x}+\sum_{k\neq n}\int_{-l_k}^{l_k}\left[g_k'(t)K_{nk}(t,x)+\overline{g_k'(t)}L_{nk}(t,x)\right]dt=\pi p_n(x),$$
$$|x|<l_n,\qquad n=1,\,2,\,\ldots,\,N \tag{3.106}$$

The integral equations (3.104) and (3.106) differ solely by the known right-hand sides. If we obtain the solution of system (3.106) under arbitrary loads $p_n(x)$, then we, in fact, also determine the solution of system (3.104). This is why, in what follows, we mainly consider the problems with self-balanced loads acting upon the cracks ($q_k(x_k) = 0$) because this type of loading is used most extensively. In particular, the problems in which the plane is loaded at infinity or at internal points are reduced to the case of self-balanced loads. It should be emphasized that if we have the solution of the problem in the case where concentrated forces act at the inner points of the domain, then the solution of the problem in the case of concentrated forces applied to a single face of the crack can be found by the limit transition. Thus, by using this solution obtained for the self-balanced loads acting upon the cracks, we can construct the fundamental solution of the problem for the general case of loading.

Note that singular integral equations for a system of arbitrarily located cracks whose faces are subjected to the action of a self-balanced load ($q_n(x_n) = 0$) were first deduced in [55] (see also [9–11]). The general case of loading was considered in the monograph [44]. We can also mention that the integral equations (3.102) can be obtained from Eqs. (3.91) under the assumption that rectilinear cracks are located on the $O_n x_n$-axes. Thus, we arrive at the integral equations (3.102) given on the rectilinear segments $t_n = x_n$.

3.5 Method of Mechanical Quadratures

The standard method aimed at the solution of singular integral equations is based on their regularization with subsequent numerical solution of obtained Fredholm integral equations of the second kind. However, this approach is quite cumbersome. In recent years, the direct methods for the solution of singular integral equations leading, without regularization, to the solution of systems of algebraic equations are especially extensively used for the numerical analyses. Among these methods, it is necessary to mention the method of mechanical quadratures based on certain formulas for interpolating polynomials and the quadrature formulas for singular integrals [4, 20, 28, 34, 60].

Numerical solution of singular integral equations. Consider a singular integral equation:

$$\int_{-1}^{1} \left[K(\xi, \eta)\, g'(\xi) + L(\xi, \eta)\overline{g'(\xi)}\right] d\xi = \pi P(\eta), \qquad |\eta| < 1, \tag{3.107}$$

where

$$K(\xi, \eta) = \frac{1}{\xi - \eta} + k(\xi, \eta). \tag{3.108}$$

The kernels $k(\xi, \eta)$ and $L(\xi, \eta)$ and the free term $P(\eta)$ are continuous functions of their arguments given on the segment $[-1, 1]$. The two-dimensional problems of the theory of cracks are reduced to Eq. (3.107) (or a system of equations of this kind). The solution of the integral equation (3.107) in the class of functions unbounded for $\xi = \pm 1$, i.e., for

$$g'(\xi) = \frac{u(\xi)}{\sqrt{1-\xi^2}}, \tag{3.109}$$

where $u(\xi)$ is a continuous function on the segment $[-1, 1]$, and satisfying the additional condition

$$\int_{-1}^{1} g'(\xi)\,d\xi = B, \tag{3.110}$$

where B is a known constant, exists and is unique [36].

The numerical solution of Eq. (3.107) is obtained by using the Gauss quadrature formulas. For the singular integral, we get [4]

$$\int_{-1}^{1} \frac{u(\xi)d\xi}{\sqrt{1-\xi^2}(\xi-\eta)} = \frac{\pi}{n}\sum_{k=1}^{n} \frac{u(\xi_k)}{\xi_k-\eta} + \pi u(\eta)\frac{U_{n-1}(\eta)}{T_n(\eta)}, \tag{3.111}$$

where $T_n(\eta) = \cos(n \arccos \eta)$ and $U_{n-1}(\eta) = \sin(n \arccos \eta)/\sqrt{1-\eta^2}$ are Chebyshev polynomials of the first and second kinds and the nodes ξ_k, which are the zeros of the polynomial $T_n(\xi)$, are given by the formula

$$\xi_k = \cos\frac{2k-1}{2n}\pi, \quad k = 1, 2, \ldots, n. \tag{3.112}$$

At the points

$$\eta_m = \cos\frac{\pi m}{n}, \quad m = 1, 2, \ldots, n-1, \tag{3.113}$$

as the roots of the equation $U_{n-1}(\eta) = 0$, the quadrature formula (3.111) takes a simpler form

$$\int_{-1}^{1} \frac{u(\xi)d\xi}{\sqrt{1-\xi^2}(\xi-\eta_m)} = \frac{\pi}{n}\sum_{k=1}^{n} \frac{u(\xi_k)}{\xi_k-\eta_m}. \tag{3.114}$$

The ordinary Gauss quadrature formula for function (3.110) has the form [38]

$$\int_{-1}^{1} \frac{u(\xi)d\xi}{\sqrt{1-\xi^2}} = \frac{\pi}{n}\sum_{k=1}^{n} u(\xi_k). \tag{3.115}$$

Thus, formula (3.114) for a singular integral with Cauchy kernel valid in the discrete set of points $\eta = \eta_m$ (3.113), coincides with the ordinary quadrature formula (3.115). Relations (3.111) and (3.115) are exact if $u(\xi)$ is a polynomial of degree not greater than $2n$ and $2n - 1$, respectively.

We now apply the quadrature formulas (3.114) and (3.115) to Eq. (3.107) and integral (3.110). As a result, we arrive at a system of n linear algebraic equations

$$\left.\begin{aligned} &\frac{1}{n}\sum_{k=1}^{n}\left[K(\xi_k, \eta_m)u(\xi_k) + L(\xi_k, \eta_m)\overline{u(\xi_k)}\right] = P(\eta_m), \quad m = 1,\ 2,\ \ldots,\ n-1 \\ &\frac{\pi}{n}\sum_{k=1}^{n} u(\xi_k) = B \end{aligned}\right\} \tag{3.116}$$

for the n unknown quantities $u(\xi_k)$ ($k = 1, 2, \ldots, n$).

The system of algebraic equations (3.116), which is a discrete analog of the integral equation (3.107) and condition (3.110) was deduced for the first time in [20].

By using the Lagrange interpolating polynomial for the required function $u(\xi)$ at nodes (3.112) (see [38], p. 527)

$$u(\xi) = \frac{1}{n}\sum_{k=1}^{n}(-1)^{k+1}u(\xi_k)\frac{T_n(\xi)\sqrt{1-\xi_k^2}}{\xi - \xi_k}, \tag{3.117}$$

we find the values of the function $u(\xi)$ at the points $\xi = \pm 1$

$$\begin{aligned} u(1) &= -\frac{1}{n}\sum_{k=1}^{n}(-1)^k u(\xi_k)\operatorname{ctg}\frac{2k-1}{4n}\pi; \\ u(-1) &= \frac{1}{n}\sum_{k=1}^{n}(-1)^{k+n}u(\xi_k)\operatorname{tg}\frac{2k-1}{4n}\pi, \end{aligned} \tag{3.118}$$

Note that the stress intensity factors are expressed via these quantities [see relation (3.81)];

$$K_{\mathrm{I}}^{\pm} - iK_{\mathrm{II}}^{\pm} = \mp p\sqrt{\pi l}\sqrt{|\omega'(\pm 1)|}\frac{u(\pm 1)}{\omega'(\pm 1)} = p\sqrt{\pi l}\left(F_{\mathrm{I}}^{\pm} - iF_{\mathrm{II}}^{\pm}\right); \tag{3.119}$$

here $F_{\mathrm{I}}^{\pm}$, $F_{\mathrm{II}}^{\pm}$ are nondimensional (relative) SIF.

Thus, for the singular integral equation (3.107) or a system of equations of this kind (the generalization of the presented results to the system of integral equations is obvious), one can easily construct the corresponding system of linear algebraic equations (3.116) and, on the basis of its solution, the numerical (approximate) solution of the initial equation (3.107) can be found by interpolation. Note that the convergence of the process for smooth functions $k(\xi, \eta)$, $L(\xi, \eta)$, and $P(\eta)$ follows from the convergence of the quadrature formulas (3.114) and (3.115) and the uniqueness of the solution of Eq. (3.107) under condition (3.110). The high efficiency of the proposed method is demonstrated by specific examples.

3.6 System of Curvilinear Cracks in Elastic Half-Plane. Singular Integral Equations (Crack Faces Are Not in Contact)

Some model contact loads. The processes of sliding and rolling take place in all contact zones in the units of machines and mechanisms. In this case, in the nearsurface areas of the elements of these units, parallel with other defects, we observe the initiation of cracks leading either to the degradation of the contact surface or to the development of the main crack and complete loss of the serviceability of joints. This is especially well visible [30, 47, 59, 63] for the forms of contact interaction accompanied by high levels of friction, namely, fretting fatigue, rolling under the conditions of dry friction, friction fatigue, etc.

As a rule, elements of tribojoints damaged by cracks are characterized by the formation of complex stressed state, and the solution of the problems of its determination encounters significant mathematical difficulties. Hence, in the construction of two-dimensional models, it is reasonable to use certain simplifications. Thus, in two-dimensional models, the damaged body is modeled by an elastic half-plane with cracks and the action of counterbody is replaced by the action of normal $p(x)$ and tangential $q(x)$ forces distributed in certain way over the boundary of the half-plane (Fig. 3.7). These forces reflect the shape and sizes of the counterbody, the type of the contact interaction, the mechanical characteristics of materials and contact surfaces, etc. In Fig. 3.8, some most extensively used schemes of model contact loadings are shown. In Fig. 3.8a–d, the tangential and normal forces obey the Coulomb-Amonton law with a friction coefficient f: $q(x) = fp(x)$. This enables us to model the conditions of slip between bodies. According to Fig. 3.8e, f, the normal forces $p(x)$ are distributed according to the elliptic (Hertz) law, whereas the tangential forces $q(x)$ have a more complicated structure (this is discussed in what follows). These schemes enable us to model the conditions under which the contact zone of the bodies contains both the regions of slip and sticking (see, e.g., [26]). The specific feature of the scheme depicted in Fig. 3.8d is that this scheme enables one to take into account the presence of the line (or band for cylindrical bodies) of previous contact and the shape of the edges of the counterbody (degree of rounding) (see [1]). It is clear that, parallel with contact loads, the elements of units are often subjected to the action of

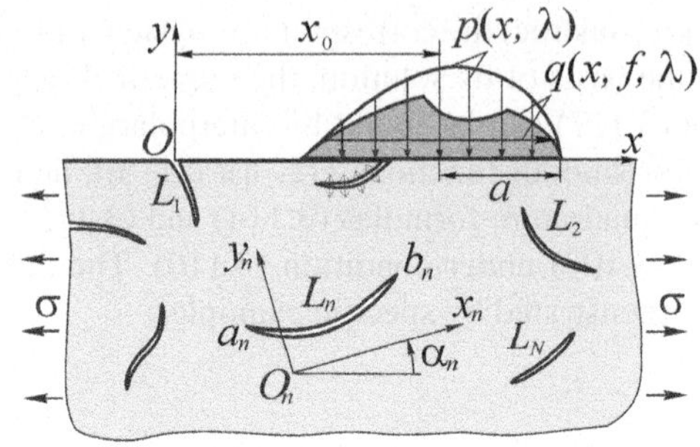

Fig. 3.7 General scheme of the problem

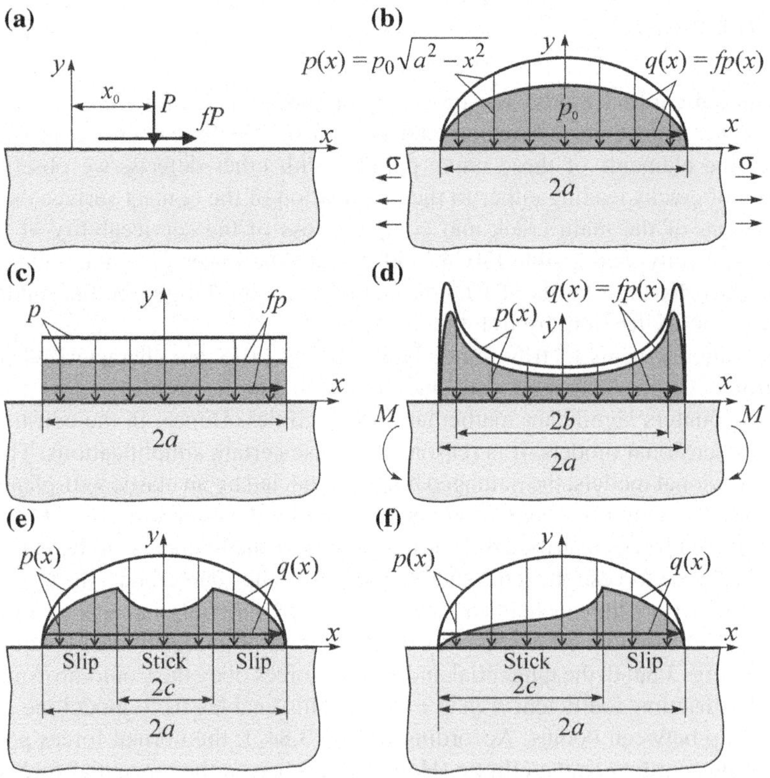

Fig. 3.8 Most extensively used schemes of model contact loading

nominal bearing tensile, compression, bending, and other loads (see Fig. 3.8b, d), which should also be taken into account.

As already indicated, the problem of finding the stress intensity factors and other parameters of fracture mechanics in a half-plane with cracks subjected to the action of model contact loads is one of the most important problems in the evaluation of lifetime. The solutions of problems of this kind were analyzed in Chap. 1 and the surveys of the accumulated results can be also found in [15, 17, 19, 24, 29, 30, 42, 48]. Note that, in the literature, the cases where the half-plane contains a crack perpendicular to its boundary and subjected to the action of a concentrated force, constant pressure, or elliptic loading on this boundary, have been extensively investigated (Fig. 3.8a–c). The number of investigations devoted to the case of inclined crack is smaller and the number works dealing with curvilinear cracks is even smaller. The data on the SIF for the last three loading schemes are practically absent (Fig. 3.8d–f).

In the present section, by using the relations of the main problems of the plane theory of elasticity for a continuous half-plane [37] and the Muskhelishvili complex potentials for a half-plane weakened by a system of arbitrarily located rectilinear [44] or curvilinear [53] cracks, we construct singular integral equations for a half-plane weakened by a system of curvilinear cracks subjected to the action of model contact loads applied to the boundary of the half-plane. In particular, we deduce singular integral equations for numerous important cases used in Chaps. 4 and 5 to construct the paths of crack propagation under the action of cyclic contact loads. We also present numerous examples of evaluation of the stress intensity factors (SIF) for curvilinear and rectilinear cracks in the half-plane.

Main problems of the theory of elasticity for a half-plane. Assume that the infinite plane contains a cut L covering the entire real axis Ox. Suppose that the stresses are continuous in passing through the contour L,, i.e.,

$$\sigma_y^{\pm} - i\tau_{xy}^{\pm} = p(x), \qquad |x| < \infty, \tag{3.120}$$

whereas the displacements have a jump $g(x)$ (3.51). We also assume that the stresses $p(x)$ on L are known and the function $p(x)$ belongs to the class H and has the order $o(1/x)$ for large $|x|$. Thus, in view of Eq. (3.52), we arrive at a singular integral equation of the first main problem for the half-plane [44]

$$\int_{-\infty}^{\infty} \frac{g'(t)dt}{t-x} = \pi p(x), \qquad |x| < \infty. \tag{3.121}$$

The solution of Eq. (3.121) is given by the Hilbert representation (see [2], p. 171) as follows:

$$g'(x) = -\frac{1}{\pi} \int_{-\infty}^{\infty} \frac{p(t)dt}{t-x}. \tag{3.122}$$

If the function $g'(x)$ is known, then the complex stress potentials of are given by the following formulas:

$$\Phi(z) = \frac{1}{2\pi} \int_{-\infty}^{\infty} \frac{g'(t)dt}{t-z}; \qquad \Psi(z) = \frac{1}{2\pi} \int_{-\infty}^{\infty} \left[\frac{\overline{g'(t)}}{t-z} - \frac{tg'(t)}{(t-z)^2} \right] dt. \tag{3.123}$$

Suppose that a body occupies the lower half-plane ($y \leq 0$). Substituting solution (3.122) in (3.123), changing the order of integration, and taking the integrals, we get

$$\Phi(z) = -\frac{1}{2\pi i}\int_{-\infty}^{\infty}\frac{p(t)dt}{t-z}; \qquad \Psi(z) = -\frac{1}{2\pi i}\int_{-\infty}^{\infty}\left[\frac{\overline{p(t)}}{t-z} - \frac{tp(t)}{(t-z)^2}\right]dt. \tag{3.124}$$

This result coincides with the well-known solution of the first main problem of the theory of elasticity for the half-plane (see [37], p. 408).

Consider the second main problem for the case where the displacements are specified on the edge of the half-plane. Assume that the displacements are continuous in passing through the contour L ($|x| < \infty,\ y = 0$) in the infinite plane, i.e.,

$$2G\frac{d}{dx}\left(u^{\pm} + iv^{\pm}\right) = f'(x), \quad |x| < \infty, \tag{3.125}$$

and that the stresses have a discontinuity $q(x)$ (3.50). If the function $f\,'(x)$ is known, then equality (3.57) yields the following singular integral equation:

$$\frac{2æ}{æ+1}\int_{-\infty}^{\infty}\frac{q(t)}{t-x}dt = \pi i f'(x), \quad |x| < \infty, \tag{3.126}$$

Here, the function $f(x)$ belongs to the class H and, for large $|x|$, has the order $o(1/x)$. The solution of Eq. (3.126) takes the form

$$q(x) = \frac{æ+1}{2æ}\frac{1}{\pi i}\int_{-\infty}^{\infty}\frac{f'(t)dt}{t-x}, \quad |x| < \infty. \tag{3.127}$$

Then the complex stress potentials of are given by relations (3.45) and (3.49). In the analyzed case, these relations take the form

$$\Phi(z) = \frac{1}{\pi i(æ+1)}\int_{-\infty}^{\infty}\frac{q(t)dt}{t-z}; \qquad \Psi(z) = \frac{1}{\pi i(æ+1)}\int_{-\infty}^{\infty}\left[\frac{æ\overline{q(t)}}{t-z} - \frac{tq(t)}{(t-z)^2}\right]dt. \tag{3.128}$$

By using relations (3.127) and (3.128), we find the solution of the second main problem for the lower half-plane:

$$\Phi(z) = -\frac{1}{2\pi iæ}\int_{-\infty}^{\infty}\frac{f'(t)dt}{t-z}; \qquad \Psi(z) = \frac{1}{2\pi iæ}\int_{-\infty}^{\infty}\left[\frac{æ\overline{f'(t)}}{t-z} + \frac{tf'(t)}{(t-z)^2}\right]dt, \tag{3.129}$$

which coincides with the well-known result (see [37], p. 409).

Complex potentials for a system of curvilinear cracks in the half-plane. Consider an infinite elastic plane containing $N+1$ cracks (cuts) L_n $(n = 0, 1, \ldots, N)$ referred to a local coordinate system $x_nO_ny_n$ (see Fig. 3.7). Assume that the contour L_0 is coincides with the real axis Ox. We also assume that the local coordinate system $x_0O_0y_0$ of the contour L_0 coincides with the principal coordinate system xOy $(\alpha_0 = 0,\ z_0^0 = 0)$ and the other cracks are located in the lower half-plane $(y < 0)$. A self-balanced load $p_0(x)$ $(q_0(x) = 0)$ is specified on the contour L_0 and the jumps of displacements $g_n(t_n)$ and stresses $q_n(t_n)$ $(n = 1, 2, \ldots, N)$ are specified on the other contours. For the considered problem, the complex potentials $\Phi_1(z)$ and $\Psi_1(z)$ have the form

$$\Phi_1(z) = \frac{1}{2\pi}\int_{-\infty}^{\infty}\frac{g_0'(t)dt}{t-z} + \Phi(z);$$
$$\Psi_1(z) = \frac{1}{2\pi}\int_{-\infty}^{\infty}\left[\frac{\overline{g_0'(t)}}{t-z} - \frac{t g_0'(t)}{(t-z)^2}\right]dt + \Psi(z), \tag{3.130}$$

where the functions $\Phi(z)$ and $\Psi(z)$ are given by relations (3.88). To determine the unknown function $g_0'(x)$, we use Eq. (3.121) in which

$$p(x) = p_0(x) - \frac{1}{2\pi}\sum_{k=1}^{N}\int_{L_k}\left\{\left[\frac{Q_k(t_k)}{T_k - x} + \frac{Q_k(t_k) + 2iq_k(t_k)}{\bar{T}_k - x}\right]e^{i\alpha_k}dt_k + \right.$$
$$\left. + \left[\frac{1}{\bar{T}_k - x} - \frac{T_k - x}{(\bar{T}_k - x)^2}\right]\overline{Q_k(t_k)\,e^{-i\alpha_k}dt_k}\right\}; \qquad \mathrm{Im}T_k < 0. \tag{3.131}$$

Thus, by virtue of relations (3.124), (3.130), and (3.131), we find [52]

$$\Phi_1(z) = \Phi(z) + \Phi_0(z)$$
$$+\frac{1}{2\pi}\sum_{k=1}^{N}\int_{L_k}\left[\frac{Q_k(t_k) + 2iq_k(t_k)}{\bar{T}_k - z}e^{i\alpha_k}dt_k + \frac{T_k - \bar{T}_k}{(\bar{T}_k - z)^2}\overline{Q_k(t_k)\,e^{-i\alpha_k}dt_k}\right];$$
$$\Psi_1(z) = \Psi(z) + \Psi_0(z)$$
$$+\frac{1}{2\pi}\sum_{k=1}^{N}\int_{L_k}\left\{\frac{Q_k(t_k) + 2iq_k(t_k)}{(\bar{T}_k - z)^2}\bar{T}_k e^{i\alpha_k}dt_k + \right.$$
$$\left. + \left[\frac{(\bar{T}_k - T_k)(\bar{T}_k + z)}{(\bar{T}_k - z)^3} - \frac{1}{\bar{T}_k - z}\right]\overline{Q_k(t_k)\,e^{-i\alpha_k}dt_k}\right], \tag{3.132}$$

where the functions $\Phi_0(z)$ and $\Psi_0(z)$ specify the solution for a continuous half-plane and the forces $p_0(x)$ given on the boundary

$$\Phi_0(z) = -\frac{1}{2\pi i}\int_{-\infty}^{\infty}\frac{p_0(t)dt}{t-z}; \qquad \Psi_0(z) = -\frac{1}{2\pi i}\int_{-\infty}^{\infty}\left[\frac{\overline{p_0(t)}}{t-z} - \frac{tp_0(t)}{(t-z)^2}\right]dt. \tag{3.133}$$

In a similar way, we obtain the integral representation for the potentials $\Phi_1(z)$ and $\Psi_1(z)$:

$$\Phi_1(z) = \Phi(z) + \Phi_0(z)$$
$$+\frac{1}{2\pi æ}\sum_{k=1}^{N}\int_{L_k}\left[\frac{Q_k(t_k)+2iq_k(t_k)}{\bar{T}_k - z}e^{i\alpha_k}dt_k - \frac{T_k-\bar{T}_k}{(\bar{T}_k-z)^2}\overline{Q_k(t_k)\,e^{-i\alpha_k}dt_k}\right];$$
$$\Psi_1(z) = \Psi(z)$$
$$+\Psi_0(z) + \frac{1}{2\pi æ}\sum_{k=1}^{N}\int_{L_k}\left\{\left[\frac{æ^2}{\bar{T}_k - z} + \frac{(T_k-\bar{T}_k)(\bar{T}_k+z)}{(\bar{T}_k-z)^3}\right]\times\right.$$
$$\left.\times\overline{Q_k(t_k)\,e^{-i\alpha_k}dt_k} - \frac{Q_k(t_k)+2iq_k(t_k)}{(\bar{T}_k-z)^2}\bar{T}_k e^{i\alpha_k}dt_k\right]. \tag{3.134}$$

in the case where displacements are given on the edge of the half-plane. Here, $\Phi_0(z)$ and $\Psi_0(z)$ specify the solution of the problem for a continuous half-plane with displacements $f_0(x)$ given on its edge

$$\Phi_0(z) = -\frac{1}{2\pi i æ}\int_{-\infty}^{\infty}\frac{f_0'(t)dt}{t-z}; \qquad \Psi_0(z) = \frac{1}{2\pi i æ}\int_{-\infty}^{\infty}\left[\frac{æ\overline{f_0'(t)}}{t-z} + \frac{tf_0'(t)}{(t-z)^2}\right]dt \tag{3.135}$$

and the functions $\Phi(z)$ and $\Psi(z)$ are given by relations (3.88).

Singular integral equations for curvilinear cracks in a half-plane [8]. Assume that a semi-infinite elastic isotropic plane occupying the lower part of the plane xOy ($y \le 0$) contain N smooth curvilinear cracks (cuts) L_n ($n = 1, \ldots, N$) (Fig. 3.7). We refer the contours of the cracks L_n to local coordinate systems $x_nO_ny_n$ connected with the principal coordinate system xOy by the formula $z = z_ne^{i\alpha_n} + z_n^0$ (z_n^0 is a complex coordinate of the point O_n in the principal coordinate system and α_n is the angle of inclination of the axis O_nx_n to Ox, $z_n = x_n + iy_n$).

The faces of the cracks are subjected to the action of nonself-balanced forces

$$N_n^{\pm}(t_n) + iT_n^{\pm}(t_n) = p_n(t_n) \pm q_n(t_n), \qquad t_n \in L_n, \qquad n = 1, \ldots, N, \tag{3.136}$$

and the loads $p_0(x)$ are applied to the edge of the half-plane. The functions $p_n(t_n)$, $q_n(t_n)$, and $p_0(x)$ belong to the class H.

Assume that the cracks faces are not in contact. This enables to represent the complex stress potentials of the problem in the form (3.132). Note that if the plate at the inner points and at infinity is subjected to the action of finitely many force factors S_j ($j = 1, 2, \ldots$), then, in (3.132), it is necessary to add the complex potentials $\Phi_0^s(z)$ and $\Psi_0^s(z)$ corresponding to the stress-strain state of the half-plane without cracks under the action of these factors. In the case where the half-plane is stretched at infinity by constant forces of intensity p parallel to the half-plane boundary, the complex potentials take the form [37].

$$\Phi_0^s(z) = \frac{p}{4}; \qquad \Psi_0^s(z) = -\frac{p}{2}$$

Satisfying the boundary conditions (3.136) on each cut with the help of potentials (3.132), we reduce the problem to N singular integral equations for the unknown functions $g'_n(t_n)$ ($n = 1, \ldots, N$):

$$\sum_{k=1}^{N} \int_{L_k} \left[R_{nk}\left(t_k, t_n'\right) Q_k(t_k)dt_k + S_{nk}\left(t_k, t_n'\right) \overline{Q_k(t_k)}d\bar{t}_k + E_{nk}\left(t_k, t_n'\right) q_k(t_k)dt_k + \right.$$
$$\left. + F_{nk}\left(t_k, t_n'\right) \overline{q_k(t_k)}d\bar{t}_k\right] + G\left(t_n'\right) = \pi p_n^*\left(t_n'\right), \quad t_n' \in L_n, \quad n = 1, \ldots, N, \tag{3.137}$$

where

$$R_{nk}\left(t_k, t_n'\right) = K_{nk}\left(t_k, t_n'\right) + \frac{e^{i\alpha_k}}{2}\left\{\frac{1}{T_n' - \bar{T}_k} + \frac{\bar{T}_k - T_k}{\left(T_n' - \bar{T}_k\right)^2} + \right.$$
$$\left. + \left[\frac{1}{\bar{T}'_n - T_k} + \frac{\bar{T}_k - T_k}{\left(\bar{T}'_n - T_k\right)^2} + 2\frac{\left(T_k - \bar{T}_k\right)\left(T_n' - \bar{T}_k\right)}{\left(\bar{T}'_n - T_k\right)^3}\right]\frac{d\bar{t}_n'}{dt_n'}e^{-2i\alpha_n}\right\};$$

$$S_{nk}\left(t_k, t_n'\right) = L_{nk}\left(t_k, t_n'\right)$$
$$+ \frac{e^{-i\alpha_k}}{2}\left[\frac{1}{\bar{T}'_n - T_k} + \frac{T_k - \bar{T}_k}{\left(\bar{T}'_n - T_k\right)^2} - \frac{T_n' - T_k}{\left(\bar{T}'_n - T_k\right)^2}\frac{d\bar{t}_n'}{dt_n'}e^{-2i\alpha_n}\right];$$

$$E_{nk}\left(t_k, t_n'\right) = M_{nk}\left(t_k, t_n'\right) + \frac{i e^{i\alpha_k}}{T_n' - \bar{T}_k};$$

$$F_{nk}\left(t_k, t_n'\right) = i e^{-i\alpha_k}\left[\frac{T_n' - T_k}{\left(\bar{T}'_n - T_k\right)^2}\frac{d\bar{t}_n'}{dt_n'}e^{-2i\alpha_n} - \frac{1}{\bar{T}'_n - T_k}\right];$$

$$G\left(t_n'\right) = \frac{1}{2i}\int_{-\infty}^{\infty}\left\{\frac{\overline{p_0(t)}}{t - \bar{T}'_n} - \frac{p_0(t)}{t - \bar{T}'_n}\right.$$

$$+\left[\frac{p_0(t)}{t-\bar{T}'_n}-\frac{t-T'_n}{\left(t-\bar{T}'_n\right)^2}\overline{p_0(t)}\right]\frac{d\bar{t}'_n}{dt'_n}e^{-2i\alpha_n}\Bigg\}dt;$$

$$K_{nk}\left(t_k,t'_n\right)=\frac{e^{i\alpha_k}}{2}\left(\frac{1}{T_k-T'_n}+\frac{e^{-2i\alpha_n}}{\bar{T}_k-\bar{T}'_n}\frac{d\bar{t}'_n}{dt'_n}\right);$$

$$L_{nk}\left(t_k,t'_n\right)=\frac{e^{-i\alpha_k}}{2}\left[\frac{1}{\bar{T}_k-\bar{T}'_n}-\frac{T_k-T'_n}{\left(\bar{T}_k-\bar{T}'_n\right)^2}\frac{d\bar{t}'_n}{dt'_n}e^{-2i\alpha_n}\right];$$

$$M_{nk}\left(t_k,t'_n\right)=\frac{ie^{i(\alpha_k-2\alpha_n)}}{\bar{T}_k-\bar{T}'_n}\frac{d\bar{t}'_n}{dt'_n};$$

$$T'_n=t'_ne^{i\alpha_n}+z^0_n \qquad p^*_n\left(t'_n\right)=p_n\left(t'_n\right)+p^s_n\left(t'_n\right).$$

The expressions $K_{nk}\left(t_k,t'_n\right)$, $L_{nk}\left(t_k,t'_n\right)$, and $M_{nk}\left(t_k,t'_n\right)$ coincide with the kernels of the singular integral equations for an infinite plane with curvilinear cracks [53] and the other components of the kernels in system (3.137) and the term $G\left(t'_n\right)$ characterize the influence of the edge of the half-plane. The functions $p^s_n\left(t'_n\right)$ are determined by the force factors S_j.

The solution of system (3.137) must satisfy the conditions

$$\int\limits_{L_n} g'_n(t_n)dt_n=0, \qquad n=1,\ldots,N, \tag{3.138}$$

guaranteeing the single-valuedness of displacements in traversing the contours L_n.

We establish the functions $g'_n(t_n)$ from the system of equations (3.137) and substitute them in relations (3.132). This gives the solution of the posed problem.

We also write the SIE (3.137) in the normalized form for a simpler case, which is used in what follow for specific calculations. Assume that an arbitrary model load acts upon the edge of the half-plane and corresponds to the following boundary conditions:

$$\begin{aligned}&\sigma_y(x)-i\tau_{xy}(x)=-p(x)-iq(x), \qquad |x-x_0|\le a, \qquad y=0;\\&\sigma_y(x)-i\tau_{xy}(x)=0, \qquad |x-x_0|>a, \qquad y=0.\end{aligned} \tag{3.139}$$

The crack faces are not in contact. They are loaded by the self-balanced forces $p_n(t_n)$, i.e.,

$$N^{\pm}_n(t_n)+iT^{\pm}_n(t_n)=p_n(t_n), \qquad t_n\in L_n, \qquad n=1,\ldots,N. \tag{3.140}$$

The corresponding system of singular integral equations (SIE) of the problem in the normalized form can be written as follows:

$$\sum_{k=1}^{N}\int_{-1}^{1}\left[R_{nk}(\xi\,,\eta)\,\varphi_k(\xi)+S_{nk}(\xi,\,\eta)\,\overline{\varphi_k(\xi)}\,\right]d\xi \\ =\pi\,P_n(\eta),\quad |\eta|<1, n=1,\,\ldots\,,\,N, \tag{3.141}$$

where $\varphi_n(\xi)=g_n'(t)\,\omega_n'(\xi)$, and $t_n=l_n\omega_n(\xi)\,(-1\le\xi\le1)$ are parametric equations of the crack contours.

The kernels $R_{nk}(\xi, \eta)$ and $S_{nk}(\xi, \eta)$ of the SIE (3.141) have the form

$$\begin{aligned} R_{nk}(\xi\,,\eta)&=K_{nk}(\xi\,,\eta)+\frac{e^{i\alpha_k}}{2}\left\{\frac{1}{T_n'-\bar{T}_k}+\frac{\bar{T}_k-T_k}{\left(\bar{T}_n'-T_k\right)^2}+\right.\\ &\left.+\left[\frac{1}{\bar{T}_n'-T_k}+\frac{\bar{T}_k-T_k}{\left(\bar{T}_n'-T_n\right)^2}+2\frac{\left(T_k-\bar{T}_k\right)\left(T_n'-T_k\right)}{\left(\bar{T}_n'-T_k\right)^3}\right]\frac{\overline{\omega_n'(\eta)}}{\omega_n'(\eta)}e^{-2i\alpha_n}\right\};\\ S_{nk}(\xi\,,\eta)&=L_{nk}(\xi\,,\eta)\\ &+\frac{e^{-i\alpha_k}}{2}\left[\frac{1}{\bar{T}_n'-T_k}+\frac{T_k-\bar{T}_k}{\left(\bar{T'}_n-\bar{T}_k'\right)^2}-\frac{T_n'-T_k}{\left(\bar{T'}_n-T_k\right)^2}\cdot\frac{\overline{\omega_n'(\eta)}}{\omega_n'(\eta)}e^{-2i\alpha_n}\right];\\ K_{nk}(\xi\,,\eta)&=\frac{e^{i\alpha_k}}{2}\left(\frac{1}{T_k-T_n'}+\frac{e^{-2i\alpha_n}}{\bar{T}_k-\bar{T}_n'}\cdot\frac{\overline{\omega_n'(\eta)}}{\omega_n'(\eta)}\right);\\ L_{nk}(\xi\,,\eta)&=\frac{e^{-i\alpha_k}}{2}\left(\frac{1}{\bar{T}_k-\bar{T}_n'}-\frac{T_k-T_n'}{\left(\bar{T}_k-\bar{T}_n'\right)^2}\cdot\frac{\overline{\omega_n'(\eta)}}{\omega_n'(\eta)}e^{-2i\alpha_n}\right);\\ T_k&=\omega_k(\xi)\,e^{i\alpha_k}+z_k^0;\qquad T_n'=\omega_n(\eta)\,e^{i\alpha_n}+z_n^0; \end{aligned} \tag{3.142}$$

$P_n(\eta)$ are known functions determined by the loads on the edge of the half-plane, on the crack faces, and at infinity, z_n^0 are the complex coordinates of the origins of local coordinate systems $x_nO_ny_n$ in the principal coordinate system xOy.

The solution of the system of SIE (3.141) must satisfy the additional conditions

$$\int_{-1}^{1}\varphi_n(\xi)d\xi=0,\qquad n=1,\,\ldots\,,\,N, \tag{3.143}$$

guaranteeing the single-valuedness of displacements in traversing the N contours of internal cracks.

The obtained system of the singular integral equations (3.142), (3.143) can be efficiently solved by the Gauss-Chebyshev method of mechanical quadratures (see Sect. 3.5). To realize this method, the unknown functions $\varphi_n(\eta)$ are represented in the form

$$\varphi_n\,(\eta)=u_n(\eta)\Big/\sqrt{1-\eta^2},\qquad n=1\,,\ldots,N, \tag{3.144}$$

where $u_n(\eta)$ are functions continuous on the segment $[-1;\,1]$.

Applying the Gauss quadrature formulas (3.111), (3.114), and (3.115) to the SIE (3.141) and relations (3.143), we reduce the problem to the solution of the following system of linear algebraic equations:

$$\sum_{k=1}^{N}\sum_{j=1}^{M}\left[R_{nk}\left(\xi_j,\eta_m\right)u_k\left(\xi_j\right)+S_{nk}\left(\xi_j,\eta_m\right)\overline{u_k}\left(\xi_j\right)\right]=MP_n(\eta_m),$$
$$\sum_{j=1}^{M}u_n\left(\xi_j\right)=0,\qquad n=1,\ldots,N,\qquad m=1,\ldots,M-1,\tag{3.145}$$

where $\xi_j=\cos\left[(2j-1)\pi/(2M)\right]$, $j=1,\ldots,M$; $\eta_m=\cos\left(\pi m/M\right)$, $m=1,\ldots,M-1$—are the roots of the Chebyshev polynomials of the first and second kinds, respectively.

By using the solution of system (3.145), we determine the stress intensity factors at the cracks tips by the following formula:

$$K_{\mathrm{I}n}^{\pm}-iK_{\mathrm{II}n}^{\pm}=\mp p\sqrt{\pi}\sqrt{\left|\omega_n'(\pm 1)\right|}\frac{u_n(\pm 1)}{\omega_n'(\pm 1)},\qquad n=1,\ldots,N.\tag{3.146}$$

In general, in the presence of $\tilde{N}$ ($\tilde{N}\le N$) edge cracks, in the numerical solution of the system of SIE (3.137) or (3.141), instead of the additional condition (3.138), we introduce the following conditions:

$$u_n(-1)=0,\qquad n=1,\ldots,\tilde{N},\tag{3.147}$$

decreasing the square-root singularity of the unknown function $\varphi_n(\eta)$ at the points where the cracks appear on the edge of the half-plane.

Singular integral equations for a system of rectilinear cracks in the half-plane. We obtain singular integral equations for a half-plane weakened by a system of arbitrarily located rectilinear cracks (Fig. 3.9) as a partial case of the SIE (3.137). Note that these equations were obtained in [11, 54] by the same method as the SIE (3.137) obtained for curvilinear cracks. In the case where the edge of the half-plane ($p_0(x)=0$) is free of loads and self-balanced loads are given on the faces of the rectilinear cracks

$$N_k^{\pm}(x_k)+iT_k^{\pm}(x_k)=p_k(x_k),\qquad |x_k|\le l_k,\qquad k=1,2,\ldots,N,\tag{3.148}$$

the system of SIE takes the form

$$\int_{-l_n}^{l_n}\frac{g_n'(t)dt}{t-x}+\sum_{k=1}^{N}\int_{-l_k}^{l_k}\left[g_k'(t)R_{nk}(t,x)+\overline{g_k'(t)}S_{nk}(t,x)\right]dt=\pi p_n(x),$$
$$|x|<l_n,\quad n=1,2,\ldots,N.\tag{3.149}$$

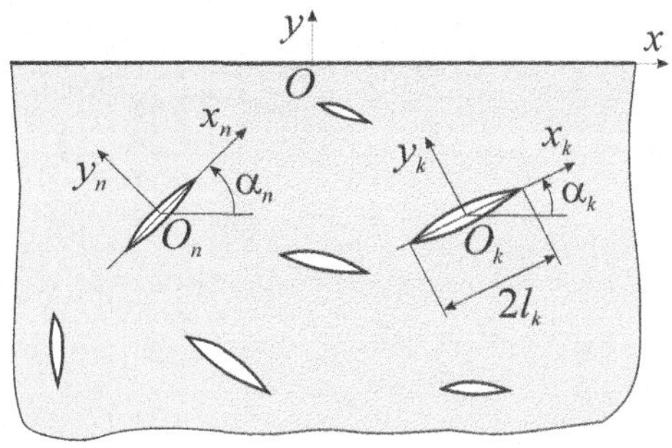

Fig. 3.9 Scheme of location of rectilinear cracks in the half-plane

Here,

$$R_{nk}(t,x) = (1-\delta_{nk})K_{nk}(t,x) + \frac{e^{i\alpha_k}}{2}\left\{\frac{1}{X_n - \bar{T}_k} + \frac{e^{-2i\alpha_n}}{\bar{X}_n - T_k} + \right.$$

$$\left. + \left(\bar{T}_k - T_k\right)\left[\frac{1+e^{-2i\alpha_n}}{\left(\bar{X}_n - T_k\right)^2} - \frac{2e^{-2i\alpha_n}(X_n - T_k)}{\left(\bar{X}_n - T_k\right)^3}\right]\right\};$$

$$S_{nk}(t,x) = (1-\delta_{nk})L_{nk}(t,x)$$

$$+ \frac{e^{-i\alpha_k}}{2}\left[\frac{T_k - \bar{T}_k}{\left(X_n - \bar{T}_k\right)^2} + \frac{1}{\bar{X}_n - T_k} - e^{-2i\alpha_n}\frac{X_n - T_k}{\left(\bar{X}_n - T_k\right)^2}\right], \quad (3.150)$$

where

$$\delta_{nk} = \begin{cases} 0 \text{ for } n \neq k; \\ 1 \text{ for } n = k. \end{cases}$$

The functions $K_{nk}(t, x)$ and $L_{nk}(t, x)$ coincide with kernels (3.103) of the system of equations for the infinite plane with cracks. It is obvious that the other components in (3.150) characterize the influence of the free edge of the half-plane.

3.7 Singular Integral Equation for Single Curvilinear Crack in Elastic Half-Plane

The case of single curvilinear crack in a half-plane is of significant practical importance. We now consider this case. We refer the half-plane to the principal coordinate system xOy and the contour of the crack L to a local coordinate system $x_1O_1y_1$ (Fig. 3.10). The system $x_1O_1y_1$ is connected with the system xOy by the formula $z = z_1e^{i\alpha} + z_1^0$, where $z_1 = x_1 + iy_1$ is the complex variable, z_1^0 the affix of the point O_1 in the system xOy, and α is the angle of inclination of the O_1x_1-axis to the Ox-axis. The shape of the contour L in the coordinate system $x_1O_1y_1$ is described by the following parametric equation:

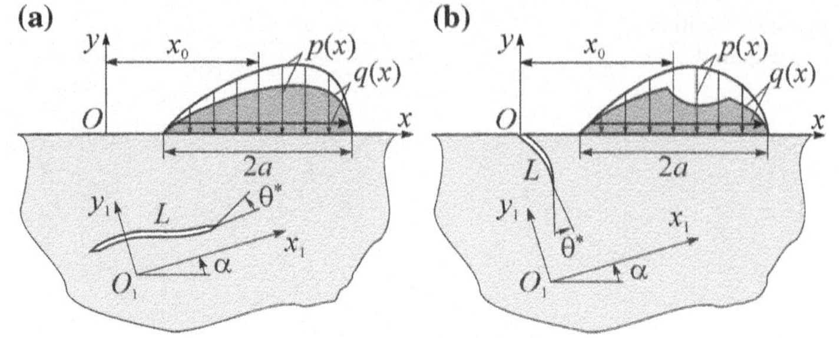

Fig. 3.10 General schemes of the problem; **a** internal (subsurface) crack; **b** edge crack

$$t = x_1(\xi) + iy_1(\xi) = l\omega(\xi) = w(\xi), \qquad t \in L, \qquad |\xi| \leq 1. \tag{3.151}$$

The boundary conditions of the problem are formulated as follows: In a segment of the boundary of the half-plane of length $2a$, we specify an arbitrary normal pressure $p(x)$ and tangential forces $q(x)$ such that

$$\begin{aligned} N(x) + iT(x) &= -p(x) - iq(x), \qquad |x - x_0| \leq a, \qquad y = 0; \\ N(x) + iT(x) &= 0, \qquad |x - x_0| > a, \qquad y = 0; \end{aligned} \tag{3.152}$$

here, $N(x)$ and $T(x)$ are, respectively, the normal and tangential components of the external forces acting on the boundary of the half-plane ($y = 0$) and x_0 is the abscissa of the middle of the segment of action of the external load. We also assume that a complex self-balanced load acts on the faces of the crack, i.e.,

$$N^{\pm}(t) + iT^{\pm}(t) = p_1(t), \qquad t \in L. \tag{3.153}$$

In relation (3.153), the superscripts "+" and "−" correspond to the limiting values of stresses on the crack faces on approaching the crack contour L from the left and from the right, respectively.

In view of relations (3.132), the Kolosov-Muskhelishvili complex potentials of the posed problem have the form

$$\begin{aligned} \Phi(z) &= \frac{1}{2\pi}\int_L \left\{\left[\frac{1}{T - z} - \frac{1}{\bar{T} - z}\right] e^{i\alpha} g'(t)\,dt + \frac{T - \bar{T}}{(\bar{T} - z)^2} e^{-i\alpha}\,\overline{g'(t)\,dt}\right\} \\ &\quad + \Phi_0(z); \\ \Psi(z) &= \frac{1}{2\pi}\int_L \left\{\left[\frac{1}{(\overline{T} - z)^2} - \frac{1}{(T - z)^2}\right] \overline{T} e^{i\alpha} g'(t)\,dt + \right. \\ &\quad \left. + \left[\frac{1}{T - z} - \frac{1}{\overline{T} - z} + \frac{(\overline{T} - T)(\overline{T} + z)}{(\overline{T} - z)^3}\right] e^{-i\alpha}\overline{g'(t)\,d\overline{t}}\right\} + \Psi_0(z); \\ T &= te^{i\alpha} + z_1^0; \end{aligned} \tag{3.154}$$

where $g'(t)$ is an unknown density of potentials, which expresses the derivative of the jump of displacements along the crack contour. The functions $\Phi_0(z)$ and $\Psi_0(z)$ determine the stressed state of the half-plane without cracks under the action of a certain model contact load (3.152) on the edge of the half-plane and a possible normal load at infinity (tension, bending, etc.).

Satisfying the boundary condition (3.153) with the help of potentials (3.154), we arrive at a singular integral equation [6, 8, 42] for the function $g'(t)$, which can be represented in the normalized form as follows:

$$\int_{-1}^{1} [R(\xi, \eta)\, \varphi(\xi) + S(\xi, \eta)\, \overline{\varphi}(\xi)]\, d\xi = \pi P(\eta), \qquad |\eta| < 1, \tag{3.155}$$

where

$$R(\xi, \eta) = \mathrm{Re}\left[\frac{w'(\eta)}{w(\xi) - w(\eta)} + \frac{\overline{w'}(\eta)}{W(\xi, \eta)}\right] +$$
$$+\frac{1}{2}\frac{W(\xi, \xi)}{W^2(\xi, \eta)}\left[\overline{w'}(\eta) + w'(\eta)\, e^{2i\alpha} + 2e^{2i\alpha}\,\overline{w'}(\eta)\,\frac{w(\xi) - w(\eta)}{W(\xi, \eta)}\right];$$
$$S(\xi, \eta) = -\frac{1}{2}\frac{\partial}{\partial\eta}\frac{w(\xi) - w(\eta)}{\overline{w}(\xi) - \overline{w}(\eta)} + \tag{3.156}$$
$$+\frac{w'(\eta)}{2}\left[W^{-1}(\xi, \eta) + \frac{w(\xi) - w(\eta)}{W^2(\xi, \eta)}\frac{\overline{w'}(\eta)}{w'(\eta)} + e^{-2i\alpha}\frac{\overline{W}(\xi, \xi)}{\overline{W}^2(\xi, \eta)}\right];$$

$$W(\xi, \eta) = \overline{w}(\eta) - w(\xi)e^{2i\alpha} - 2ie^{i\alpha}\,\mathrm{Im}\, z_1^0\,; \qquad \varphi(\xi) = g'(t)\, w'(\xi)\,;$$
$$P(\eta) = P_0(\eta) + P_1(\eta) = [p_0(t) + p_1(t)]\, w'(\eta)\,;$$
$$p_0(t) = -\Phi_0(t) - \overline{\Phi_0(t)} - \left[t\overline{\Phi_0'(t)} + \overline{\Psi_0(t)}\right]\frac{d\bar{t}}{dt}\,. \tag{3.157}$$

If, e.g., the plane is subjected to tension by uniformly distributed forces $\sigma_x^\infty = \sigma$ at infinity, then we observe the appearance the following component of the function $P_0(\eta)$:

$$P_0^\infty(\eta) = \sigma\left[\bar{w}'(\eta)e^{-2i\alpha} - w'(\eta)\right]/2. \tag{3.158}$$

We determine this term by using the second relation in (3.157) and the known Kolosov-Muskhelishvili complex potentials $\Phi_0(z)$ and $\Psi_0(z)$ for the case of tension of the half-plane without cracks at infinity [37].

Since, in what follows, the integral equation (3.155) is solved by the Gauss-Chebyshev method of mechanical quadratures, we represent, as earlier, the unknown function in the form

$$\varphi(\eta) = \frac{u(\eta)}{\sqrt{1 - \eta^2}}. \tag{3.159}$$

In the case of internal crack, we supplement the SIE (3.155) with condition (3.143) by setting $n = 1$. This guarantees the single-valuedness of displacements in traversing the crack contour. By using the quadrature formulas, we arrive at a system of algebraic equations of type (3.116), where $B = 0$. Solving this system, we compute the SIF with the help of relation (3.119).

In the case of edge crack, since the singularity of stresses at the point $\eta = -1$ is not a square-root singularity, we assume [44, 53] that $u(\eta)$ is a function continuous on the segment $[-1; 1]$ and satisfies the additional condition

$$u(-1) = 0 \tag{3.160}$$

guaranteeing the boundedness of the solution at the point $\eta = -1$, where the crack appears on the boundary of the half-plane. Applying the Gauss quadrature formulas to Eq. (3.155), representing the function $u(\eta)$ via the Lagrange interpolating polynomial at the Chebyshev nodes, and using Eq. (3.155) and condition (3.160) we arrive at the following system of M linear algebraic equations for M the unknowns $u(\xi_k)$

$$\begin{gathered}\sum_{k=1}^{M}\left[R(\xi_k,\eta_m)\,u(\xi_k)+S(\xi_k,\eta_m)\,\overline{u}(\xi_k)\right]=MP(\eta_m),\qquad m=1,\ \dots\ ,M-1;\\ \sum_{k=1}^{M}\left[(-1)^k\mathrm{tg}\left(\frac{2k-1}{4M}\pi\right)u(\xi_k)\right]=0,\end{gathered} \tag{3.161}$$

where $\xi_k = \cos\,[(2k-1)\pi/(2M)]$, $\quad \eta_m = \cos\,(\pi m/M)$ are the roots of Chebyshev polynomials of the first and second kinds, respectively. With the help of the solution of system (3.161) and the formulas (3.118) i (3.119) we determine the stress intensity factors K_{I} and K_{II} at the tip of an edge curvilinear crack.

$$K_{\mathrm{I}} - iK_{\mathrm{II}} = \frac{p\sqrt{\pi l|\omega'(1)|}}{M\omega'(1)}\sum_{k=1}^{M}\left[(-1)^k\mathrm{ctg}\left(\frac{2k-1}{4M}\pi\right)u(\xi_k)\right]. \tag{3.162}$$

3.8 Stress Intensity Factors for Curvilinear Cracks (Internal and Edge) in Half-Plane

Internal cracks. On the basis of the numerical solution of the SIE (3.141), for $N = 1$ or (3.155), the problem of stress intensity factors at the tips of a curvilinear crack located along an arc of a circle parabola, or semiellipse was solved in the following cases [8]: stretching of a half-plane in the direction parallel to the Ox-axis by constant forces $\sigma_x^\infty = p$ at infinity (Fig. 3.11) and under a constant pressure normal to the crack faces $N_1^\pm(t_1) = p_1(t) = -p$.

The parametric equations of the crack contour were as follows [53]:
$w(\xi) = l(\xi - i\varepsilon)/(1 - i\varepsilon\xi)$ for a circular arc;

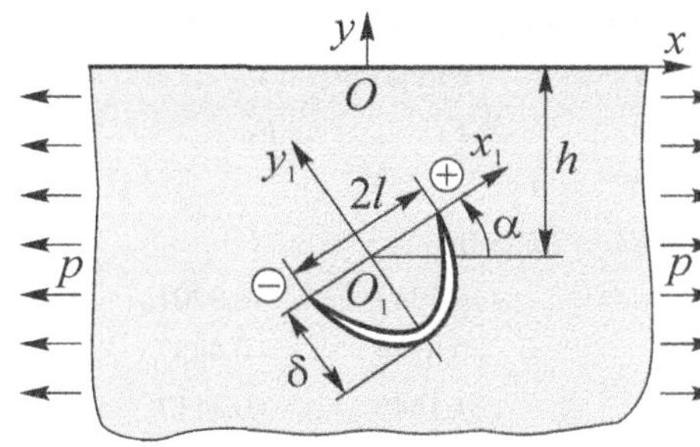

Fig. 3.11 Scheme of location of a curvilinear crack in the half-plane

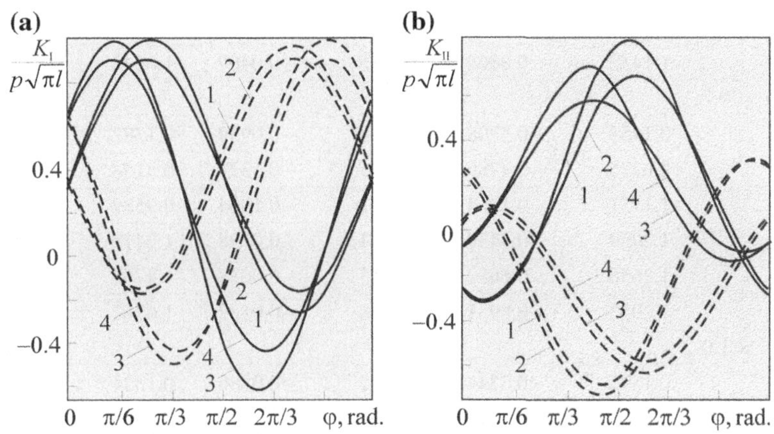

Fig. 3.12 Dependences of the stress intensity factors for a crack made along the circle on the angle of its orientation in the half-plane stretched at infinity. The solid curves correspond to the right tip of the crack ("+"), while the dashed curves correspond to its left tip ("–"). (1) $\lambda = l/h = 0.7$, $\varepsilon = \delta/l = 0.5$; (2) $\lambda = 0$, $\varepsilon = 0.5$; (3) $\lambda = 0.7$, $\varepsilon = 1.0$; (4) $\lambda = 0$, $\varepsilon = 1.0$

$w(\xi) = l[\xi + i\varepsilon(\xi^2 - 1)]$ for a parabolic arc;

$w(\xi) = l[2\xi - i\varepsilon(1 - \xi^2)]/(1 + \xi^2)$ for a semiellipse;

here, $\varepsilon = \delta/l$ is a dimensionless parameter, $2l$ distance between tips of curvilinear crack; and $z_1^0 = ih$ (Fig. 3.11). The parameter $\lambda = l/h$ specifies the location of the crack relative to the edge of the half-plane. In the case of uniaxial tension of a half-plane weakened by a crack, the function $P(\eta)$ in Eq. (3.155) has the following structure: $P(\eta) = -p\left[w'(\eta) - \overline{w'(\eta)}e^{-2i\alpha}\right]/2$. At the same time, under the action of pressure upon the crack faces, it has the following structure: $P(\eta) = -pw'(\eta)$.

By analyzing the numerical data for the case of symmetric location of the crack ($\alpha = 0$ and $\alpha = \pi$) (see Table 3.1), we arrive at the conclusion that, for $\alpha = 0$, the stress intensity factor K_1 always increases as the crack approaches the edge. This regularity is violated only for a crack made along a circle in the case where $p_1(t) = -p$ and $\varepsilon = \delta/l > 1$. This, in particular, follows from the comparison of our results with the data of [44] for the infinite plane containing a crack ($\lambda = 0$). Note, in Tables 3.1 and 3.2

Table 3.1 Normalized SIF for internal curvilinear cracks in elastic half-plane

$\varepsilon = \delta/l$	Circle		Ellipse		Parabola	
	F_{I}	F_{II}	F_{I}	F_{II}	F_{I}	F_{II}
Load $\sigma_x^\infty = p$						
$\alpha = 0$; $\lambda = l/h = 0{,}5$						
0.2	0.1383	−0.3201	0.6523	−0.0993	0.1312	−0.3173
0.5	0.6150	−0.4857	0.9412	−0.1418	0.5034	−0.5176
1.0	1.1645	−0.1117	1.1645	−0.1117	0.9447	−0.5034
1.5	1.2726	0.3439	1.3559	−0.0798	1.2283	−0.3793
2.0	1.2230	0.6664	1.5314	0.0585	1.4463	−0.3722
2.5	1.1432	0.8862	1.6945	−0.0407	1.6342	−0.3263
$\alpha = 0$; $\lambda = 0.7$						
0.2	0.1453	−0.3208	0.6558	−0.0939	0.1382	−0.3172
0.5	0.6275	−0.4786	0.9465	−0.1329	0.5158	−0.5140
1.0	1.1751	−0.1064	1.1751	−0.1064	0.9589	−0.4981
1.5	1.2939	0.3439	1.3754	−0.0798	1.2478	−0.4325
2.0	1.2638	0.6682	1.5615	−0.0638	1.4765	−0.3775
2.5	1.2053	0.8951	1.7388	−0.0532	1.6767	−0.3368
$\alpha = 0$; $\lambda = 1.0$						
0.2	0.1595	−0.3244	0.6682	−0.0886	0.1524	−0.3208
0.5	0.6523	−0.4715	0.9571	−0.1170	0.5406	−0.5087
1.0	1.1929	−0.0904	1.1929	−0.0904	0.9890	−0.4874
1.5	1.3240	0.3563	1.4073	−0.0691	1.2868	−0.4236
2.0	1.3134	0.6824	1.6112	−0.0567	1.5296	−0.3740
2.5	1.2709	0.919	1.8079	−0.0496	1.7459	−0.3385
$\alpha = 0$; $\lambda = 2.0$						
0.2	0.2056	−0.3421	0.7214	−0.0709	0.1967	−0.3385
0.5	0.7267	−0.4573	1.001	−0.0762	0.6168	−0.5034
1.0	1.2549	−0.0479	1.2549	−0.0479	1.0794	−0.4644
1.5	1.4073	0.4094	1.5137	−0.0319	1.4109	−0.4024
2.0	1.4304	0.7622	1.7671	−0.0230	1.6980	−0.3580
2.5	1.4144	1.0333	2.0153	−0.0195	0.1932	−0.3261
$\alpha = 0$; $\lambda = 4.0$						
0.2	0.2694	−0.3704	0.8118	−0.0514	0.2625	−0.3669
0.5	0.7941	−0.4449	1.0404	−0.0425	0.6859	−0.4998
1.0	1.3240	−0.0213	1.3240	−0.0213	1.1662	−0.4538
1.5	1.5261	−0.4555	1.6413	−0.0106	1.5491	−0.4005
2.0	1.5934	0.8508	1.9586	−0.0071	1.8965	−0.3633

(continued)

Table 3.1 (continued)

$\varepsilon = \delta/l$	Circle		Ellipse		Parabola	
	F_{I}	F_{II}	F_{I}	F_{II}	F_{I}	F_{II}
Load $\sigma_x^{\infty} = p$						
2.5	1.6059	1.1681	2.2687	−0.0035	2.2280	−0.3385
Load $\sigma_x^{\infty} = p$						
$\alpha = \pi$; $\lambda = 0.5$						
0.2	0.1294	−0.3261	0.6540	−0.1046	0.1223	−0.3226
0.5	0.6168	−0.5034	0.9554	−0.1524	0.4998	−0.5371
1.0	1.2141	−0.1064	1.2141	−0.1064	0.9890	−0.5229
$\alpha = \pi$; $\lambda = 0.7$						
0.2	0.1276	−0.3315	0.6576	−0.1064	0.1205	−0.3279
0.5	0.6301	−0.5158	0.9748	−0.1524	0.5122	−0.5512
1.0	1.2850	−0.0798	1.2850	−0.0798	1.0723	−0.5211
Load $p_1(t) = -p$						
$\alpha = 0$; $\lambda = 0.5$						
0.2	1.9550	0.3828	0.6239	0.9411	1.9621	0.6386
0.5	1.4853	0.8171	0.6345	0.9713	1.6430	0.7178
1.0	0.8366	1.0156	0.8366	1.0156	1.4321	8454
1.5	0.4980	1.0315	1.0492	1.0085	1.4498	0.8614
2.0	0.3066	1.0244	1.2354	1.0032	1.5367	0.8667
2.5	0.1790	1.0138	1.4073	1.0014	1.6483	0.8720
$\alpha = 0$; $\lambda = 0.7$						
0.2	2.1854	0.3881	0.7160	1.0563	2.1943	0.3704
0.5	1.6359	0.9181	0.6646	1.1237	1.8150	0.8011
1.0	0.8596	1.1591	0.8596	1.1591	1.5455	0.9677
1.5	0.4466	1.1751	1.0812	1.1538	1.5455	0.9908
2.0	0.2056	1.1591	1.2814	1.1467	1.6288	0.9978
2.5	0.0407	1.1609	1.4675	1.1432	1.7458	1.0032
$\alpha = \pi$; $\lambda = 0.5$						
0.2	1.8894	0.4874	0.5051	0.9181	1.9000	0.4732
0.5	1.3931	0.8383	0.5671	0.9429	1.5526	0.7586
1.0	0.7922	0.9659	0.7922	0.9659	1.3701	0.8383
$\alpha = \pi$; $\lambda = 0.7$						
0.2	2.0436	0.6292	0.4431	1.0103	2.0578	0.6150
0.5	1.4321	0.9819	0.5051	1.0315	1.6218	0.9075
1.0	0.7479	1.0617	0.7479	1.0617	1.4197	0.9766

Table 3.2 Normalized SIF for edge curvilinear cracks in elastic half-plane

$\varepsilon = \delta/l$	Circle		Ellipse		Parabola	
	F_{I}	F_{II}	F_{I}	F_{II}	F_{I}	F_{II}
Load $p_1(t) = -p$						
0	1.121	0	1.121	0	1.121	0
0.2	1.167	0	0.728	0.455	1.166	−0.007
0.5	1.402	0	1.238	0.345	1.394	−0.079
1.0	2.242	0	2.242	0	2.113	−0.390
1.5	3.640	0	3.490	−0.540	3.160	−0.900
2.0	5.590	0	5.000	−1.290	4.500	−1.660
2.5	8.110	0	6.780	−2.250	6.130	−2.610
Load $\sigma_x^\infty = p$						
0	$\frac{1.121}{1}$	$\frac{0}{0}$	$\frac{1.121}{1}$	$\frac{0}{0}$	$\frac{1.121}{1}$	$\frac{0}{0}$
0.2	$\frac{1.045}{0.869}$	$\frac{0.257}{0.370}$	$\frac{0.253}{-0.048}$	$\frac{0.516}{0.503}$	$\frac{1.050}{0.878}$	$\frac{0.249}{0.361}$
0.5	$\frac{0.741}{0.403}$	$\frac{0.479}{0.649}$	$\frac{0.263}{-0.171}$	$\frac{0.494}{0.543}$	$\frac{0.838}{0.537}$	$\frac{0.438}{0.620}$
1.0	$\frac{0.280}{-0.177}$	$\frac{0.447}{0.530}$	$\frac{0.280}{-0.177}$	$\frac{0.447}{0.530}$	$\frac{0.611}{0.215}$	$\frac{0.487}{0.661}$
1.5	$\frac{0.072}{-0.362}$	$\frac{0.297}{0.289}$	$\frac{0.282}{-0.164}$	$\frac{0.418}{0.502}$	$\frac{0.505}{0.085}$	$\frac{0.472}{0.623}$
2.0	$\frac{-0.004}{-0.387}$	$\frac{0.185}{0.119}$	$\frac{0.281}{-0.157}$	$\frac{0.400}{0.482}$	$\frac{0.447}{0.022}$	$\frac{0.453}{0.587}$
2.5	$\frac{-0.030}{-0.367}$	$\frac{0.116}{0.011}$	$\frac{0.279}{-0.152}$	$\frac{0.388}{0.467}$	$\frac{0.411}{-0.014}$	$\frac{0.437}{0.560}$

Comment: The results obtained for an infinite plane containing a crack are presented in the denominators

values of normalized SIF $F_{\mathrm{I}} = K_{\mathrm{I}}\Big/\left(p\sqrt{\pi l}\right)$ and $F_{\mathrm{II}} = K_{\mathrm{II}}\Big/\left(p\sqrt{\pi l}\right)$ have been present.

We now compare the stress intensity factors for $\alpha = 0$ and $\alpha = \pi$ in the case where the other values of the parameters of the problem (λ, ε, $p_1(t)$, $\omega(\xi)$) are identical. In both cases, the ends of the cracks are located at the same distance from the boundary and the cracks are located on different sides of the straight line that passes through their ends. For $p_1(t) = -p$, we have $K_{\mathrm{I}}|_{\alpha=\pi} < K_{\mathrm{I}}|_{\alpha=0}$, while for $\sigma_x^\infty = p$, this is not always true.

The influence of the edge of the half-plane in the case of an arbitrary orientation of the crack ($0 \leq \alpha \leq \pi$) was studied by analyzing an example of a crack made along a circular arc in the case of a half-plane stretched at infinity. The character of the curves $K_{\mathrm{I}}(\alpha)\Big/\left(p\sqrt{\pi l}\right)$ (Fig. 3.12a) and $K_{\mathrm{II}}(\alpha)\Big/\left(p\sqrt{\pi l}\right)$ (Fig. 3.12b) for the analyzed values of the parameters λ and ε in the half-plane (curves 1 and 3) is the same as for the infinite plane (curves 2 and 4). The influence of the edge is maximum in the vicinity of the values of α at which the stress intensity factors are extreme.

Note that the numerical results obtained for $\lambda = 0$ are in good agreement with the data obtained on the basis of the exact analytic solution [40]. Indeed, even for $M = 10$, their absolute error in the entire range of α does not exceed 0.3×10^{-4} for $\varepsilon = 0.5$ and 0.4×10^{-2} for $\varepsilon = 1.0$.

Edge cracks [7]. Assume that a half-plane is weakened by a smooth edge curvilinear crack (cut) appearing on its boundary (Fig. 3.13).

We refer the half-plane to the coordinate system xOy and the contour of the cut L to a system $x_1O_1y_1$ connected with the system xOy by the formula $z = z_1e^{i\alpha} - ih$ ($z_1^0 = -ih$). The parametric equation of the contour L has the form

$$t = w(\xi), \qquad t \in L, \qquad |\xi| \leq 1. \tag{3.163}$$

The faces of the crack are subjected to the action of self-balanced forces

$$N_1^{\pm}(t) + iT_1^{\pm}(t) = p_1(t_1), \qquad t_1 \in L. \tag{3.164}$$

As in the case of internal curvilinear cracks, this problem is reduced to the solutions of the SIE (3.155) with additional condition (3.160) or to the system of algebraic equations (3.161).

The influence of configuration of curvilinear edge cracks on the SIF was investigated for two types of external loads: constant pressure upon the crack ($p_1(t) = -p$) and stretching of the half-plane at infinity by the forces $\sigma_x^\infty = p$; $p_0(t) = p\left(e^{-2i\alpha}d\bar{t}/dt - 1\right)/2$; see (3.137).

We determined the SIF (Table 3.2) for cracks located along the arcs of semicircles, semiparabolas, and quarters of ellipses and appearing on the boundary of the half-plane at the right angle ($h = 0$, $\alpha = -\pi/2$; Fig. 3.14a–c); the parametric equations of the contours of these cracks have, respectively, the following forms [53]:

$$w(\xi) = l((\xi + 1)/2 - i\varepsilon)/(1 - i\varepsilon(\xi + 1)/2),$$

$$w(\xi) = l[(\xi + 1)/2 + i\varepsilon((\xi + 1)^2/4 - 1)],$$

$$w(\xi) = l[\xi + 1 - i\varepsilon(1 - (\xi + 1)^2/4)/[1 + (\xi + 1)^2/4],$$

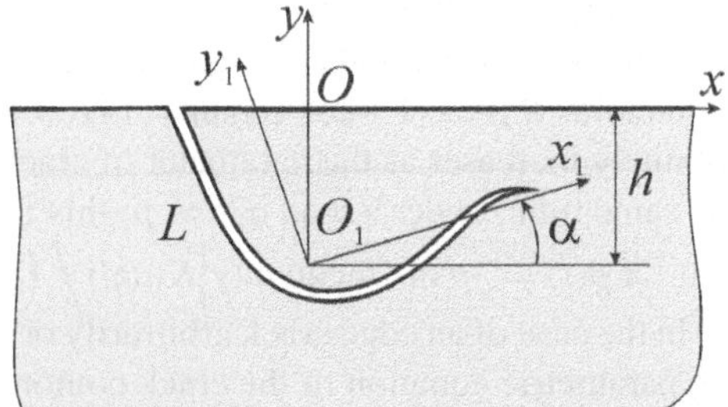

Fig. 3.13 General scheme of location of the crack in the half-plane

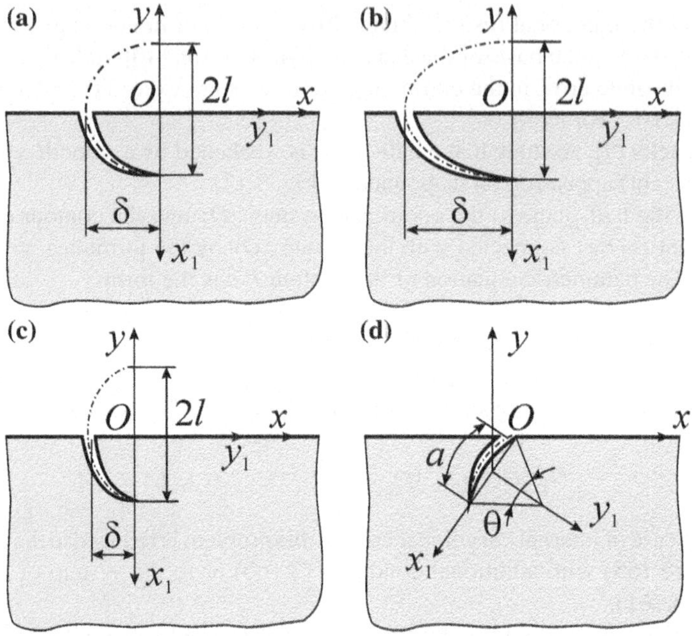

Fig. 3.14 Schemes of curvilinear cracks in the half-plane corresponding to the results presented in Table 3.2 (**a–c**) and Table 3.3 (**d**)

where $\varepsilon = \delta/l$.

Comparing these results with the data accumulated for the infinite plane obtained by complementing the half-plane containing a crack by its mirror image about to the Ox-axis, we conclude that, under the load $p_1(t) = -p$, the crack made in the plane containing a crack changes the character of the dependence $K_{\rm I}(\varepsilon)\Big/\left(p\sqrt{\pi l}\right)$: For $0 \le \varepsilon \le 2.5$, in the case of the plane, we have $K_{\rm I}(\varepsilon)\Big/\left(p\sqrt{\pi l}\right) \le 1$ [53]. At the same time, for the half-plane, this quantity rapidly increases. Under a load $\sigma_x^\infty = p$, the crack made in the plane mainly does not affect the character of dependences of the SIF.

The analysis of the dependences $K_{\rm I,II}(\varphi)\Big/\left(p\sqrt{\pi R}\right)$ shows that, for both types of loading of the half-plane containing a crack located along the arc of a semicircle with radius R ($\varepsilon = (1 - \cos\varphi)/\sin\varphi$; Fig. 3.15) under a load $p_1(t) = -p$, the factor $K_{\rm I}$ infinitely increases as the parameter of crack length φ approaches π (Fig. 3.15a). At the same time, under a load $\sigma_x^\infty = p$, this factor approaches zero (Fig. 3.15b). Note that for $p(t) = -p$ the inequality $\left|K_{\rm II}(\varphi)\Big/\left(p\sqrt{\pi R}\right)\right| < 0,01$ is realized for all φ.

In the case of an edge crack arbitrarily oriented along an arc of a circle (Fig. 3.16), the parametric equation of the crack contour takes the form

$$w(\xi) = l/2[(\xi - i\varepsilon)/(1 - i\xi\varepsilon)], \quad |\xi| \le 1, \tag{3.165}$$

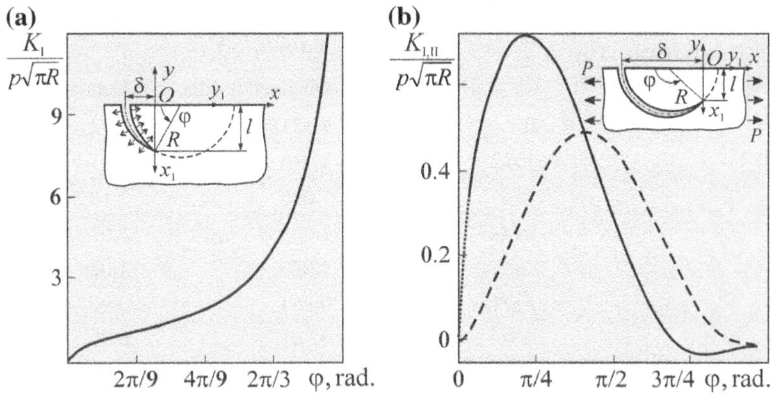

Fig. 3.15 Dependences of the stress intensity factors on the length of a crack located along an arc of a semicircle: the solid lines correspond to $K_{\rm I}$ and the dashed lines correspond to $K_{\rm II}$

Fig. 3.16 Dependences of the SIF for a crack arbitrarily oriented along the arc of a circle on the angle β of orientation and the curvature ε of the circle: the solid lines correspond to $K_{\rm I}$; the dashed lines correspond to $K_{\rm II}$; (1) ε = 0 ($\beta_1^*= \pi$); (2) ε = 0.2 ($\beta_2^*\approx 2.75$); (3) ε = 0.5 ($\beta_3^*\approx 2.21$)

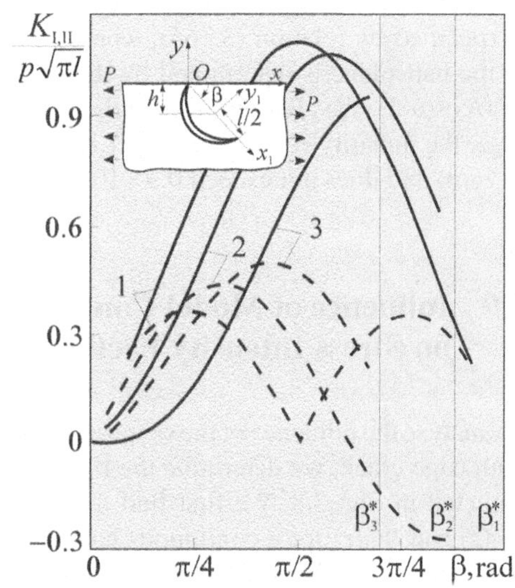

and its location in the half-plane is determined by the parameters $\beta = -\alpha$, $\lambda = l/(2h) = 1/\sin\beta$; $0 < \beta < \beta^*$, $\beta^* = \pi - \arcsin(2\varepsilon/(1 + \varepsilon^2))$, and $\varepsilon = \delta/a$ (the possibility of contact of the crack contour with the boundary of the half-plane is excluded). It was established that, under a load $\sigma_x^\infty = p$, the value of the factor $K_{\rm I}$ for a curvilinear crack (curves 2 and 3) subtended by a chord of length l may slightly exceed $K_{\rm I}$ obtained for the rectilinear crack of length l (curve 1) but never exceeds $K_{\rm I}$ for the rectilinear crack perpendicular to the direction of tension.

In the case of an arbitrarily oriented rectilinear crack, the obtained values of the SIF are in good agreement with the data of [25], while the angles of initial crack propagation are in good agreement with the results obtained in [41].

Table 3.3 SIF comparative table with results obtained by authors and in paper [21] for edge crack along of circle arc

θ	$K_{\mathrm{I}}(\theta)/\, p\sqrt{\pi a}$	
	Obtained results	Data of [21]
0	1.1215	1.12
π/12	1.115	1.11
π/6	1.096	1.10
π/4	1.065	1.07
π/3	1.020	1.02
5π/12	0.964	0.96
π/2	0.901	0.90

We also consider the case of tension of a half-plane containing a crack of length a along a circular arc located so that the tangent to its contour at the end is perpendicular to the direction of tension (Fig. 3.14d). The parametric equation of the crack contour is specified by relation (3.165), where $\varepsilon = (1 - \cos(\theta/2))/\sin(\theta/2)$ and the position in the half-plane is determined by the parameters $\alpha = -(\pi + \theta)/2$ and $\lambda = l/(2h) = 1/\cos(\theta/2)$. We also determined the values of $K_{\mathrm{I}}(\theta)/\left(p\sqrt{\pi a}\right)$ (Table 3.3). In this case, the quantity $K_{\mathrm{II}}(\theta)/\left(p\sqrt{\pi a}\right)$ is negative (for the rectilinear crack, it is equal to zero) and does not exceed 0.9×10^{-2} in modulus.

3.9 Influence of Model Contact Loads Shape (Distribution) on Stress Intensity Factors for Edge Rectilinear Crack

To analyze the influence of the shape of model contact loads on the SIF for a half-plane with edge crack, we determine the right-hand sides of the SIE (3.155) for the loads depicted in Fig. 3.8. We first find the Kolosov-Muskhelishvili complex potentials $\Phi_0(z)$ and $\Psi_0(z)$ for a continuous half-plane under the action of the corresponding model loads and then, on the basis of relations (3.157), determine the functions $P(\eta)$. For simplicity, it is assumed that the crack is free of loads ($p_1(t) = 0$) and the stresses at infinity are also absent ($P_0^{\infty}(\eta) = 0$), i.e., $P(\eta) = P_0(\eta)$.

Concentrated force. For a concentrated force with the normal component P and tangential component $Q = fP$ applied at a point x_0 on the boundary of the half-plane (Fig. 3.8a), the first boundary condition in (3.152) takes the form

$$N(x) + iT(x) = -(P + iQ)\delta(x - x_0) = -P(1 + if)\delta(x - x_0), \qquad y = 0, \tag{3.166}$$

where $\delta(x)$ is the Dirac delta-function.

In this case, the complex potentials $\Phi_0(z)$ and $\Psi_0(z)$ are given by the following formulas [6, 37, 42]:

$$\Phi_0(z) = \frac{P}{2\pi}\frac{i-f}{z-x_0}; \qquad \Psi_0(z) = \frac{P}{2\pi}\left[\frac{i+f}{z-x_0} + \frac{i-f}{(z-x_0)^2}x_0\right]. \tag{3.167}$$

On the basis of relations (3.157), the right-hand side of Eq. (3.155) takes the form [6, 42]

$$P(\eta) = \frac{P}{\pi}\left\{\mathrm{Re}\left[\frac{(f-i)e^{-i\alpha}}{a(\eta)}\right]w'(\eta) - \frac{\mathrm{Re}\left[e^{i\alpha}(f+i)a(\eta)\right]}{\overline{a^2(\eta)}}\overline{w'(\eta)}\right\}, \tag{3.168}$$

where $t = w(\xi)$ is a parametric equation of the crack contour (see (3.151)); $a(\eta) = w(\eta) + \left(z_1^0 - x_0\right)e^{-i\alpha}$.

Constant pressure. In the case where a uniformly distributed pressure with normal component $p(x) = p$ and constant component $q(x) = fp$ acts on a segment of the boundary of half-plane of length $2a$, (Fig. 3.8c, 3.10), we represent the first boundary condition in (3.152) in the form

$$N(x) + iT(x) = -p(1+if), \qquad |x - x_0| \le a, \qquad y = 0, \tag{3.169}$$

and find the complex potentials $\Phi_0(z)$ and $\Psi_0(z)$ as follows [16, 37]:

$$\begin{aligned}\Phi_0(z) &= \frac{p(1+if)}{2\pi i}\ln\frac{z-x_0-a}{z-x_0+a};\\ \Psi_0(z) &= -\frac{pf}{\pi}\ln\frac{z-x_0-a}{z-x_0+a} + \frac{p(i-f)az}{\pi\left[(z-x_0)^2-a^2\right]}.\end{aligned} \tag{3.170}$$

For the right-hand side of the SIE (3.155), we can write [16, 17]:

$$\begin{aligned}P(\eta) = \frac{p}{\pi}\Bigg\{&w'(\eta)\mathrm{Re}\left[(i-f)\ln\frac{b_1(\eta)+1}{b_1(\eta)-1}\right]\\ &+\overline{w'(\eta)}e^{-2i\alpha}\left[(i+f)\frac{b_1(\eta)-\overline{b_1(\eta)}}{b_1^2(\eta)-1} + f\ln\frac{\overline{b_1(\eta)}+1}{\overline{b_1(\eta)}-1}\right]\Bigg\},\end{aligned} \tag{3.171}$$

where $b_1(\eta) = \lambda - \gamma - \varepsilon\omega(\eta)\, e^{i\alpha}$; $\varepsilon = l/a$; $\lambda = x_0/a$; $\gamma = z_1^0/a$; $\omega(\eta) = w(\eta)/l$. Here, l is a parameter characterizing the length of the curvilinear crack and λ is the relative distance from the middle of the contact segment to the crack mouth.

Elliptic distribution: total slip. If a segment of the boundary of the half-plane is subjected to the action of a load distributed according to the elliptic law (Hertz contact forces with tangential component; Figs. 3.8b and 3.10), then the first boundary condition in (3.148) takes the form

$$N(x) + iT(x) = -p_0(1+if)\sqrt{a^2-(x-x_0)^2}\Big/a, \qquad |x-x_0| \le a, \qquad y = 0, \tag{3.172}$$

where p_0 is the maximum value of pressure in the contact zone.

Under this load, the complex potentials can be found in the form [16, 37]

$$\Phi_0(z) = \frac{p_0(i-f)}{2a}\left\{(z-x_0) + i\sqrt{a^2-(z-x_0)^2}\right\};$$

$$\Psi_0(z) = \frac{p_0}{2a}\left\{(i+f)\left((z-x_0) + i\sqrt{a^2-(z-x_0)^2}\right) - \right.$$

$$\left. -(i-f)\left[\left((z-x_0) + i\sqrt{a^2-(z-x_0)^2}\right) + z\left(1 + \frac{(z-x_0)}{i\sqrt{a^2-(z-x_0)^2}}\right)\right]\right\}. \quad (3.173)$$

On the basis of relations (3.157) and (3.173), the right-hand side of the integral equation takes the form [16]

$$P(\eta) = p_0\left\{\mathrm{Re}[(1+if)(a(\eta)-ib(\eta))]w'(\eta) - \left[(1-if)\left(\overline{b(\eta)}/\overline{a(\eta)} - i\right)\times\right.\right.$$
$$\left.\left.\times i\mathrm{Im}\left(\varepsilon\omega(\eta)e^{i\alpha}+\gamma\right) - if\left(\overline{a(\eta)} + i\overline{b(\eta)}\right)\right]e^{-2i\alpha}\overline{w'(\eta)}\right\}, \quad (3.174)$$

where

$$a(\eta) = \sqrt{1-b^2(\eta)}; \quad b(\eta) = \varepsilon\omega(\eta)e^{i\alpha} + \gamma - \lambda; \quad \gamma = z_1^0/a;$$
$$\omega(\eta) = w(\eta)/l. \quad (3.175)$$

Note that this formula was first established by Datsyshyn and Terlets'kyi in the paper "Stress-strain state near edge curvilinear cracks in the half plane under elliptic loading applied on its boundary", Fiz.-Khim. Mekh. Mater, Deposited in the State Scientific-Technical Library of Ukraine 9.01.1997, No. 57 UK 97.

Distribution of forces taking into account the degree of rounding of edges of the counterbody. If the load acts on a segment of the boundary of half-plane according to the scheme presented in Fig. 3.8d, i.e., with regard for the presence of the initial contact line (band) and the degree of rounding of the edges of counterbody, then, by using the solution of the corresponding contact problem, [1], we represent the boundary condition (3.152) in the form

$$N(x) + iT(x) = -p(x)(1+if), \qquad |x-x_0| \le a, \qquad y=0, \quad (3.176)$$

where

$$p(x) = \frac{P(1+\delta)}{b\pi F(\delta)}\left[\left(\pi - 2\varphi_0 + \sin\varphi_0 \ln\left|tg\frac{\varphi+\varphi_0}{2}tg\frac{\varphi-\varphi_0}{2}\right|\right]\right)\cos\varphi$$
$$+ \sin\varphi\ln\left|\frac{\sin(\varphi+\varphi_0)}{\sin(\varphi-\varphi_0)}\right|, \quad (3.177)$$

$$\varphi = \arcsin \frac{x}{a}, \qquad F(\delta) = \frac{\pi - 2\varphi_0}{2\sin^2 \varphi_0} - ctg\varphi_0,$$
$$\varphi_0 = \arcsin \frac{1}{1+\delta}, \delta = (a-b)/b. \tag{3.178}$$

Here, P is the resultant vector of normal forces of the contact load, $2b$ is the length of the initial contact line, $2a$ is its total length, which increases as a result of contact of the bodies. In this case, the parameter δ characterizes the degree of rounding of the edges of counterbody whose radii are determined by the formula

$$R = \frac{b^2 E F(\delta)}{4P(1-\nu^2)}, \tag{3.179}$$

where E and ν are the elastic constants of the materials (we consider the case where the materials of the contacting bodies are identical).

Note that, as in all previous cases, it is now assumed that the conditions of total slip between the contacting bodies are satisfied.

As above, to determine the complex potentials $\Phi_0(z)$ and $\Psi_0(z)$, we use their representations (3.124) [37, 44, 53] in the first main problem of the theory of elasticity for a half-plane. By using these representations, we get

$$\Phi_0(z) = \frac{1+if}{2\pi i} \int\limits_{x_0-a}^{x_0+a} \frac{p(x)}{x-z} dx;$$
$$\Psi_0(z) = \frac{1}{2\pi i} \int\limits_{x_0-a}^{x_0+a} \left[\frac{(1-if)p(x)}{x-z} - \frac{(1+if)xp(x)}{(x-z)^2} \right] dx, \tag{3.180}$$

where $p(x)$ is given by relation (3.177). In view of of (3.177), (3.157), and (3.180), we represent the right-hand side of the SIE (3.155) in the integral form. In this case, for the numerical solution of the problem, we apply the Simpson quadrature formula to the right-hand side of the SIE [61].

Elliptic distribution: slipping-sticking (symmetric case). In the case where the boundary of a half-plane suffers the action of a contact load used to model the presence not only of the sections of slip but also the section of sticking (Figs. 3.8e and 3.10b) in contacting bodies [24, 26], the normal and tangential forces in the entire contact section are distributed as follows:

$$p(x) = \frac{p_0}{a}\sqrt{a^2 - (x-x_0)^2}, \qquad |x - x_0| \le a, \qquad y = 0$$

$$q(x) = \begin{cases} f\dfrac{p_0}{a}\sqrt{a^2-(x-x_0)^2}, & -a \le x - x_0 \le -c, \quad c \le x - x_0 \le a\,, \\ y = 0; & \\ f\dfrac{p_0}{a}\left[\sqrt{a^2-(x-x_0)^2} - \sqrt{c^2-(x-x_0)^2}\right], & |x - x_0| \le c, \\ y = 0. & \end{cases} \tag{3.181}$$

Here, $2c$ is the length of the section of sticking symmetric about the center of the contact section. For this distribution of the model contact load, by virtue of relation (3.124), we find the complex potentials in the form

$$\begin{aligned} \Phi_0(z) &= \frac{p_0}{2a}\left\{(i-f)\left[Z + i\sqrt{a^2-Z^2}\right] + f\left[Z + i\sqrt{c^2-Z^2}\right]\right\}; \\ \Psi_0(z) &= \frac{p_0}{a}\left\{\frac{f-i}{2}z\left(1 - \frac{iZ}{\sqrt{a^2-Z^2}}\right) + f\left[i\sqrt{a^2-Z^2} + Z\right] - \right. \\ &\left. - \frac{f}{2}z\left(1 - \frac{iZ}{\sqrt{c^2-Z^2}}\right) - f\left[Z + i\sqrt{c^2-Z^2}\right]\right\}, \end{aligned} \tag{3.182}$$

where $Z = z - x_0$.

As above, the right-hand side of the integral equation (3.155) takes the form

$$\begin{aligned} &P(\eta) = p_0\left\{\mathrm{Re}[(1+if)(a(\eta) - ib(\eta))]w'(\eta) - \right. \\ &- \left[(1-if)\left(\overline{b(\eta)}/\overline{a(\eta)} - i\right)i\,\mathrm{Im}\left(\varepsilon\omega(\eta)e^{i\alpha} + \gamma\right)\right. \\ &\left.\left. -if\left(\overline{a(\eta)} + i\overline{b(\eta)}\right)\right]e^{-2i\alpha}\overline{w'(\eta)}\right\} + \\ &+ fp_0\varepsilon_1 c/a\left\{-\mathrm{Re}(ia_1(\eta) + b_1(\eta))w'(\eta) - \left[\left(i\overline{b_1(\eta)}/\overline{a_1(\eta)} + 1\right)\times\right.\right. \\ &\left.\left.\times\ \mathrm{Im}\left(\varepsilon_1\omega(\eta)e^{i\alpha} + \gamma_1\right) - \left(i\overline{a_1(\eta)} - \overline{b_1(\eta)}\right)\right]e^{-2i\alpha}\overline{w'(\eta)}\right\} \end{aligned} \tag{3.183}$$

where the notation $a(\eta)$, $b(\eta)$, ε, λ, γ, and $\omega(\eta)$ is the same as in relations (3.175);

$$\begin{aligned} a_1(\eta) &= \sqrt{1 - b_1^2(\eta)}, \qquad b_1(\eta) = \varepsilon_1\omega(\eta)e^{i\alpha} + \gamma_1 - \lambda_1; \\ \varepsilon_1 &= l/c; \qquad \lambda_1 = x_0/c; \qquad \gamma_1 = z_1^0/c. \end{aligned} \tag{3.184}$$

It should be emphasized that, according to [26], the relative half length of the section of sticking c/a (see Fig. 3.8f) is given by the formula

$$c/a = \sqrt{1 - |Q/fP|}, \tag{3.185}$$

where Q and P are, respectively, the tangential and normal components of the vector of the external load.

Elliptic distribution: slipping-sticking (asymmetric case). If the section of sticking is located asymmetrically, e.g., is adjacent to the left edge of the contact section, then the distribution of normal and tangential forces in this section (Fig. 3.8f, 3.10) can be represented as follows [26]:

$$
p(x) = \frac{p_0}{a}\sqrt{a^2 - (x - x_0)^2}, \qquad |x - x_0| \le a, \qquad y = 0;
$$
$$
q(x) = \begin{cases} f\dfrac{p_0}{a}\sqrt{a^2 - (x - x_0)^2}, & c - d \le x - x_0 \le a, \qquad y = 0; \\ f\dfrac{p_0}{a}\left[\sqrt{a^2 - (x - x_0)^2} - \sqrt{c^2 - ((x + d) - x_0)^2}\right], \\ -a \le x - x_0 \le c - d, \quad y = 0, \end{cases} \tag{3.186}
$$

where $2c$ is the length of the section of sticking, $c/a = \sqrt{1 - |Q/fP|}$, and $d/a = 1 - c/a$.

For the indicated distribution of the contact load, the complex potentials of the problem take the form:

$$
\Phi_0(z) = \frac{p_0}{2a}\left\{(i - f)\left[Z + i\sqrt{a^2 - Z^2}\right] + f\left[i\sqrt{c^2 - (z - D)^2} + (z - D)\right]\right\};
$$
$$
\Psi_0(z) = \frac{p_0}{a}\left\{\frac{i - f}{2}z\left(1 - \frac{iZ}{\sqrt{a^2 - Z^2}}\right) + f\left[i\sqrt{a^2 - Z^2} + Z\right] - \right.
$$
$$
\left. - \frac{f}{2}z\left(1 - i\frac{z - D}{\sqrt{c^2 - (z - D)^2}}\right) - f\left[i\sqrt{c^2 - (z - D)^2} + (z - D)\right]\right\}, \tag{3.187}
$$

where $Z = z - x_0$ and $D = x_0 - d$.

In this case, the right-hand side of the integral equation (3.155) is given by formula (3.183) with $\lambda_1 = (x_0 - d)/c$.

Numerical results. We perform the numerical analysis of the stress intensity factors K_{I} and K_{II} for the edge rectilinear arbitrarily oriented crack in the half-plane loaded by model contact forces according to the schemes presented in Fig. 3.8a–f. In what follows, as already indicated, the crack length, the angles of its inclination to the edge, the parameters of the external load, and the friction coefficient between the contacting bodies are mainly chosen with regard for the conditions of fretting fatigue and rolling. Consider a crack of length l inclined at an angle β to the edge of the half-plane (Fig. 3.17). We locate the origins of the principal and local coordinate systems at the crack mouth and direct the O_1x_1-axis along the crack. Then, in the main relations of the previous subsection, we have $\alpha = -\beta$ and $z_1^0 = 0$, and the parametric equation of the crack contour takes the form

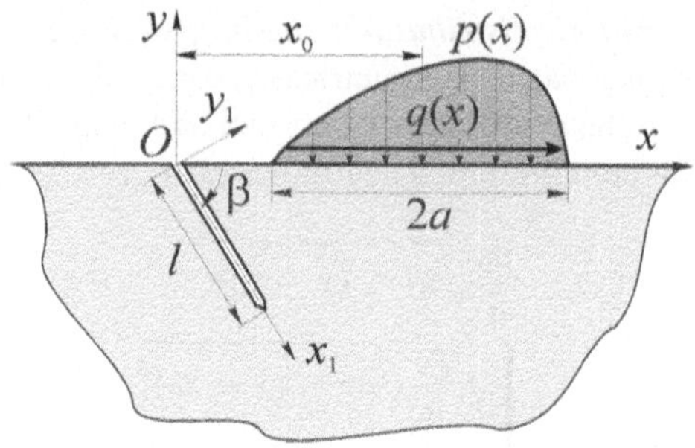

Fig. 3.17 Computational scheme of the problem

$$t = x_1(\xi) = w(\xi) = l\omega(\xi) = l(\xi + 1)/2, \quad 0 \le x_1 \le l, \quad |\xi| \le 1. \tag{3.188}$$

As already indicated, we solve the singular integral equations obtained from Eq. (3.155) with its right-hand sides (3.168), (3.171), (3.173), and (3.183) by the method of mechanical quadratures. It should also be emphasized that all numerical results, i.e., the dependences $K_{\mathrm{I}}(\lambda, \varepsilon, \beta)$ and $K_{\mathrm{II}}(\lambda, \varepsilon, \beta)$, are presented for the locations of contact loads guaranteeing, for a given direction of tangential forces, the validity of the condition $K_{\mathrm{I}}(\lambda) > 0$, i.e., the condition of absence of the contact of crack faces in the vicinity of the crack tip. For the direction of tangential forces that coincides with the Ox-axis, the cracks with completely noncontacting faces appear for $\lambda > 1.0$, i.e., when the contact load is located to the right of the crack mouth. We also note that, in analyzing the SIF, we focus our attention on their maximum values because all predictions of possible mechanisms of the fracture processes in contacting bodies made in what follows are performed by using these values.

It is known [29, 30, 48] that *concentrated forces* and *elliptic distributions of forces* are most often used to model the action of counterbody in the case of contact rolling interaction. For this interaction, the contact zone of the bodies is small as compared with their sizes and the initial contact is a point contact (linear for cylindrical bodies).

For these cases, the data on the dependence of the SIF on the parameters of the geometric-force scheme are available from the literature. In order to establish the limits of applicability of the scheme of loading by concentrated forces instead of the elliptic distribution (see the schemes in Figs. 3.8a, b), we present the plots of the normalized stress intensity factors $F_{\mathrm{I,II}} = K_{\mathrm{I,II}}(\lambda, \varepsilon_0, \beta)\sqrt{\pi l}\big/P$ for both these types of loads in Fig. 3.18. The parameter $\lambda = x_0/l$ characterizes the location of the load on the boundary of the half-plane (Fig. 3.17) and $\varepsilon_0 = a/l$ is the relative half length of the contact section. The numerical analysis was performed for the friction coefficient $f = 0.3$ and two angle of inclination of the crack to the boundary ($\beta = \pi/6$ and $\beta = 5\pi/6$) at which the edge cracks are most often initiated in rolling bodies.

On the basis of the data from [6, 24, 42] and the plots depicted in Fig. 3.18, we can make the following conclusions:

- the characters of the curves $K_{\mathrm{I}}(\lambda, \varepsilon_0, \beta)$ and $K_{\mathrm{II}}(\lambda, \varepsilon_0, \beta)$ obtained for the elliptic loads and concentrated forces are identical;

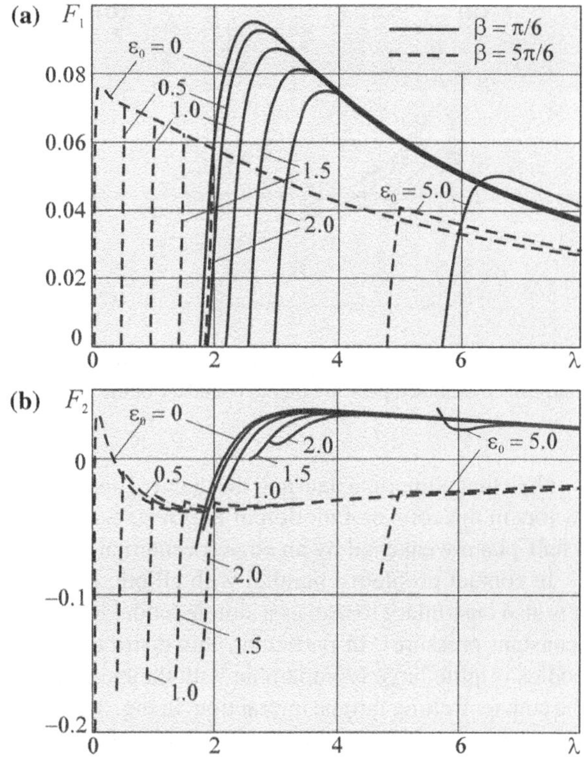

Fig. 3.18 Normalized SIF F_{I} and F_{II} for the edge rectilinear inclined crack under the action of: **a** a concentrated force ($\varepsilon_0 = 0$) or **b** forces distributed according to the elliptic law ($\varepsilon_0 > 0$) on the boundary of the half-plane

- the smaller the contact section for the elliptic load (the smaller ε_0), the faster the corresponding curves (both for K_{I} and K_{II}) merge with the curve for the concentrated force ($\varepsilon_0 = 0$);
- the concentrated force induces the highest maximum values of K_{I} and $|K_{\mathrm{II}}|$. As the contact zone enlarges, the maximum values of the SIF decrease.
- the maximum values of $K_{\mathrm{I}}(\lambda, \varepsilon_0, \beta)$ for the angle $\beta = \pi/6$ are much higher than for the angle $\beta = 5\pi/6$ and the same ε_0;
- for the angle $\beta = 5\pi/6$, the contact of the crack faces is terminated (i.e., K_{I} becomes positive) immediately after opening of the crack mouth by the load moving in the positive direction of the Ox-axis; for the angle $\beta = \pi/6$, this happens for much larger λ; hence, K_{I} and $|K_{\mathrm{II}}|$ also attain their maximum values for the angle $\beta = 5\pi/6$ faster; it is clear that, for both these angles, the larger the contact section, the later the crack mouth is opened;
- $K_{\mathrm{II}}(\lambda, \varepsilon_0, \beta)$ are predominantly positive for $\beta = \pi/6$ and negative for $\beta = 5\pi/6$; $\max |K_{\mathrm{II}}(\lambda, \varepsilon_0, \beta)|$ for the angle $\beta = 5\pi/6$ are much larger than for the angle $\beta = \pi/6$ for the same ε_0.

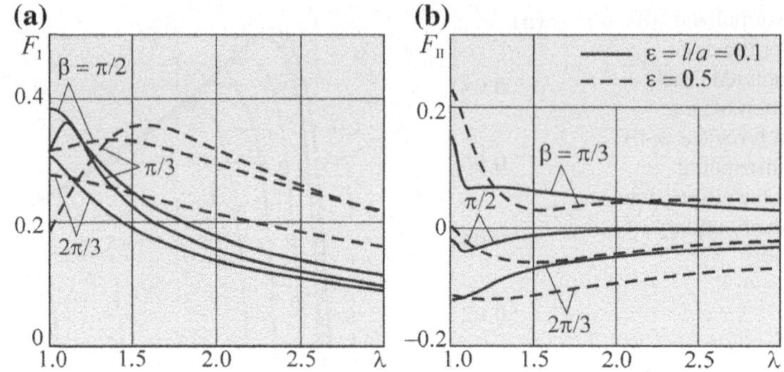

Fig. 3.19 Normalized SIF $F_{\rm I}$ and $F_{\rm II}$ for a rectilinear inclined edge crack under the action of uniformly distributed pressure on the boundary of the half-plane

Note that numerical data and the detailed analysis of changes in the stress intensity factors in the course of motion of the *Hertz (elliptic) contact load* along the edge of a half-plane weakened by an edge (or internal) crack are presented in Chap. 4.

In contact problems, parallel with elliptic distributions and concentrated forces, it is also customary to use, as a simple model load, the uniform distribution of forces (constant pressure). In particular, this distribution is used if the contact section of bodies is quite large (comparable with their sizes), which is typical, in particular, of the contact fretting fatigue interaction. In Fig. 3.19, we present the plots of normalized stress intensity factors $F_{\rm I,II}(\lambda, \varepsilon, \beta) = K_{\rm I,II}/\left(p\sqrt{\pi a}\right)$ *for the case of contact loading* of the edge of a half-plane by *normal and tangential forces of the same intensity* (scheme in Fig. 3.8c). The parameter $\lambda = x_0/a$ characterizes the location of the load on the boundary of the half-plane (Fig. 3.17) and $\varepsilon = l/a$ characterizes the relative crack length. The plots were constructed for the friction coefficient $f = 0.75$, fixed values of the crack length ($\varepsilon = 0.1$ and $\varepsilon = 0.5$), and different angles of inclination of the crack to the boundary of the half-plane ($\beta = \pi/3$, $\beta = \pi/2$, and $\beta = 2\pi/3$) observed in contacting bodies under the conditions of fretting fatigue.

It was established that:

- in the case where the boundary of the loading zone is close to the crack mouth ($1.0 \le \lambda \le 1.5$), the character of the curves $K_{\rm I}(\lambda, \varepsilon, \beta)$ and $|K_{\rm II}(\lambda, \varepsilon, \beta)|$ strongly depends on the orientation of the crack and its length; as the distance from the load to the mouth increases (as a function of λ), the values of $K_{\rm I}(\lambda, \varepsilon, \beta)$ and $|K_{\rm II}(\lambda, \varepsilon, \beta)|$ decrease and, hence, the curves $\varepsilon = 0.5$ are, in this case, located "higher";
- the maximum values of $K_{\rm I}(\lambda, \varepsilon, \beta)$ are higher; moreover, for short cracks, they are attained faster (for smaller λ); the maxima of $K_{\rm I}(\lambda, \varepsilon, \beta)$ appear within the range of angles $\pi/3 \le \beta \le \pi/2$;
- the dependences of $K_{\rm II}(\lambda, \varepsilon, \beta)$ have a well-pronounced maximum at the angle $\beta = \pi/3$ for $\lambda = 1.0$; in this case, the value of $K_{\rm II}$ for $\varepsilon = 0.5$ exceeds the corresponding value for $\varepsilon = 0.1$;

- for large angles of inclination of the cracks to the boundary of the half-plane ($\beta \geq \pi/2$), the SIF $K_{II}(\lambda, \varepsilon, \beta)$ are predominantly negative, i.e., the sign of tangential stresses in the vicinity of the crack tip differs from they sign observed for small angles.

It should be emphasized that a large body of numerical data is presented in Chap. 5 devoted to the contact fretting fatigue interaction. Moreover, the same chapter also contains the analysis of the data obtained under the action *constant pressure* applied to the edge of the half-plane weakened by a crack.

A *contact load that takes into account the degree of rounding of the edges of the counterbody* is described by relations (3.176–3.179) (see the scheme in Fig. 3.8d) is used, parallel with constant pressure, as model loading in the cases where the contact section is relatively large. Moreover, this load takes into account the shape of the edges of counterbody more accurately. It can be used to model the contact interaction both of fixed joints (press fits, riveted, splined, and bolted joints) often subjected to fretting fatigue and of the gearings operating under the conditions of rolling.

The detailed numerical analysis of this case is presented in Chap. 5. Here, we restrict ourselves to the evident conclusion: The smaller the radius of rounding of the edges of the counterbody, the higher the SIF and, in particular, their maximum values at the tip of an edge crack located near the end of the zone of contact loading (Fig. 3.17).

Complicated elliptic contact load (the boundary conditions (3.181); Fig. 3.8e). This type of loading is used to *model* the case where the *conditions of sticking* are realized in the central part of the contact zone between the bodies of length $2c$ and the conditions of slip are preserved on the edges of this region. This type of contact conditions is typical of the fretting fatigue contact interaction.

The performed calculations demonstrate, as the section of sticking becomes larger, the SIF $K_I(\lambda)$ and $K_{II}(\lambda)$ decrease; in particular, the maximum value of $K_I(\lambda)$ becomes 4.4 time lower, whereas $K_{II}(\lambda)$ decreases by a factor of 2.8 as the relative length of the section of sticking c/a increases from 0 up to 0.85. A more detailed analysis of this case is carried out in Chap. 5.

In Fig. 3.20, we show the dependences $F_{I,II} = K_{I,II}(\lambda, \varepsilon, \beta)/\left(p_0\sqrt{\pi a}\right)$ plotted for a complex elliptic contact load simulating the case where the conditions of sticking between the contacting bodies are realized in a part of the contact zone on the side of the crack, while the conditions of slip are preserved in the remaining part of this zone. This type of contact load is described by the boundary conditions (3.186) and schematically depicted in Fig. 3.8f. It corresponds to the transient stage at the beginning of the process of rolling. The calculations were performed for the friction coefficient $f = 0.3$, the angles of inclination of the crack to the edge of the half-plane $\beta = \pi/6$ and $\beta = 5\pi/6$, and the relative crack length $\varepsilon = 0.2$. As already indicated, these combinations of the parameters are characteristic and dangerous from the viewpoint of propagation of cracks in rolling. The analysis of the obtained curves yields the following conclusions:

- the character of the curves of $K_I(\lambda)$ and $K_{II}(\lambda)$, as in the case of simple elliptic loading (Fig. 3.18), strongly depends on the orientation of the crack;

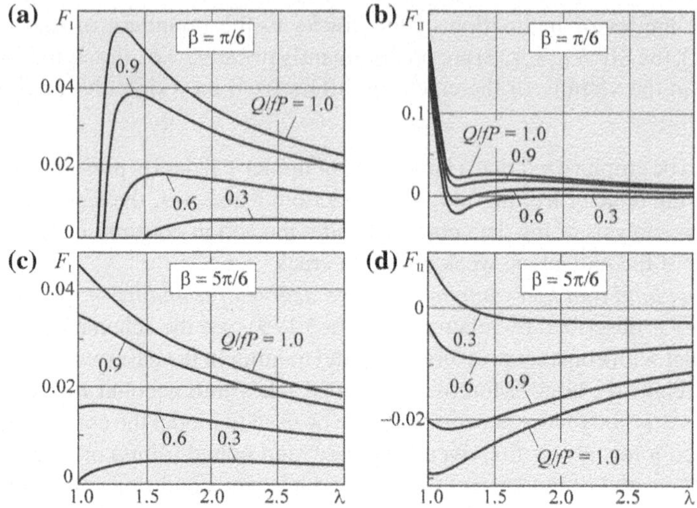

Fig. 3.20 Normalized SIF F_{I} and F_{II} for the rectilinear inclined edge crack under the action of a complex elliptic load applied to the boundary of the half-plane simulating the presence of an asymmetric section of sticking in the contact zone

Table 3.4 Dependence of maximum values of normalized SIF on relative length c/a of sticking section

Q/fP	c/a	$\beta = \pi/6$		$\beta = 5\pi/6$	
		max F_{I}	max $\lvert F_{\text{II}}\rvert$	max F_{I}	max $\lvert F_{\text{II}}\rvert$
1.00	0.00	0.05582	0.19423	0.04500	0.02988
0.90	0.32	0.03862	0.19340	0.03501	0.02168
0.60	0.63	0.01704	0.17513	0.01606	0.00954
0.30	0.84	0.00514	0.16149	0.00493	0.00850

- $\max K_{\text{I}}(\lambda)$ and $\max\lvert K_{\text{II}}(\lambda)\rvert$ for the angle $\beta = \pi/6$ are higher than for the angle $\beta = 5\pi/6$ for the identical lengths of the sections of sticking;
- as the section of sticking becomes larger, the values of $K_{\text{I}}(\lambda)$ and $\lvert K_{\text{II}}(\lambda)\rvert$ decrease; in particular, for the angle $\beta = \pi/6$, the quantity max $K_{\text{I}}(\lambda)$ becomes 10.9 times lower, while the quantity $\max\lvert K_{\text{II}}(\lambda)\rvert$ decreases by a factor of 1.2 as the relative length of the section of sticking c/a increases from 0 to 0.84; for the angle $\beta = 5\pi/6$, $\max K_{\text{I}}(\lambda)$ becomes 9.13 times lower, while the quantity max $\lvert K_{\text{II}}(\lambda)\rvert$ decreases by a factor of 3.5 (see Table 3.4).

We also mention that the numerical results obtained for the cases of action of concentrated forces and constant pressure are in good agreement with data of Rooke and Jones [30, 48], while the results obtained for elliptic loading agree with data of Keer et al. [29]. The character of the curves obtained for the case of complex symmetric elliptic loading agrees with data obtained by Hills and Nowell [24] and, for the case of contact loading taking into account the degree of rounding of edges of the counterbody, with the data of Dubourg and Villechaise [19].

3.10 Edge Crack with Contacting Faces in Half-Plane Under Action of Moving Hertz Load

Singular integral equations [12, 13]. Much attention is given to the determination of the stressed state in the vicinity of edge cracks whose faces may overlap under the action of an external contact load [14]. The theoretical works are mainly devoted to the analysis of two-dimensional problems [3, 18, 23, 32, 33, 56] and, in particular, to the evaluation of the stress intensity factors (SIF) at the tips of a rectilinear crack in the half-plane under the action of a given load applied to its boundary (concentrated forces, Hertz and uniformly distributed contact forces) for various boundary conditions imposed on the crack faces and modeling their contact interaction (see the bibliography in [44, 51, 53]).

In this section, by the method of singular integral equations [44, 53], we consider a more general case, namely, the case of a curvilinear crack in a half-plane for various conditions of interaction between the crack faces (contact in the presence of the sections of sticking and slip with friction).

Curvilinear crack under general conditions of contact of the faces. Assume that an elastic isotropic half-plane referred to the coordinate system xOy is weakened by a curvilinear crack appearing on its edge and is referred to a local coordinate system $x_1O_1y_1$ (Fig. 3.21). The local coordinate system is connected with the principal coordinate system by the formula $z = z_1e^{i\alpha} + z_1^0$, where α is the inclination angle of the axis O_1x_1 to Ox and z_1^0 is the affix of the point O_1 in the system xOy. It is assumed that the Hertz contact forces

$$s(x) = -p_0(1 + if)\sqrt{a^2 - (x - x_0)^2}/a, \qquad |x - x_0| \le a, \qquad (3.189)$$

where p_0 is the maximum pressure at the center of the contact section of length $2a$ and f is the Coulomb friction coefficient in contact, are applied to the boundary of the half-plane.

The boundary conditions of the problem on the boundary of the half-plane take the form

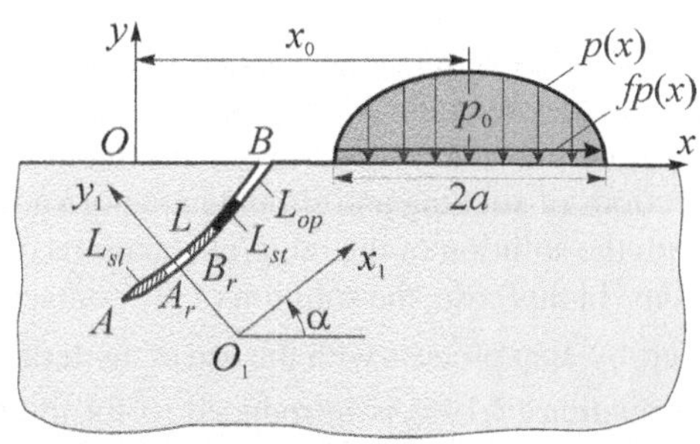

Fig. 3.21 General scheme of the problem

$$\sigma_y(x) - i\tau_{xy}(x) = s(x), \qquad |x - x_0| \leq a; \tag{3.190}$$

$$\sigma_y(x) - i\tau_{xy}(x) = 0, \qquad |x - x_0| > a. \tag{3.191}$$

By L_{op} we denote a collection of sections A_rB_r ($r = 1, \ldots, R$) of the crack contour L along which the crack faces are not in contact. These sections are enumerated in the direction from the crack tip A to its mouth B, namely,

$$L_{op} = \bigcup_{r=1}^{R} A_r B_r. \tag{3.192}$$

Assume that, under the action of contact forces $s(x)$, the crack faces in the zones of compression may overlap, which promotes the formation of contact stresses in these sections. Beyond the contact sections, the crack faces are free of loads.

Assume that the crack faces contact with friction and, therefore, the contact sections $L\backslash L_{op}$ can be split into the sections of slipping L_{sl} and sticking L_{st}, namely,

$$L\backslash L_{op} = L_{sl} \bigcup L_{st}. \tag{3.193}$$

Thus, the boundary conditions on the crack faces take the form

$$N^{\pm}(\tau) + iT^{\pm}(\tau) = 0, \qquad \tau \in L_{op}; \tag{3.194}$$

$$v_n^+(\tau) - v_n^-(\tau) = 0, \qquad \tau \in L_{sl}; \tag{3.195}$$

$$T^{\pm}(\tau) = f_c \, sign\left[\tilde{T}^{\pm}(\tau)\right] \, \left|N^{\pm}(\tau)\right|, \qquad \tau \in L_{sl}; \tag{3.196}$$

$$u^+(\tau) - u^-(\tau) + i\left[v^+(\tau) - v^-(\tau)\right] = 0, \qquad \tau \in L_{st}; \tag{3.197}$$

$$N^+(\tau) - N^-(\tau) + i\left[T^+(\tau) - T^-(\tau)\right] = 0, \qquad \tau \in L_{st}. \tag{3.198}$$

Here, $v_n(\tau)$ are the normal components of displacements of the crack faces, u and v are the components of the vector of displacements in the directions of the O_1x_1 and O_1y_1-axes, respectively, and f_c is the Amonton-Coulomb friction coefficient in contact between the crack faces.

Note that the boundaries of the sections of contact of the crack faces and the sections of sticking and slipping are not known in advance and can be found parallel with the solution of the integral equations of the problem from the additional conditions. In this case, the solution of the contact problem is constructed as a result of the step-by-step process with the use of an iterative procedure. The quantity $sign\left[\tilde{T}(\tau)\right]$ in relation (3.196) is introduced to fix the sign of contact stresses in the iterative procedure of finding the boundaries of the contact sections on the crack faces. To determine the contact sections $L \backslash L_{op}$ of the crack, we use the condition of vanishing of the mode I stress intensity factor at the ends of the intervals $A_r B_r$, namely,

$$K_{\rm I}(A_r) = 0, \quad K_{\rm I} = (B_r) = 0, \quad r = 1, \ldots, R, \tag{3.199}$$

and find the sections of sticking and slipping in the case of contact of the crack faces from the condition

$$f_c \left| N^{\pm}(\tau) \right| > \left| T^{\pm}(\tau) \right|, \qquad \tau \in L_{st}. \tag{3.200}$$

We now represent the derivative of the jump of the unknown vector of displacements on the crack line L in the form of a sum of two functions as follows:

$$g'(\tau) = g_1'(\tau) + g_2'(\tau), \tag{3.201}$$

where

$$g_1(\tau) = \frac{2G}{1+æ} \left[v_n^+(\tau) - v_n^-(\tau) \right] \frac{d\tau}{ds}; \tag{3.202}$$

$$g_2(\tau) = -\frac{2iG}{1+æ} \left[u_n^+(\tau) - u_n^-(\tau) \right] \frac{d\tau}{ds}. \tag{3.203}$$

Here, G is the shear modulus, $æ = 3 - 4\mu$ for plane deformation, $æ = (3-\mu)/(1+\mu)$ for the generalized plane stressed state, μ is Poisson's ratio, and $u_n(\tau)$, $v_n(\tau)$ are, respectively, the tangential and normal components of the vector of displacements of the crack faces.

Satisfying the boundary conditions of problem (3.194)–(3.198) with the help of the integral representations (3.154) of the Kolosov-Muskhelishvili complex potentials expressed via the functions $g_1'(t)$ and $g_2'(t)$, we arrive at the following system of singular integral equations:

$$\mathrm{Re}\, D(\tau) = \pi \mathrm{Re} P(\tau), \qquad \tau \subset L_{op}; \tag{3.204}$$

$$\mathrm{Im}\, D(\tau) + f_0 \mathrm{Re} D(\tau) = \pi [\mathrm{Im} P(\tau) + f_0 \mathrm{Re} P(\tau)], \qquad \tau \in L_{op} \bigcup L_{sl}. \tag{3.205}$$

Here,

$$D(\tau) = I\{L_{op}\}\, g_1'(\tau) + I\left\{ L_{op} \bigcup L_{sl} \right\} g_2'(\tau); \tag{3.206}$$

$$f_0 = \begin{cases} 0, & \tau \in L_{op}, \\ -f_c sign\, [\mathrm{Im}(P(\tau))], & \tau \in L_{sl}; \end{cases} \tag{3.207}$$

$$P(\tau) = p_0 \left\{ \mathrm{Re}\left[(1+if)\,c(\tau)\right] - \left[(1-if)\frac{\overline{c(\tau)}}{a(\tau)}\mathrm{Im}b(\tau) - if\overline{c(\tau)}\right]\frac{d\bar{\tau}}{d\tau}e^{-2i\alpha}\right\}; \tag{3.208}$$

$$a(\tau) = \sqrt{1-b^2(\tau)}; \quad b(\tau) = (\tau/a)e^{i\alpha} + \gamma - \lambda; \tag{3.209}$$

$$c(\tau) = a(\tau) + ib(\tau); \tag{3.210}$$

$$\tau_1 = \tau/a, \qquad \gamma = z_1^0/a, \qquad \lambda = x_0/a. \tag{3.211}$$

The operator $\mathrm{I}\{L\}$ is given by the formula

$$\mathrm{I}\{L\}\psi(t) = \int_L \left[R(t,\tau)\psi(t)dt + S(t,\tau)\overline{\psi(t)}d\bar{t}\right], \tag{3.212}$$

where

$$R(t,\tau) = \frac{e^{i\alpha}}{2}\left\{\frac{1}{T-T'} - \frac{1}{\overline{T}-\overline{T'}} - 2i\frac{\mathrm{Im}T}{(T-\overline{T'})^2} - \right.$$
$$\left. -\left[\frac{1}{T-\overline{T'}} - \frac{1}{\overline{T}-\overline{T'}} - 2i\,\mathrm{Im}T\frac{T+\overline{T'}-2T'}{(T-\overline{T'})^3}\right]\frac{d\bar{t}}{dt}e^{-2i\alpha}\right\};$$
$$S(t,\tau) = \frac{e^{-i\alpha}}{2}\left\{\frac{1}{\overline{T}-\overline{T'}} - \frac{1}{T-\overline{T'}} + 2i\frac{\mathrm{Im}T}{(\overline{T}-T')^2} - \right.$$
$$\left. -\left[\frac{T-T'}{(T-\overline{T'})^2} + \frac{T-T'}{(\overline{T}-\overline{T'})^2}\right]\frac{d\bar{t}}{dt}e^{-2i\alpha}\right\}; \tag{3.213}$$

$$T = te^{i\alpha} + z_1^0, \qquad T' = \tau e^{i\alpha} + z_1^0. \tag{3.214}$$

In order that system (3.204), (3.205) be complete, it is necessary to complement it with the following equations [53]:

$$\mathrm{Im}\left[g_1(\tau)\frac{d\bar{\tau}}{ds}\right] = 0, \qquad \tau \in L_{op}; \tag{3.215}$$

$$\mathrm{Re}\left[g_2(\tau)\frac{d\bar{\tau}}{ds}\right] = 0, \qquad \tau \in L_{op}\bigcup L_{sl}, \tag{3.216}$$

To solve the system of equations (3.204), (3.205), (3.215), and (3.216), in view of the fact that the limits of integration (the sizes of the sections of contact of the crack faces) are unknown, we use the method of successive approximations. As the zero-order approximation of coordinates of the points A_r and B_r separating the free sections of the crack from the contact sections, we take points at which the normal stresses on the crack line in a continuous half-plane with loaded boundary are equal to zero, i.e.,

$$\operatorname{Re} P(\tau) = 0, \qquad \tau \in L. \tag{3.217}$$

Further, on the basis of the solution of the system, we seek the values of the factors K_{I} that may differ from zero at these points. Thus, in each next stage of construction of the solution, we gradually change the boundaries of the open sections of the crack until conditions (3.199) are satisfied with a given accuracy. Simultaneously, by using the iterative procedure and taking into account condition (3.200), we determine the slipping and sticking sections in the contact zones of the crack faces.

Note that, in the open regions L_{op} of the crack, the following condition must be satisfied by the final result:

$$\delta(\tau) = v_n^+(\tau) - v_n^+(\tau) = \frac{1+æ}{2G}\operatorname{Re}\left[g(\tau)\frac{d\overline{\tau}}{d\tau}\right] > 0, \qquad \tau \in L_{op}. \tag{3.218}$$

This condition reflects the absence of contact of the crack faces and is deduced from relations (3.200)–(3.203).

Rectilinear crack under the conditions of contact of its faces with friction. The case of edge rectilinear crack in the half-plane subjected to the action of a Hertz contact load on its boundary is thoroughly investigated in the next chapters of the book. In the present section, in order to illustrate the results presented above, we consider two special cases. Assume that a crack of length l is inclined at an angle $\beta = -\alpha$ to the boundary of the half-plane (Fig. 3.22). The location of the contact load relative to the crack mouth is specified by the parameter $\lambda = x_0/a$. In the course of motion of the contact load, there exist the ranges of the parameter λ for which the crack is either closed or partially closed. For the sake of simplicity, we assume that, in contact, the crack faces slip with friction. At the same time, in the case of partial contact, the contact section is located near the crack tip (see Fig. 3.22).

Then the boundary conditions of the problem on the boundary of the half-plane take the form (3.190) and (3.191) and, for the crack faces, the boundary conditions can be written as follows:

$$N^{\pm}(x_1) + iT^{\pm}(x_1) = 0, \qquad 0 < x_1 < l^*; \tag{3.219}$$

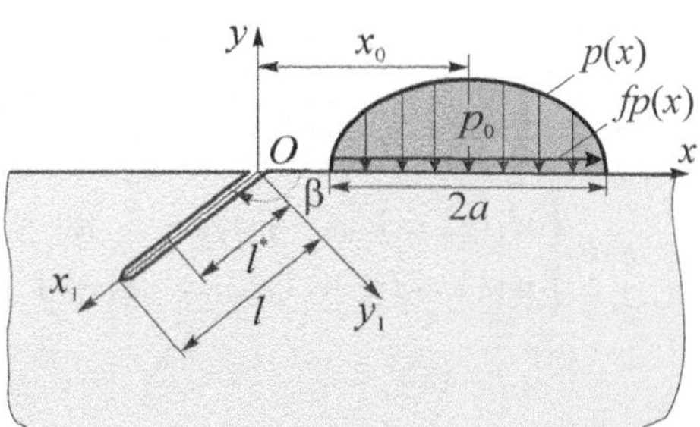

Fig. 3.22 Edge rectilinear crack in the half-plane under the action of the Hertz load

$$v^{+}(x_1) - v^{-}(x_1) = 0, \qquad l^{*} < x_1 < l; \tag{3.220}$$

$$T^{\pm}(x_1) = f_c \operatorname{sign}\left[\tilde{T}^{\pm}(x_1)\right] \left|N^{\pm}(x_1)\right|, \qquad l^{*} < x_1 < l, \tag{3.221}$$

where l^{*} is the open part of the crack.

According to relations (3.204) and (3.205), we get the following singular integral equations of the problem:

$$\operatorname{Re}\int_{0}^{l^{*}} E(t, x_1)\varphi_1(t)dt - \operatorname{Im}\int_{0}^{l} F(t, x_1)\varphi_2(t)dt = \pi \operatorname{Re} P(x_1), \qquad 0 < x_1 < l^{*}; \tag{3.222}$$

$$\begin{aligned} &\int_{0}^{l^{*}} [\operatorname{Im} E(t, x_1) + f_0 \operatorname{Re} E(t, x_1)]\varphi_1(t)dt \\ &+ \int_{0}^{l} [\operatorname{Re} F(t, x_1) - f_0 \operatorname{Im} F(t, x_1)]\varphi_2(t)dt = \\ &= \pi[\operatorname{Im} P(x_1) + f_0 \operatorname{Re} P(x_1)], \qquad 0 < x_1 < l \end{aligned} \tag{3.223}$$

for the following unknown functions:

$$\varphi_1(x_1) = \frac{2G}{1+æ}\frac{d}{dx_1}\left[v^{+}(x_1) - v^{-}(x_1)\right], \qquad 0 < x_1 < l^{*}; \tag{3.224}$$

$$\varphi_2(x_1) = -\frac{2Gi}{1+æ}\frac{d}{dx_1}\left[u^{+}(x_1) - u^{-}(x_1)\right], \qquad 0 < x_1 < l. \tag{3.225}$$

The kernels of the SIE (3.222) and (3.223) are given by the formulas

$$E(t, x_1) = R(t, x_1) + S(t, x_1); \tag{3.226}$$

$$F(t, x_1) = R(t, x_1) - S(t, x_1). \tag{3.227}$$

In this case, the functions $P(x_1)$, $R(t, x_1)$, and $S(t, x_1)$ are given by relations (3.208), (3.213).

By the change of variables

$$t = \begin{cases} w_1(\xi) = k^{*}l(\xi+1)/2, & |\xi| \le 1 \quad \text{for} \quad 0 < t < l^{*} \quad (k^{*} = l^{*}/l) \\ w(\xi) = l(\xi+1)/2, & |\xi| \le 1 \quad \text{for} \quad 0 < t < l \end{cases} \tag{3.228}$$

and

$$x_1 = \begin{cases} w_1(\eta) = k^* l(\eta + 1)/2, & |\eta| < 1 \quad \text{for} \quad 0 < x_1 < l^* \\ w(\eta) = l(\eta + 1)/2, & |\eta| < 1 \quad \text{for} \quad 0 < x_1 < l \end{cases} \tag{3.229}$$

the integral equations of the problem (in the normalized form) are as follows:

$$\int_{-1}^{1} \left[\mathrm{Re}E(\xi, \eta)\varphi_1(\xi) - \mathrm{Im}F'(\xi, \eta)\varphi_2(\xi)\right] d\xi = \pi \mathrm{Re}P_1(\eta), \qquad |\eta| < 1; \tag{3.230}$$

$$\begin{gathered} \int_{-1}^{1} \left\{ \left[k^* \mathrm{Im}E''(\xi, \eta) + f_0 \mathrm{Re}E(\xi, \eta)\right]\varphi_1(\xi) \right. \\ \left. +\left[\mathrm{Re}F(\xi, \eta) - f_0 \mathrm{Im}F'(\xi, \eta)\right]\varphi_2(\xi)\right\} d\xi = \\ -\pi[\mathrm{Im}P(\eta) + f_0 \mathrm{Re}P_1(\eta)], \qquad |\eta| < 1. \end{gathered} \tag{3.231}$$

Here, we have introduced the notation

$$\begin{gathered} E(\xi, \eta) = E(w(\xi), w(\eta)); \qquad F(\xi, \eta) = F(w(\xi), w(\eta)); \\ E'(\xi, \eta) = E(w(\xi), w_1(\eta)); \qquad F'(\xi, \eta) = F(w(\xi), w_1(\eta)); \\ E''(\xi, \eta) = E(w_1(\xi), w(\eta)); \end{gathered} \tag{3.232}$$

$$\varphi_1(\xi) = \varphi_1(t) w'(\xi); \tag{3.233}$$

$$\varphi_2(\xi) = \varphi_2(t) w'(\xi); \tag{3.234}$$

$$P_1(\eta) = P(w_1(\eta)); \qquad P(\eta) = P(w(\eta)). \tag{3.235}$$

The integral equations (3.230) and (3.231) are solved numerically by the method of mechanical quadratures under the assumption assuming that the functions $\varphi_1(\xi)$ and $\varphi_2(\xi)$ have square-root singularities, i.e. $\varphi_1(\xi) = v_1(\xi)/\sqrt{1-\xi^2}$, $\varphi_2(\xi) = v_2(\xi)/\sqrt{1-\xi^2}$, $v_1(\xi)$ and $v_2(\xi)$ are continuous on the segment $|\xi| \le 1$. As a result of application of the Gauss quadrature formulas to these equations, we arrive at the system of $2n$ real linear algebraic equations for $2n$ unknowns $v_1(\xi_k)$ and $v_2(\xi_k)$:

$$\frac{1}{n}\sum_{k=1}^{n} \left\{\mathrm{Re}E(\xi_k, \eta_m)\, v_1(\xi_k) - \mathrm{Im}F'(\xi_k, \eta_m)\, v_2(\xi_k)\right\} = \mathrm{Re}P_1(\eta_m) + f_0 \mathrm{Re}P_1(\eta_m),$$

$$m = 1, \ldots, n-1; \tag{3.236}$$

$$\begin{gathered} \tfrac{1}{n}\sum_{k=1}^{n} \left\{ \left[k^* \mathrm{Im}E''(\xi_k, \eta_m) + f_0 \mathrm{Re}E(\xi_k, \eta_m)\right] v_1(\xi_k) + \right. \\ \left. +\left[\mathrm{Re}F(\xi_k, \eta_m) - f_0 \mathrm{Im}F'(\xi_k, \eta_m)\right] v_2(\xi)\right\} = \mathrm{Im}P(\eta_m) + f_0 \mathrm{Re}P_1(\eta_m), \\ m = 1, \ldots, n-1; \end{gathered} \tag{3.237}$$

$$\sum_{k=1}^{n}\left[(-1)^k tg\left(\frac{2k-1}{4n}\pi\right)v_1(\xi_k)\right]=0; \tag{3.238}$$

$$\sum_{k=1}^{n}\left[(-1)^k tg\left(\frac{2k-1}{4n}\pi\right)v_2(\xi_k)\right]=0, \tag{3.239}$$

where n is the number of nodes in the segment of integration $|\xi| \leq 1$.

By using the solution of the system of equations (3.236)–(3.239), we find the mode I stress intensity factor K_{I} at the end of the open section of the crack and the mode II stress intensity factor K_{II} at the crack tip

$$K_{\mathrm{I}}=p_0\sqrt{2\pi a}\frac{1}{n}\sum_{k=1}^{n}\left[(-1)^k ctg\left(\frac{2k-1}{4n}\pi\right)v_1(\xi_k)\right]; \tag{3.240}$$

$$K_{\mathrm{II}}=-p_0\sqrt{2\pi a}\frac{1}{n}\sum_{k=1}^{n}\left[(-1)^k ctg\left(\frac{2k-1}{4n}\pi\right)v_2(\xi_k)\right]. \tag{3.241}$$

The absence of contact of the crack faces is controlled by condition (3.218). By the change of variables, this condition is reduced to the form [53]

$$\frac{1+æ}{2Gn}\sum_{k=1}^{n}\left\{v_1(\xi_k)\left[\arccos\eta+2\sqrt{1-\eta^2}+\sum_{m=1}^{n-1}\frac{T_m(\xi_k)U_{m-1}(\eta)}{m}\right]\right\}<0, \quad |\eta|<1. \tag{3.242}$$

If condition (3.242) is not satisfied on some segments of the crack contour, then this means that the crack faces are in contact. Hence, the solution of the integral equations can be found by the method of successive approximations.

Stress intensity factors. The numerical analyses were carried out for the following types of contact of the faces of edge rectilinear cracks (Fig. 3.22): (1) smooth contact under the assumption that, for some locations of load on the boundary of the half-plane in the vicinity of the crack mouth, the crack can be partially or completely closed; (2) contact with friction under the assumption that contact is realized over the entire crack length.

Smooth contact of crack faces. The numerical analysis was performed for different values of the angles of inclination of the crack to the edge of the half-plane (β = $\pi/6$, $\pi/4$, $3\pi/4$, $5\pi/6$, $8\pi/6$), for two values of the relative crack length (ε = l/a = 0,5; 1,0), and the friction coefficient in contact between the bodies f = 0.1. We established the ranges of the parameter $\lambda = x_0/a$ characterizing the location of the contact load (counterbody) for which the crack is completely or partially closed. The plots of the dependences of the normalized stress intensity factors $F_{\mathrm{II}}(\lambda,\beta,f,\varepsilon)=K_{\mathrm{II}}(\lambda,\ \beta,\ f,\ \varepsilon)/\left(p_0\sqrt{\pi a}\right)$ on the location of the contact load (on λ) are shown in Figs. 3.23 and 3.24.

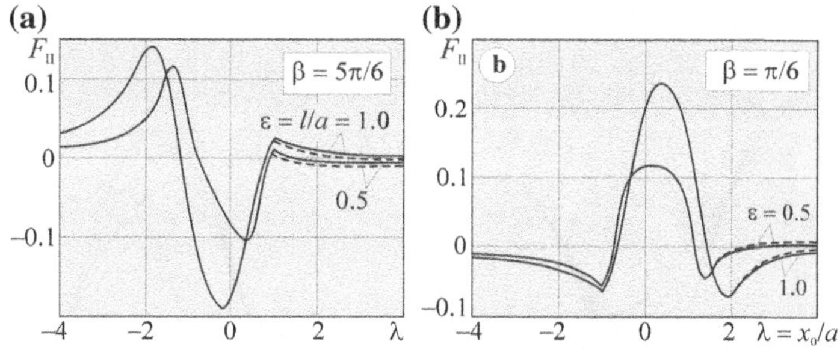

Fig. 3.23 Dependence of the normalized SIF $F_{II}(\lambda)$ on the location of the contact load relative to the crack mouth for different angles between the crack and the boundary of the half-plane: **a** $\beta = 5\pi/6$, **b** $\beta = \pi/6$; $f = 0.1$; (——) contact of faces over the entire length; (– – –) partial contact

Fig. 3.24 Dependence of the normalized SIF $F_{II}(\lambda)$ on the orientation (angle β) of the edge crack

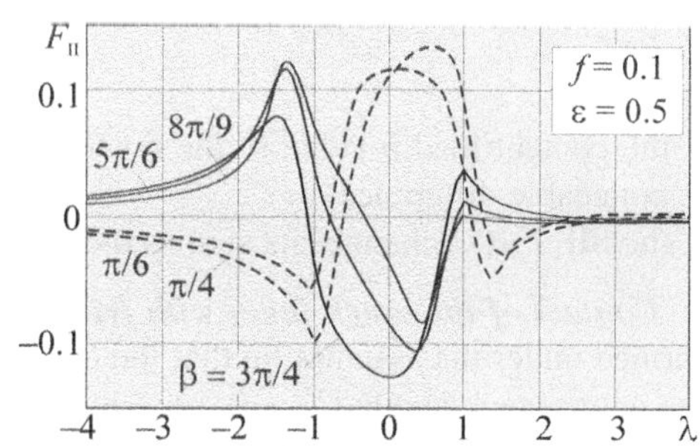

The analysis of the curves in Fig. 3.23 enables us to make the following conclusions:

- the increase in the relative length $\varepsilon = l/a$ of the crack from 0.5 to 1.0 leads, in general, to the increase in the SIF $|K_{II}|$; at the same time, for both inclination angles of the crack, the quantity $\max|K_{II}|$ becomes almost twice larger;
- for given values of the parameters f, ε, and β, in the case of motion of the contact load from the right to the left, the crack is initially open and then partially open; at the same time, in the case of motion of the load over the crack mouth and behind it, the crack is completely closed; moreover, the cracks oriented at acute angles ($\beta < \pi/2$) close faster (for larger λ) than the cracks making blunt angles ($\beta > \pi/2$) with the direction of the Ox-axis (to the direction of tangential forces in contact between the bodies).

Note that, in both cases ($\beta = \pi/6$ and $\beta = 5\pi/6$), in the zero-order approximation of the iterative process, it was assumed that, as the load moves from the right to the left, the crack closes from its tip.

The analysis of the curves presented in Fig. 3.24 demonstrates that:

- for acute ($\beta < \pi/2$) and blunt ($\beta > \pi/2$) angles of inclination of the crack to the edge of the half-plane, $\max\,|K_{II}|$ is attained for absolutely different locations of

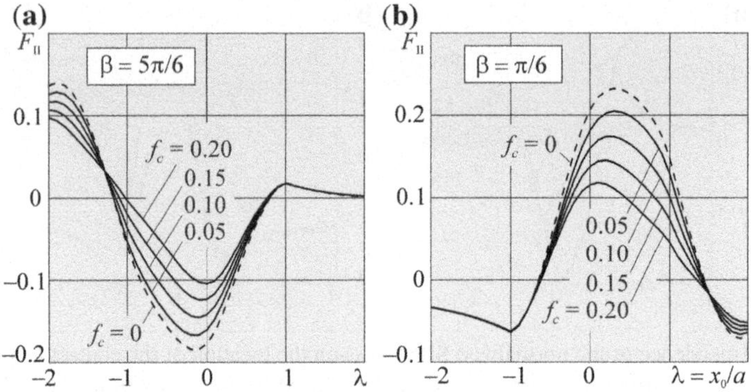

Fig. 3.25 Dependence of the normalized SIF F_{II} on the location of the contact load (λ) for various coefficients of friction f_c between the crack faces; $f = 0.1$; $\varepsilon = l/a = 1.0$

the contact load relative to the crack mouth but their values do not exhibit any noticeable difference;

- the SIF $F_{II}(\lambda)$ changes its sign as the contact load moves over the crack mouth.

Contact of the crack faces with friction. The numerical calculations were performed under the assumption that the crack faces are in contact with friction along the entire crack length ($l^* = 0$; see Fig. 3.22) if the contact load is located near the crack mouth ($|\lambda| \leq 2$). In this case, the boundary conditions of the problem take the form (3.190), (3.191), and (3.219)–(3.221).

The problem is reduced to the solution of the singular integral equations (3.230) and (3.231). The discrete analog of these equations has the form of the system of linear algebraic equations (3.236), (3.237).

We now seek the mode II stress intensity factor K_{II} at the crack tip on the basis of the solution of this system given by relation (3.241).

In Figs. 3.25, 3.26, 3.27 and 3.28, we present the dependences of the normalized stress intensity factors $F_{II} = K_{II}/\left(p_0\sqrt{\pi a}\right)$ on the parameter λ specifying the location of the contact load for various values of the friction coefficients in contact between the bodies ($f = 0, 0.05, 0.1, 0.15$, and 0.2), relative crack lengths ($\varepsilon = l/a = 0.3, 0.5, 0.7, 1.0$), angles made by the crack with the boundary of the half-plane ($\beta = 5\pi/6, 3\pi/4, 2\pi/3, \pi/4, \pi/6$), and friction coefficients in contact between the crack faces ($f_c = 0, 0.05, 0.1, 0.15, 0.2$).

The analysis of the curves presented in Fig. 3.25 shows that the presence of friction between the crack faces strongly decreases the SIF at the crack tip independently of its orientation. Thus, for the friction coefficient $f_c = 0.1$, the quantity max $|K_{II}|$ decreases, on the average (for $\beta = \pi/6$ and $\beta = 5\pi/6$), by 22%, as compared with the case $f_c = 0$ and by 48% if the friction coefficient increases to $f_c = 0.2$. At the same time, the increase in the friction coefficient f under contact loading (in the case of contact between the bodies) weakly affects the SIF $|K_{II}|$ (Fig. 3.26).

The dependences of the SIF F_{II} on the crack length for different locations of the contact load (Fig. 3.27) indicate that F_{II} strongly increases as the crack length

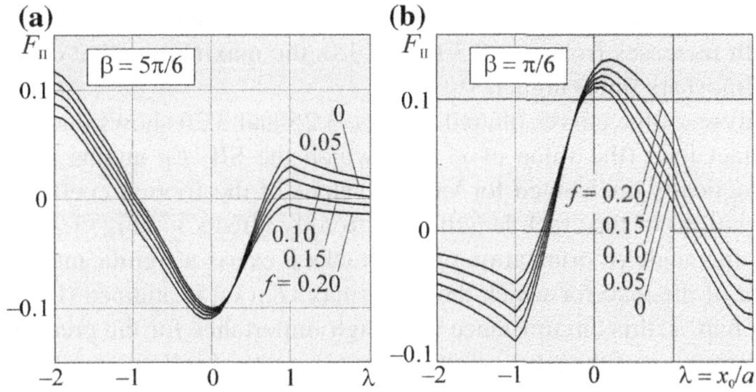

Fig. 3.26 Dependence of the normalized SIF F_{II} on the location of the contact load (λ) for different friction coefficients f in the case of contact between the bodies; $\varepsilon = 1.0; f_c = 0.2$

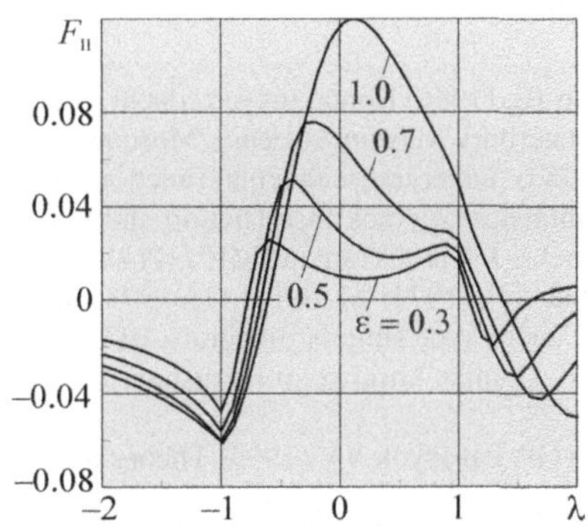

Fig. 3.27 Dependence of the normalized SIF F_{II} on the location of the contact load (λ) for different relative crack lengths $\varepsilon = l/a$; $\beta = \pi/6; f = 0.1; f_c = 0.2$

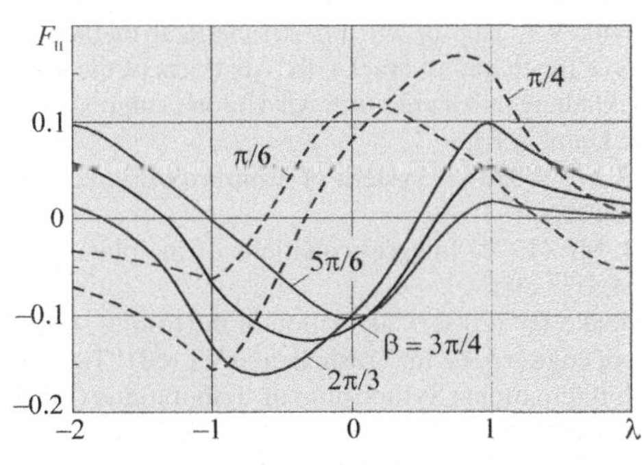

Fig. 3.28 Dependence of the normalized SIF F_{II} on the location of the contact load (λ) for different angles β between the crack and the boundary of the half-plane; $f = 0.1$; $\varepsilon = 1.0; f_c = 0.2$

increases (for fixed values of the parameters β, f, and f_c). In particular, as the relative crack length increases from $\varepsilon = 0.3$ to $\varepsilon = 1.0$, the maximum value of the SIF F_{II} becomes almost six times higher.

The analysis of the curves plotted in Figs. 3.25 and 3.26 shows that the location of the contact load (the value of λ^*) for which the SIF F_{II} attains its maximum remains practically unchanged for various values of the friction coefficients f and f_c. The variations of the crack length also strongly affects λ^* (Fig. 3.27), while the changes in the angle of orientation of the cracks β exerts a significant influence on the location of the load for which the value max $|F_{II}(\lambda)|$ is attained (Fig. 3.28). As shown in Chap. 4, this circumstance is of high importance for the prediction of the angles of formation of the edge cracks in the elements of rolling couples.

References

1. Aleksandrov VM, Romalis BL (1986) Kontaktnyye zadachi v mashinostroyenii (Contact Problems in Mechanical Engineering). Mashinostroenie, Moscow
2. Bateman H, Erdelyi A (1953) higher transcendental functions, vol 2. McGraw-Hill, New York
3. Bower AF (1988) The influence of crack face friction and trapped fluid on surface initiated rolling contact fatigue cracks. J Tribol Trans ASME 110(4):704–711
4. Chawla MM, Ramakrishan TR (1974) Modified Gauss-Jacoby quadrature formulas for the numerical evaluation of Cauchy type singular integrals. BIT (Sver) 14(1):14–21
5. Cherepanov GP (1974) Mekhanika khrupkogo razrusheniya (Mechanics of Brittle Fracture). Nauka, Moscow
6. Datsishin OP, Marchenko GP, Panasyuk VV (1993) Theory of crack growth in rolling contact. Mater Sci 29(4):373–383
7. Datsyshin AP, Marchenko HP (1985) An edge curvilinear crack in an elastic half plane. Sov Mater Sci 21(1):66–70
8. Datsyshin AP, Marchenko HP (1985) Interaction of curvilinear cracks with the boundary of an elastic half plane. Sov Mater Sci 20(5):466–473
9. Datsyshin AP, Panasyuk VV, Savruk MP (1976) General method for the solution of two-dimensional problems of the theory of cracks. In: Abstracts of the 4 Vsesoyuznogo Kongressa po teoreticheskoy i prikladnoy mekhaniki (4th All-Union Congress on Theoretical and Applied Mechanics). Naukova Dumka, Kiev
10. Datsyshin AP, Savruk MP (1973) A system of arbitrarily oriented cracks in elastic solids. J Appl Math Mech 37(2):306–313
11. Datsyshin AP, Savruk MP (1974) Integral equations of the plane problem of crack theory. J Appl Math Mech 38(4):677–686
12. Datsyshyn OP, Marchenko HP (2002) Calculation of the rolling surface durability at the stage of shear mode growth of edge cracks. In: Troshchenko VT (ed) "Trybofatyka" ("Tribofatigue"), Proceedings of the 4th International Symposium on Tribofatigue (ISTF 4), Vol 1. Ternopil', pp 420–425
13. Datsyshyn OP, Marchenko HP (2003) Estimation of mode II surface crack growth under rolling contact. Mashynoznavstvo 7:17–23
14. Datsyshyn OP, Marchenko HP (2008) Stressed state of a half plane with shallow edge crack under Hertzian loading (a survey). Mater Sci 44(1):22–34
15. Datsyshyn OP, Panasyuk VV (2017) Methods for the evaluation of the contact durability of elements of tribojoints (a survey). Mater Sci 52(4):447–459
16. Datsyshyn OP, Pryshlyak RE, Prykhods'ka SV, et al Influence of the shape of model contact load on the stress intensity factors for an edge crack. Probl Trybol 3:3–16

17. Datsyshyn OP, Shchur RB (1998) Development of edge cracks under conditions of fretting fatigue. Probl Tribol 2:7–16
18. Dawson PH (1967) Contact fatigue in soft steel with random loading. J Mech Engn Sci 9(1):79–80
19. Dubourg MC, Villechaise B (1992) Stress intensity factors in a bent crack: a model. Eur J Mech A/Solids 11(2):169–179
20. Erdogan F, Gupta GD (1972) On the numerical solution of singular integral equations. Q Appl Math 29(4):525–534
21. Frolish MF, Fletcher DI, Beynon JH (2002) A quantitative model for predicting the morphology of surface initiated rolling contact fatigue cracks in back-up roll steels. Fatigue Fract Eng Mater Struct 25:1073–1086
22. Gakhov FD (1963) Krayevyye zadachi (Boundary-Value Problems). Fizmatgiz, Moscow
23. Hills DA, Comninou M (1985) An analysis of fretting fatigue crack during loading phase. Int J Solids Struct 21(7):721–730
24. Hills DA, Nowell D (1994) Mechanics of fretting fatigue. Dordrecht: Kluwer Academic Publishers, 236 p
25. Isida M (1979) Tension of a half plane containing array cracks, branched cracks, and cracks emanating from sharp notches. Trans Jpn. Soc Mech Eng A45(392):306–314
26. Johnson KL (1985) Contact mechanics. Cambridge University Press, Cambridge
27. Kachanov LM (1974) Osnovy mekhaniki razrusheniya (Fundamentals of Fracture Mechanics). Nauka, Moscow
28. Kalandiya AI (1973) Matematicheskiye metody dvumernoy uprugosti (Mathematical Methods of Two-Dimensional Elasticity). Nauka, Moscow
29. Keer LM, Bryant MD, Haritos GK (1982) Subsurface and surface cracking due to Hertzian contact. Trans ASME J Lubric Technol 104(3):347–351
30. Kolesnikov YuV, Morozov EM (1989) Mekhanika kontaktnogo razrusheniya (Mechanics of Contact Fracture). Nauka, Moscow
31. Kolosov GV (1935) Primeneniye kompleksnoy peremennoy k teorii uprugosti (Application of Complex Variable to the Theory of Elasticity). ONTI Moscow-Leningrad
32. Komvopoulos K (1996) Subsurface crack mechanisms under indentation loading. Wear 199:9–23
33. Komvopoulos K, Cho S-S (1997) Finite element analysis of subsurface crack propagation in a half-space due to a moving asperity contact. Wear 209:57–68
34. Korneichuk AA (1964) Quadrature formulas for singular integrals. In: Chislennyye metody resheniya differentsial'nykh i integral'nykh uravneniy i kvadraturnyye formuly (Numerical Methods for the Solution of Differential and Integral Equations and Quadrature Formulas). Nauka, Moscow, pp 64–74
35. Lin'kov AM (1976) Problems of the theory of elasticity for a plane with a finitely many curvilinear cuts. Issled Uprug Plast 11:3–11
36. Muskhelishvili NI (1962) Singulyarnyye integral'nyye uravneniya (Singular Integral Equations). Fizmatgiz, Moscow
37. Muskhelishvili NI (1966) Nekotoryye osnovnyye zadachi matematicheskoy teorii uprugosti (Some Basic Problems of the Mathematical Theory of Elasticity). Nauka, Moscow
38. Natanson IP (1949) Konstruktsionnaya teoriya funktsiy (Constructive Theory of Functions). Gostekhizdat, Moscow-Leningrad
39. Panasyuk VV (1968) Predel'noye ravnovesiye khrupkikh tel s treshchinami (Limit Equilibrium of Brittle Bodies with Cracks). Naukova Dumka, Kiev
40. Panasyuk VV, Berezhnits'kyi LT (1964) Determination of the limit forces in tension of a plate with an arc-shaped crack. In: Karpenko GV (ed) Voprosy mekhaniki real'nogo tverdogo tela (Problems of Mechanics of Real Solid Body). Naukova Dumka, Kiev, pp 3–19
41. Panasyuk VV, Datsyshin AP (1974) The equilibrium limit of a half plane with an arbitrarily oriented crack at its boundary. Sov Mater Sci 7(6):751–752
42. Panasyuk VV, Datsyshyn OP, Marchenko HP (1995) To crack propagation theory under rolling contact. Eng Fract Mech 52(1):179–191

43. Panasyuk VV, Savruk MP, Datsyshin AP (1977) Application of singular integral equations to the solution of two-dimensional problems in crack theory. Sov Mater Sci 12(3):245–259
44. Panasyuk VV, Savruk MP, Datsyshin AP (1976) Raspredeleniye napryazheniy okolo treshchin v plastinakh i obolochkakh (Distribution of Stresses near Cracks in Plates and Shells). Naukova Dumka, Kiev
45. Panasyuk VV, Savruk MP, Datsyshyn AP (1977) A general method of solution of two-dimensional problems in the theory of cracks. Eng Fract Mech 9(2):481–497
46. Parton VZ, Morozov EM (1974) Mekhanika uprugo-plasticheskogo razrusheniya (Mechanics of Elastoplastic Fracture). Nauka, Moscow
47. Pinegin SV (1969) Kontaktnaya prochnost‘ i soprotivleniye kacheniyu (Contact Strength and Rolling Resistance). Mashinostroenie, Moscow
48. Rooke DP, Jones DA (1979) Stress intensity factors in fretting fatigue. J Strain Anal 14(1):1–6
49. Savruk MP (1977) Constructing integral equations for two-dimensional elasticity theory problems of a body with curvilinear cracks. Sov Mater Sci 12(6):682–683
50. Savruk MP (1980) Stress intensity factors for a curvilinear crack differing little from arc shaped or straight. Sov Mater Sci 16(2):156–161
51. Savruk MP (1988) Koefficienty intensivnosti napryazheniy v telakh s treshchinami (Stress Intensity Factors in Bodies with Cracks). In: Panasyuk VV (ed) Fracture mechanics and strength of materials: a handbook, vol 2. Naukova Dumka, Kiev
52. Savruk MP (1978) System of curvilinear cracks in an elastic body under different boundary conditions on their lips. Sov Mater Sci 14(6):641–649
53. Savruk MP (1981) Dvumernyye zadachi uprugosti dlya tel s treshchinsmi (Two-Dimensional Problems of Elasticity for Bodies with Cracks). Naukova Dumka, Kiev
54. Savruk MP, Datsyshin AP (1974) Interaction between a system of cracks and the boundaries of an elastic body. Sov Appl Mech 10(7):755–761
55. Savruk MP, Datsyshin AP (1972) Limiting equilibrium state of a body weakened by a system of randomly oriented cracks. In: Termomekhanicheskiye metody razrusheniya gornykh porod (Thermomechanical Methods of Fracture of Rocks), Part 2. Naukova Dumka, Kiev, pp 97–102
56. Sheppard SD, Hills DA, Barber JR (1986) An analysis of fretting cracks. Part 2: Unloading and reloading phases. Int J Solids Struct 22(4):387–396
57. Sih GC, Liebowitz H (1968) Mathematical theory of brittle fracture. In: Liebowitz H (ed) Fracture, vol 2. Academic Press, New York-London, pp 68–191
58. Sih GC, Paris PC, Erdogan F (1962) Crack-tip stress-intensity factors for plane extension and plate bending problems. Trans ASME J Appl Mech 29:306–312
59. Sosnovskii LA, Makhutov NA, Shurinov VA (1992) Friction-mechanical fatigue: basic regularities, Zavodskaya Laboratoriya 9:46–63
60. Theocaris PS, Ioakimidis NI (1979) A method of numerical solution of Chauchy-type singular integral equations with generalized kernels and arbitrary complex singularities. J Comput Phys 30(3):309–323
61. Turchak LI (1987) Osnovy chisel‘nykh metodov (Fundamentals of Numerical Methods): A Tutorial. Nauka, Moscow
62. Vladimirov VI (1976) Obobshchennyye funktsii v matematicheskoy fizike (Generalized Functions in Mathematical Physics). Nauka, Moscow
63. Waterhause RB (1972) Fretting corrosion. Pergamon Press, Oxford-New York
64. Weighardt R (1907) Über das Spalten und Zerreißen elastischer Körper. Z. Math. und Phys. 55(1/2):60–103
65. Williams ML (1957) On the stress distribution at the base of a stationary crack. J. Appl. Mech. 24(1):109–114

Chapter 4
Rolling Contact Fatigue

Abstract This chapter contains the results of investigations of the fracture processes (paths of crack propagation) for bodies under the conditions of rolling (rolling with sliding). These results were obtained by using the model proposed in Chap. 2, the solutions of singular integral equations for the corresponding problems, and also the results of evaluation of lifetime under the conditions of contact fatigue. We study and describe the specific features of formation of the typical defects usually observed in rolling bodies, such as pitting, spalling, squat ("dark spot"), checks depending on the operating parameters of the analyzed rolling couple and the characteristics of cyclic crack growth resistance of the corresponding materials. We present examples of evaluation of the residual contact durability of rail, wheel, and roll steels according to the criteria of pitting and spalling formation. The contact fatigue curves (i.e., the dependences of the number N_f of rolling cycles on the maximum value of contact pressure p_0) are plotted by using the criteria of pitting or spalling formation on the rolling surfaces. We also give some recommendations useful for the engineering practice.

The model proposed in the previous sections and the solutions of singular integral equations (SIE) are used for the investigation of the processes of fracture in rolling bodies and for the evaluation of their lifetimes under the conditions of cyclic contact. We now determine the specific features of formation of typical defects, such as pitting and spalling, squat ("dark spots"), and checks in rolling bodies depending on the operation parameters of rolling couples and on the characteristics of cyclic crack growth resistance of materials.

We consider the case of unidirectional rolling with slip. The body damaged by cracks is modeled by an elastic half-plane with cracks (cuts) and the contact action of the counterbody is modeled by the repeated motion of Hertzian normal forces with tangential component along the edge of half-plane (Fig. 4.1). In the contact section, these forces are described by the formula

$$\sigma_y - i\tau_{xy} = -p(x) - iq(x) = -p_0(1 + if)\frac{\sqrt{a^2 - (x - x_0)^2}}{a}, \quad |x - x_0| \leq a, \quad y = 0, \tag{4.1}$$

O. Datsyshyn and V. Panasyuk, *Structural Integrity Assessment of Engineering Components Under Cyclic Contact*, Structural Integrity 9,
https://doi.org/10.1007/978-3-030-23069-2_4

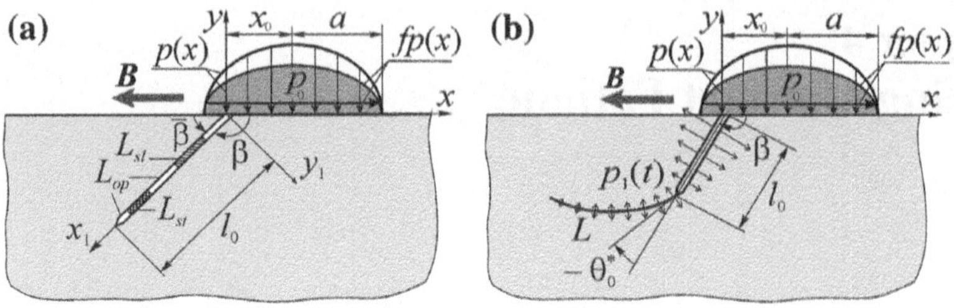

Fig. 4.1 Computational schemes of the two stages of pit formation: **a** stage of shear, **b** opening stage; **B** is the direction of the motion of the counterbody; f is the coefficient of friction between contacting bodies; L_{sl}, L_{st}, and L_{op} are the sections of slipping with friction, sticking, and opening displacement (the crack faces are not in contact), respectively

where $2a$ is the length of the contact section and p_0 is the maximum pressure in the central part of the contact section. Outside the contact section, forces are equal to zero.

4.1 Rolling Under Boundary Lubrication Conditions. Pitting Formation

Pitting appears practically on the surfaces of all elements of the rolling pairs. In the engineering practice, their contact lifetime is determined by the depth of the pits in rolling bodies and by the number of cycles prior to their appearance. As early as in 1935, Way [99] assumed that pitting appears only in the presence of lubricants in the contact zone between rolling bodies. Under this assumption, the process of growth of edge cracks in rolling bodies is studied under the conditions of boundary lubrication. In the computational model, it is assumed that the thickness of lubricating layer is equal to zero. In other words, this layer does not separate the rolling bodies but significantly decreases the friction forces between the bodies. The lubricant is regarded as an incompressible liquid. It is also assumed that the rolling body damaged by cracks is a driven body. Note that the edge cracks oriented in the direction of motion of the counterbody (contact load) are frequently initiated in this body. This promotes the penetration of lubricants into these cracks in rolling. In this case, the lubricant reduces the friction forces between the crack faces in the stage of their propagation by the mechanism of shear (Fig. 4.1a). Later, in the process of rolling, the lubricant may separate the crack faces and even wedge them out. In this case, we observe the transition from fracture by the mechanism of shear to fracture by mechanism of normal opening. The wedging action of the lubricant [or another liquid (medium)] upon the crack faces is modeled either uniformly [14, 72] or linearly [40] by distributed normal pressure acting along the entire crack length (Fig. 4.1b).

Hence, in the computational model, it is assumed that the growth of an edge rectilinear crack (inclined to the side of motion of the counterbody) by the mode II mechanism in the direction of its continuation is the first stage (onset) of the process of pitting formation. The second stage of crack propagation by the mode I mechanism takes into account the influence of lubricants penetrating into the crack and creating pressure on its faces.

4.1.1 Edge Crack in Contact Zone Under Transverse Shear Conditions

The available numerous experimental data [11, 47, 55, 65, 81, 99, 102] demonstrate that shallow surface macrocracks are frequently initiated in rolling bodies (race tracks). As a rule, they grow rectilinearly in the direction (plane) opposite to the action of tangential contact forces in the damaged body. As the main cause of development of these cracks, we can mention significant compressive contact forces induced by the action of the counterbody. The theoretical analyses [13, 15] demonstrate that the SIF $K_{\mathrm{I}} < 0$ at the tips of these cracks. Moreover, the τ_θ-criterion (formula (2.21)) implies that, in this case, the angle of initial deviation is $\theta^* = 0$. In other words, the crack grows rectilinearly. In the general case, three types of contact conditions can be realized on the crack faces in the case of motion of the counterbody along the boundary of the body (half-plane): the absence of contact (L_{op}), slip of the crack faces with friction (L_{sl}), and sticking (L_{st}) (Fig. 4.1a). In what follows, we consider these cases by using the approaches presented in Chaps. 2 and 3. The solution of the corresponding contact problems, the determination of the stress intensity factors (SIF), and the analysis of the specific features of contact of the crack faces are reduced to the solution of the system of Eqs. (3.204), (3.205) whose right-hand side $P(\eta)$ (3.208) is formed by the Hertzian load with regard for condition (3.160). We solve the system of these equations by using the already described method of mechanical quadratures with the use of iterative procedures in order to determine the ends of the sections of contact of the crack faces [25, 34, 72, 73].

Contact of crack faces along the entire crack length: slipping with friction and sticking. First, we consider the case of full contact of the faces of edge crack (the crack faces are in contact along the entire length) with friction. In this case, the possibility of sticking is not neglected, i.e., condition (2.10) may be satisfied, which could be guaranteed, in particular, by high friction between the faces of shear crack (high values of f_c).

To get the complete picture of changes in the stress-strain state (stress intensity factors) at the crack tip for a single passage of a model contact load along the boundary of the half-plane (in a single contact cycle), we plot the dependence of the SIF K_{II} on the parameter $\lambda = x_0/a$ specifying the location of the contact load (Fig. 4.1) with regard for the values of typical operating parameters of rolling couples.

Note that the SIFs K_{I} and K_{II} are normalized (here and in what follows) by the formula $F = K/(p_0\sqrt{\pi a})$.

Among the parameters that give an adequate description of the operating characteristics of rolling couples in the two-dimensional formulation of the problem,

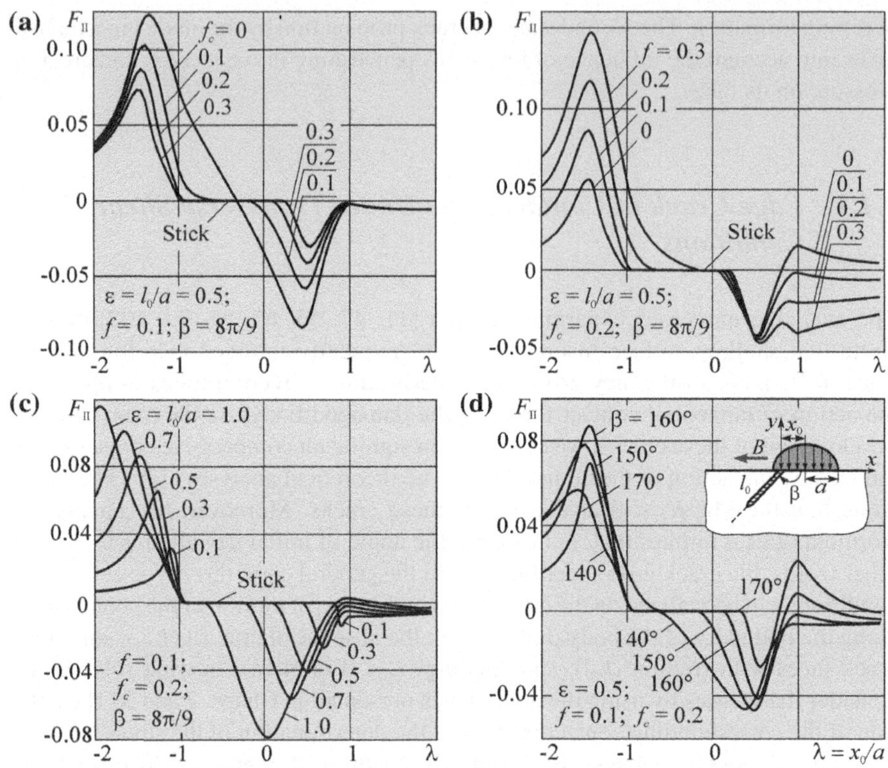

Fig. 4.2 Dependences of the normalized SIF F_{II} on the location of the counterbody for different values of the: **a** friction coefficient f_c between the crack faces; **b** friction coefficient f between the contacting bodies; **c** relative crack length $\varepsilon = l_0/a$; **d** angle β of crack orientation

we can choose the following characteristics: the maximum value of contact load p_0, the length of the contact section $2a$, the coefficient of sliding friction f between the rolling bodies; the friction coefficient f_c between the crack faces; the angle β of orientation of shallow shear edge macrocracks typical of the rolling bodies; the relative crack length $\varepsilon = l_0/a$, etc. In view of the possibility of investigation of rolling couples, such as wheel-rail couples and rolls of cold-rolling mills, the indicated parameters are chosen within the following ranges: $f = 0\ldots4$; $f_c = 0\ldots0.7$; $\beta = 140°\ldots170°$ ($\bar{\beta} = 180° - \beta = 10°\ldots40°$), and $\varepsilon = 0.1\ldots3.0$.

To study the influence of these parameters on the stress strain state (SSS) at the crack tip, we plot the dependences $F_{\text{II}}(\lambda)$ (Fig. 4.2) by adding the data presented in Table 4.1 with the maximum and minimum values of SIF F_{II}, the values of its range $\Delta F_{\text{II}} = F_{\text{IImax}} - F_{\text{IImin}}$, and the values of $\lambda_{\min}$ and $\lambda_{\max}$ corresponding to the locations of the counterbody for which the maximum and minimum values of SIF F_{II} are attained within a single rolling cycle (one passage of the contact load along the boundary of the half-plane near the crack mouth).

Table 4.1 Dependences of the normalized SIF range $\Delta F_{\rm II}$, its extreme values $F_{\rm IImax}$ and $F_{\rm IImin}$, and locations of the counterbody ($\lambda_{\max}$, $\lambda_{\min}$) for which they are realized, on the parameters f, f_c, β, and ε

	$\Delta F_{\rm II}$	$F_{\rm IImin}$	$F_{\rm IImax}$	$\lambda_{\min}$	$\lambda_{\max}$
β	$f_c = 0.2, f = 0.1$, $\varepsilon = 0.5$				
140°	0.102	−0.046	0.056	0.35	−1.60
150°	0.124	−0.046	0.078	0.45	−1.50
290°	0.128	−0.041	0.087	0.50	−1.45
170°	0.096	−0.026	0.070	0.55	−1.45
ε	$f_c = 0.2, f = 0.1$, $\beta = 160°$				
0.1	0.045	−0.013	0.032	0.90	−1.10
0.3	0.092	−0.026	0.066	0.70	−1.25
0.5	0.128	−0.041	0.087	0.50	−1.45
0.7	0.157	−0.056	0.101	0.35	−1.65
1.0	0.194	−0.079	0.115	0.05	−1.95
f_c	$\varepsilon = 0.5, f = 0.1$, $\beta = 160°$				
0.0	0.206	−0.083	0.123	0.45	−1.35
0.1	0.161	−0.058	0.103	0.50	−1.40
0.2	0.128	−0.041	0.087	0.50	−1.45
0.3	0.103	−0.030	0.073	0.55	−1.50
f	$f_c = 0.2$, $\varepsilon = 0.5$, $\beta = 160°$				
0.0	0.097	−0.040	0.057	0.50	−1.45
0.1	0.128	−0.041	0.087	0.50	−1.45
0.2	0.159	−0.042	0.117	0.55	−1.45
0.3	0.191	−0.044	0.147	0.55	−1.40

We now analyze the plots presented in Fig. 4.2 and the values of SIF $F_{\rm II}$ from Table 4.1 obtained by using the solutions of Eqs. (3.204) and (3.205) for the formulated problems. It is seen that if the friction coefficient f_c (Fig. 4.2a) increases, then the maximum absolute values of the SIF ($F_{\rm II}$) monotonically decrease. As a result, the SIF range ($\Delta F_{\rm II} = F_{\rm IImax} - F_{\rm IImin}$) becomes mush smaller. This, in turn, decreases the hazard of fracture (by the shear mechanism). Thus, the subsequent increase in f_c by 0.1 causes the a decrease in the relative values of $F_{\rm IImax}$ by about 16%, and $|F_{\rm IImin}|$ becomes smaller by more than 28% (Table 4.1, Fig 4.2a). A similar increase in the friction coefficient f causes an increase in $F_{\rm IImax}$ by 27%, on the average (see Table 4.1, Fig 4.2b) but almost does not affect the values of $F_{\rm IImin}$.

The increase in the relative crack length (Table 4.1, Fig 4.2c) results in a noticeable elevation of the maximum absolute value of the SIF $F_{\rm II}$ as a result of the left shift (in the direction of motion of the contact load) of the location of counterbody for which this extremum is realized. This fact implies that the hazardous locations of the contact load always move in the direction of crack growth. The variations of the angle of inclination of the shear crack (Table 4.1, Fig. 4.2d) cause a noticeable growth of the maximum absolute values of $F_{\rm II}$. However, instead of a monotonically increasing dependence observed in the previous case, we see a jump at $\beta = 160°$, which enables us to conclude that this angle plays a predominant role in the analyzed range of the angle β.

We also add that, in the case of contact loading moving from the right to the left, the range of locations of the counterbody for which the crack faces are completely stuck (this state of the crack appears if the SIF F_{II} remains equal to zero during the motion of the counterbody, i.e., when the crack faces are rigidly connected and prevent the formation of stresses at the crack tip, if we do not take into account the loading history) enlarges as the friction coefficient f between the rolling bodies becomes lower and diminishes as the friction coefficient f_c between the crack faces decreases.

Hence, for the analyzed range of the friction coefficients f and f_c, it is possible to assert that if both parameters simultaneously increase, then they mutually compensate their influence on the range of SIF F_{II} (see Table 4.1) and on the range of the locations of the counterbody for which the crack faces can be restrained (Fig. 4.2a, b). The range of locations of the counterbody for which the crack faces are completely restrained (stuck) becomes larger as the relative crack length ε decreases and the angle of crack orientation β increases.

Maps of contact of the crack faces: slipping with friction and sticking. We now perform a comprehensive analysis of the $F_{II}(\lambda)$ dependences presented in Fig. 4.2a for a more detailed investigation of the kinetics of contact of the faces of an edge shear crack starting from slipping with friction to sticking. For this purpose, the maps of sticking and slipping are plotted for different values of the friction coefficient f_c in Fig. 4.3. These maps are plotted on the coordinates λ–η, where parameter η is a dimensionless argument in the parametric equation $t = w(\eta) = l(1 - \eta)/2$; $|\eta| \leq 1$, which describes the crack contour.

As can be expected, the elevation of the friction coefficient f_c between the crack faces leads to a significant enlargement of the sections of sticking (Fig. 4.3). In all analyzed cases, if the counterbody moves from the right to the left, then the process of sticking of faces of the shear crack starts from its mouth, as soon as the counterbody begins to cover the mouth (this location corresponds to $\lambda = x_0/a = 1.0$). The effect of sticking begins to disappear also from the crack mouth. The lower the friction coefficient f_c, the higher the rate of this process. Note that the sticking of the faces is also preserved in the case where the counterbody has already passed the mouth and is located at a certain distance from the crack. Thus, for $f_c = 0.3$, the phenomenon of sticking completely disappears only for $\lambda = -1.4$.

The problem of contact of the faces of shear edge cracks under the action of a contact load moving on the body boundary was studied by numerous researchers (in particular, by Bower [10], Keer and Bryant [67]). Hence, we now perform the results of comparative analysis of our results with the results obtained in the cited works. In Fig. 4.4, we present the dependences of $F_{II}(\lambda)$ obtained for the case of contact sliding of the faces of edge crack with friction (over the entire crack length), including the presence of sticking presented, taken from [10, 67] (dashed lines) and obtained in the present work (solid lines).

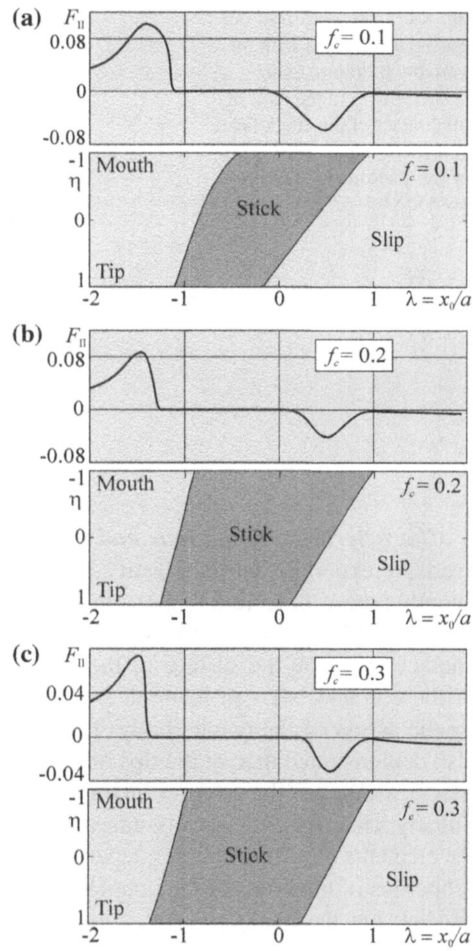

Fig. 4.3 Dependences of the SIF F_{II} and the sections of sticking on the location of the counterbody for different values of the friction coefficient f_c; $\varepsilon = l_0/a = 0.5$, $f = 0.1$, $\beta = 8\pi/9$

The comparison of our results with the data presented in Bower's work shows that they are in satisfactory agreement everywhere except the range $\lambda = -0.5\ldots1.0$. According to our calculations, the complete sticking of the crack faces is observed within the range $\lambda = 0.5\ldots1.0$. Therefore, the SIF F_{II} decreases to zero with gradual renewal for $\lambda = -0.5\ldots0.5$, where the section of sticking begins to decrease from the tip. On the contrary, Bower fixed the values of SIF F_{II} in the entire analyzed range according to its maximum value attained at the point $\lambda = -0.5$. This means that the applied mathematical model can only reveal the presence of a section with sticking but is unable to determine the SIF value in this zone. Unlike the previous case, Keer and Bryant, parallel with the determination of the section of sticking of the crack faces found the SIF F_{II} values in these sections, which almost completely coincide with our results for all locations of the contact load and all considered relative crack lengths ε.

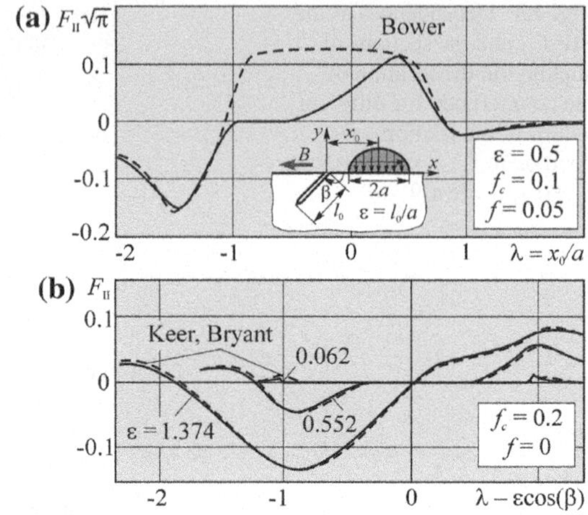

Fig. 4.4 Comparison of our results (solid lines) with the available literature data (dashed lines) in the case of full contact of the crack faces with friction and with regard for the sections of sticking; $\beta = 155°$

Characteristic orientations and critical lengths of the edge crack. Numerous available experimental data testify that shallow subsurface edge macrocracks frequently appear in rolling bodies (on the race tracks). These cracks, as a rule, propagate rectilinearly in the direction opposite to the direction of action of tangential contact forces on the surface of the damaged body. As the main cause of growth of these cracks, we can mention significant compressive contact stresses induced by the action of the counterbody. The theoretical analyses carried out in [13, 15] also demonstrated that, at the tips of these cracks in the compression zone, we have $K_{\mathrm{I}} < 0$. Moreover, the τ_θ-criterion implies that $\theta^* = 0$, i.e., the crack propagates rectilinearly. Thus, it is logical to assume that the angle of propagation of the shear crack in the elements of rolling couples depends on the operating parameters and mechanical properties of materials of the couples(friction in contact between the rolling bodies and between the crack faces; direction of rolling, presence of the medium; values and distribution of contact pressures, etc.). In turn, this angle strongly affects the contact durability, and the depth of pits (dimensions of pitting) on the rolling surface. The researchers mainly determine this angle on the basis of the available engineering or experimental data. We now try to determine this angle theoretically (numerically).

We now assume that the most favorable direction of growth of the shear crack is the direction (angle $\beta = \beta^*$) guaranteeing the maximum range of the normalized SIF F_{II} (i.e., ΔF_{II}) under the conditions of repeated unidirectional motion of the load over the surface of the body damaged by the crack. Thus, the angle β^* is given by the formula

$$\left.\frac{\partial(\Delta F_{\mathrm{II}}(\beta))}{\partial\beta}\right|_{\beta=\beta^*} = 0. \tag{4.2}$$

Hence, the problem of determination of the angle (β^*) most favorable (characteristic) for the growth of a shear edge crack is reduced to the determination of the extremum (maximum) of the function $\Delta F_{\mathrm{II}}(\beta)$.

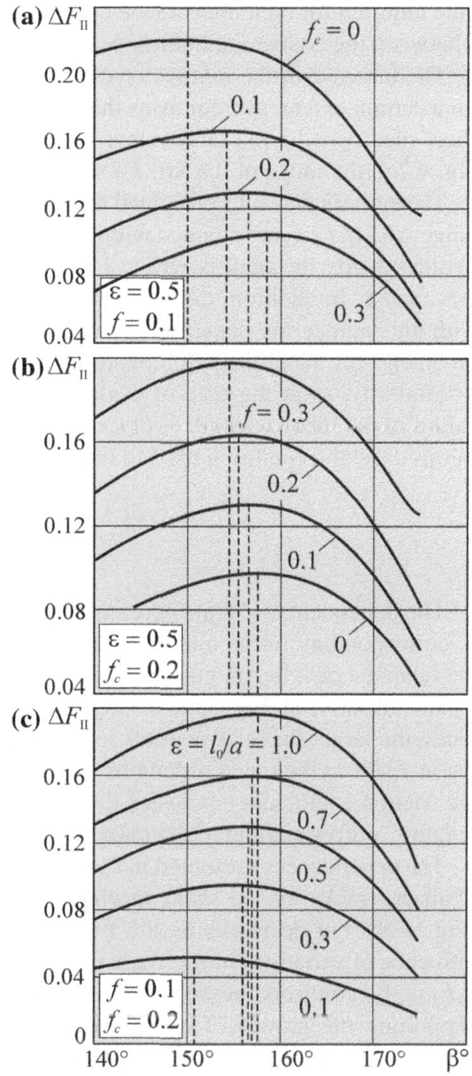

Fig. 4.5 SIF range ΔF_{II} established with regard for the sections of slipping with friction, and sticking of the crack faces depending on the angle β of crack orientation for different values of the: **a** friction coefficient f_c between the faces; **b** friction coefficient f between the rolling bodies; **c** relative crack length $\varepsilon = l_0/a$

Analyzing the dependences presented in Fig. 4.5, we conclude that the increase in the friction coefficient f_c between the crack faces promotes an increase in the angle β^* for which the maximum of ΔF_{II} is attained. Moreover, the increase of the friction coefficient f between the contacting bodies causes an insignificant decrease in the angle β for which the maximum of ΔF_{II} is attained. Further, by analyzing the specific features of the behavior of the SIF range F_{II} depending on the coefficients f and f_c (see Table 4.1), it is possible to conclude that the higher the values of the friction coefficient f_c, the smaller the sizes of the obtained defects. Moreover, the time required for their appearance becomes much longer. For the friction coefficient f between the contacting bodies, we observe the opposite situation.

The increase in the relative crack length ε has almost no effect on the angle β^* (to a certain extent, this confirms the assumption about the rectilinear growth of the shear macrocrack) and increases it only by several degrees. At the same time (see Fig. 4.2c), the range of the SIF F_{II} significantly increases.

The numerical results presented in Fig. 4.5 demonstrate that the maxima of the SIF range $\Delta F_{II}(f, f_c, \varepsilon)$ are attained within a small range of crack angles ($\beta = 150\ldots160°$). In other words, the angle $\beta = \beta^* \approx 155°$ is stable under the changes of the parameters f, f_c, and ε. In addition, the angle β^* established in this way is in good agreement with the engineering data. Hence, we assume that this angle is characteristic (most favorable) for the propagation of shear edge cracks in rolling bodies.

Similarly, as in the case of evaluation of the angle β^*, we can find the critical values of the relative length ε^* of a shear edge crack for which max ΔF_{II} is attained. In this case, the condition used to find ε^* takes the following form:

$$\left.\frac{\partial(\Delta F_{II}(\varepsilon))}{\partial\varepsilon}\right|_{\varepsilon=\varepsilon^*} = 0. \tag{4.3}$$

The dependences $\Delta F_{II}(\varepsilon)$ presented in Fig. 4.6 enable to determine the quantity ε^* corresponding to the maximum SIF range ΔF_{II}. Hence, the indicated values of the relative crack length guarantee the highest rate of its growth.

As follows from Fig. 4.6a, the higher the friction coefficient f_c between the crack faces, the larger the relative crack length ε^* guaranteeing the maximum of ΔF_{II} and the increase in the range of lengths (ε) for which ΔF_{II} monotonically increases. As the friction coefficient f between the contacting bodies increases, the relative crack length ε^* corresponding to the maximum of ΔF_{II} decreases.

The dependences presented in Fig. 4.6 demonstrate that the probability of growth of macrocracks by the shear mechanism increases with the friction coefficient f (Fig. 4.6b) but decreases as the friction coefficient f_c (Fig. 4.6a) increases. The influence of variations of the angle of inclination of a shear crack in the dependence $\Delta F_{II}(\varepsilon, \beta)$ confirms the advantage of angles within the range $\beta = 150\ldots160°$ in supporting the growth of shear cracks (this is also demonstrated in Fig. 4.5). In addition, for these angles, ΔF_{II} attains its maximum for the least possible relative length ε. This means that these values are most favorable in the shear stage of growth of the macrocrack. Hence, these angles are preferable for the numerical analyses. They are regarded as basic parameters in our investigations.

Summarizing the analysis of the curves presented in Fig. 4.6, we can make a conclusion that the maximum values of the SIF range ΔF_{II} that controls the process of growth of shear crack are attained within the range of relative crack lengths $\varepsilon^* = 2.1\ldots2.6$, i.e., in a quite narrow range in view of the considered ranges of the parameters f, f_c, and β.

General conditions of contact of the crack faces: normal opening, slipping with friction and sticking. We now describe the iterative procedure (algorithm) used in this case for the determination of the boundaries of the sections of contact of

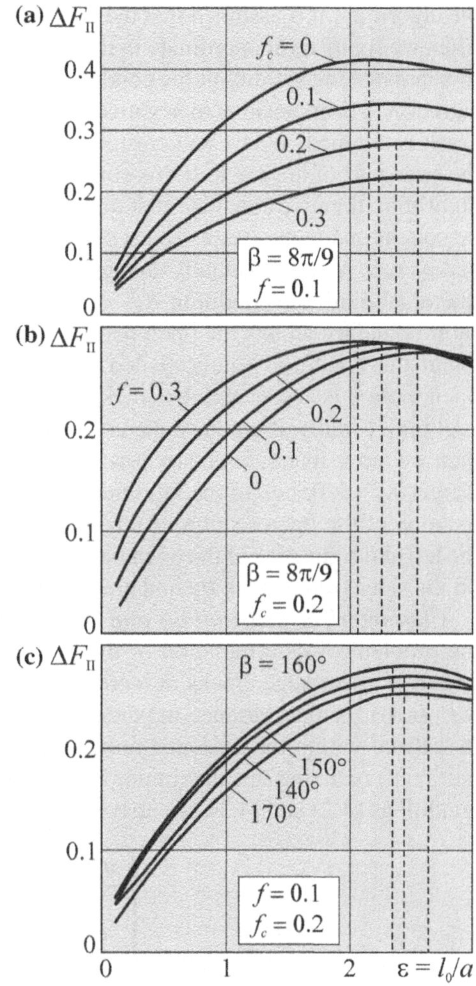

Fig. 4.6 Dependences of the SIF ΔF_{II} range on the relative crack length $\varepsilon = l_0/a$ for different values of the: **a** friction coefficient f_c between the crack faces; **b** friction coefficient f between the contacting bodies; **c** angle β of crack orientation

the crack faces. It is assumed [25, 72, 73] that, in the case of motion of the contact load along the boundary of the half-plane, the conditions of contact slipping with friction and sticking are realized on the crack faces, and only one open (inner or edge) section L_{op} arbitrarily located along the crack (the crack faces are not in contact) is formed. As a criterion for the determination of the location of this section, we use the condition according to which the SIF K_I should be positive on its ends ($K_I > 0$).

The algorithm used for the determination of the ends of an open segment of the edge crack has three parts: preliminary fixing of location of the open section of the crack; determination of its left end with regard for the sections of slipping with friction section in the case of the edge location of the open section, and the determination of its right (l^+) and left (l^-) ends in the case of internal location. In the first step of the algorithm, it is assumed that the open part is located at the right end of the crack. Then we fix its right coordinate in the mouth and change the left coordinate in order to guarantee the validity of the condition $K_{\mathrm{I}}^{-} > 0$, i.e., $K_{\mathrm{I}}(l^{-}) > 0$. The change in the sign of K_{I}^{-} from positive to negative indicates that left end of the open section of the crack is attained. We fix its coordinate. This completely determines the location of the open part of the crack. If this condition is not satisfied even in the first step of the algorithm, then we conclude that our assumption is wrong and the open part should be sought inside the crack. To do this, we fix the location of the left boundary (l^{-}) at the crack tip and gradually change the location of the right end (l^{+}) starting from the mouth until the condition $K_{\mathrm{I}}^{+} > 0$ ($K_{\mathrm{I}}(l^{+}) > 0$) is satisfied. This means that the right boundary attains the open part of the crack. We fix its coordinate. In the case where this condition is not satisfied along the entire crack length, we assume that the open section is absent and the crack is completely closed (slipping with friction and sticking). Finally, if we have the coordinate of the right boundary of the open part, then we move its left boundary toward the crack tip starting from the right end, until we get $K_{\mathrm{I}}^{-} > 0$, i.e., to the left end of the open part. In this way, we determine the location of the open section inside the crack. Note that, in the proposed algorithm, the left and right ends of the open part of the crack are understood as the ends located on the sides of the crack tip and crack mouth, respectively (see Fig. 4.1).

Characteristic orientations and critical lengths of edge crack. We now consider the problem of determination of the characteristic orientation and relative critical length of shear edge cracks in more detail. For this purpose, we study the function $\Delta F_{\mathrm{II}}(\varepsilon, \beta)$ for the presence maximum in the parameters ε and β. If this extremum is realized within the analyzed ranges of these parameters, then the corresponding pair (ε, β) realizing this maximum is just the required solution. Hence, generalizing conditions (4.2) and (4.3), we arrive at the following system of equations:

$$\begin{cases} \dfrac{\partial \Delta F_{\mathrm{II}}(\beta, \varepsilon)}{\partial \beta} = 0, \\ \dfrac{\partial \Delta F_{\mathrm{II}}(\beta, \varepsilon)}{\partial \varepsilon} = 0. \end{cases} \tag{4.4}$$

Therefore, the analyzed problem is reduced to the determination of the extremum (maximum) of a function of two variables. In Figs. 4.7 and 4.8, we show the surfaces graphically illustrating the results of numerical solution of this problem. We also note that, in order to find the values of function $F_{\mathrm{II}}(\lambda, \varepsilon, \beta, f, f_c)$, it is necessary to solve the problem posed above under the condition of contact with friction of the crack faces along the entire crack length. The absence of sticking is controlled by condition (3.197). However, in the analyzed case, it is neglected.

As follows from Fig. 4.7, all depicted surfaces have clearly pronounced extrema, which, in turn, ensures the uniqueness of a pair of parameters $(\varepsilon^{*}, \bar{\beta}^{*} = \pi - \beta^{*})$. Moreover, for relative crack length, this extremum is most often attained within the range $\varepsilon = 2.0\ldots2.5$. One more local maximum appears within the range $\varepsilon = 0.1\ldots1.0$ only for $f_c \geq 0.2$.

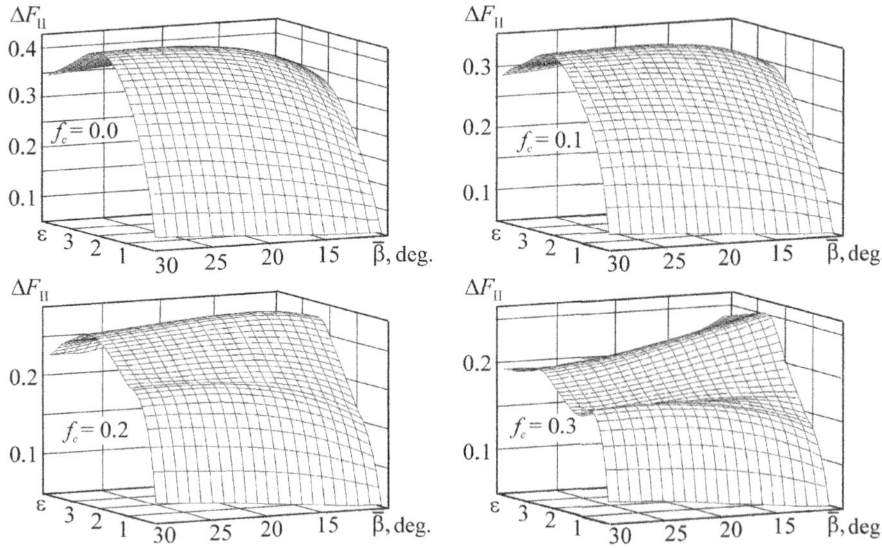

Fig. 4.7 Dependence of the SIF ΔF_{II} range on the relative crack length $\varepsilon = l_0/a$ and on the complementary angle $\bar{\beta} = \pi - \beta$ for various values of the friction coefficient $f_c; f = 0.1$

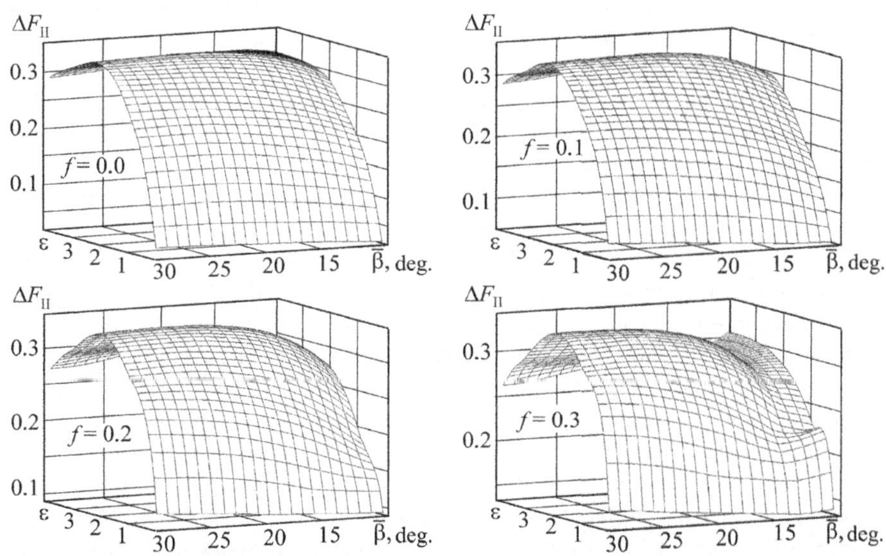

Fig. 4.8 Dependence of the SIF ΔF_{II} range on the relative crack length $\varepsilon = l_0/a$ and on the complementary angle $\bar{\beta} = \pi - \beta$ for different values of the friction coefficient $f; f_c = 0.1$

Table 4.2 The relative critical length ε^* and the complementary angle $\bar{\beta}^*$ guaranteeing the maximal value of ΔF_{II} depending on the friction coefficients f and f_c

a	$\varepsilon^* = l_0^*/a$	$\bar{\beta}^* = \pi - \beta^*$	max ΔF_{II}	**b**	$\varepsilon^* = l_0^*/a$	$\bar{\beta}^* = \pi - \beta^*$	max ΔF_{II}
f_c	$f = 0.1$			f	$f_c = 0.1$		
0.0	2.1	26°	0.4212	0.0	2.3	22°	0.3445
0.1	2.3	22°	0.3445	0.1	2.3	22°	0.3445
0.2	2.4	15°	0.2835	0.2	2.2	24°	0.3417
0.3	2.5	10°	0.2556	0.3	2.0	26°	0.3379

Based on the dependences $\Delta F_{II}(\varepsilon, \bar{\beta})$ depicted in Figs. 4.7 and 4.8, we constructed Tables 4.2a, b for a pair of values of the relative length and orientation of the shear crack (ε^*,$\bar{\beta}^*$) satisfying condition (4.4) and, hence, creating the required configuration of the parameters for the fastest (easiest) growth of shear cracks in the material. In order to study the influence of behavior of the analyzed parameters on the maximum range of normalized SIF F_{II}, the tables also contain the values of the quantity $\max\Delta F_{II} = \Delta F_{II}(\varepsilon^*, \bar{\beta}^*)$. Recalculated to $\max\Delta K_{II}$, these data determine the actual conditions of crack initiation and propagation according to the shear mechanism.

The results presented in Table 4.2 imply that, as the friction coefficient f_c between the crack faces increases, the relative crack length ε^* guaranteeing the maximum of ΔF_{II} also increases, whereas the complementary angle $\bar{\beta}^*$ decreases. As can be expected, the SIF range ΔF_{II} strongly decreases as f_c increases. Note that the angles of initial crack growth $\bar{\beta}^*$ given in the table agree with the data obtained by Way under the laboratory conditions [99].

The dependences presented in Fig. 4.8 reveal a less pronounced relationship between the maximum value of the SIF range ΔF_{II} and the coefficient of friction f between the rolling bodies than similar dependences for the coefficient f_c (Fig. 4.7). If we analyze the qualitative aspect of changes in the pair (ε^*,$\bar{\beta}^*$), then we reveal an absolutely opposite trend: as the friction coefficient f increases, the relative length ε^* decreases, while the complementary angle $\bar{\beta}^*$, on the contrary, increases and max ΔF_{II} remains almost constant.

Summarizing, we conclude that the coefficient of friction f_c between the crack faces strongly affects the maximum value of the SIF K_{II} in the case of motion of the contact load along the edge of the half-plane (over the surface of the rolling body): Its increase leads to a significant decrease in the SIF $|K_{II}|$ and in $\max\Delta K_{II}$. Moreover, f_c noticeably affects the direction (angle $\bar{\beta}^*$) of growth of the edge macrocrack. The coefficient of friction between the rolling bodies weakly affects $\max\Delta K_{II}$ and the initial direction of crack growth (see Table 4.2).

Maps of the crack faces contact: normal opening, slipping with friction and sticking. Based on the values of $\bar{\beta}^*$ presented in Table 4.2, we construct the maps of contact between the crack faces and a function $F_{II}(\lambda)$ that take into account the phenomenon of partial contact (partial opening displacement of the crack faces) for

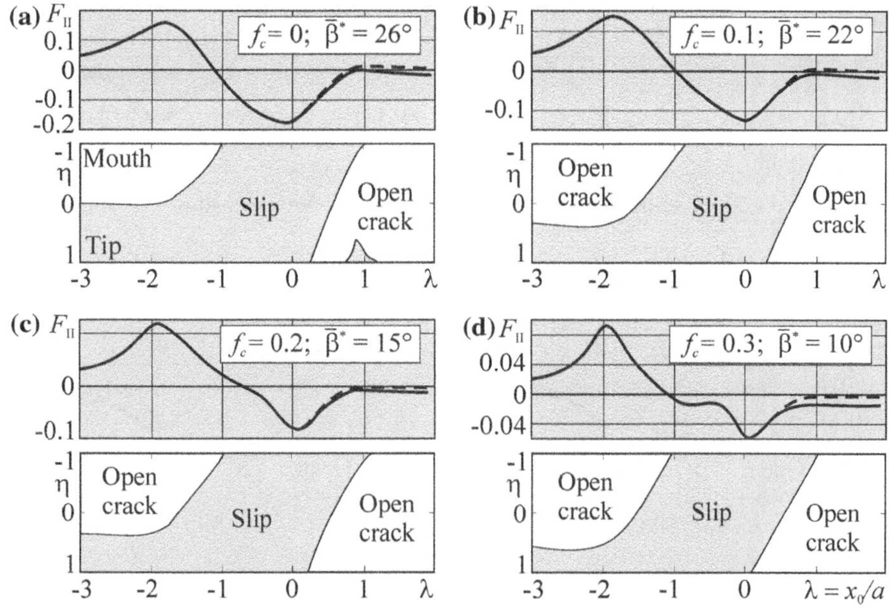

Fig. 4.9 Maps of the crack faces contact and the SIF F_{II} depending on the location of counterbody for various values of the friction coefficient f_c between the crack faces; $f = 0.1$; $\varepsilon = 1.0$; the values of the corresponding characteristic angles $\bar{\beta}^*$ are taken from Table 4.2a

different values of the friction coefficients f and f_c. Here, it is assumed that the parameter that describes the relative crack length is $\varepsilon = 1.0$.

We now analyze the specific features of the behavior of F_{II} caused by the presence open sections (partial contact) between the faces of the shear crack in the course of motion of the counterbody (the changes in λ). To this end, in Figs. 4.9 and 4.10, we present the plots of the function $F_{II}(\lambda)$ (dashed lines) for the case of full contact of the crack faces with friction (the section of sticking are neglected). The dependences depicted in Fig. 4.9 show that the increase in the friction coefficient f_c between the crack faces (in the case where the angle $\bar{\beta} = \bar{\beta}^*$ is determined from condition (4.4), and the counterbody moves from the right to the left) makes the contact section smaller for all locations of the contact load to the left of the crack mouth (the range of negative values of λ). However, in the case of changes in the coefficient of friction f between the contacting bodies (Fig. 4.10), we observe the opposite trend: As the indicated coefficient increases, we observe a noticeable enlargement of the contact section for the same locations of the counterbody; at the same time, this zone remains practically constant if the counterbody is located to the right of the crack mouth (interval of positive values of λ).

Comparing the results obtained for various types of contact of the crack faces, we conclude that the difference between the results obtained for the cases of partial and full contact of the faces (the case of full contact of the crack faces is illustrated by the

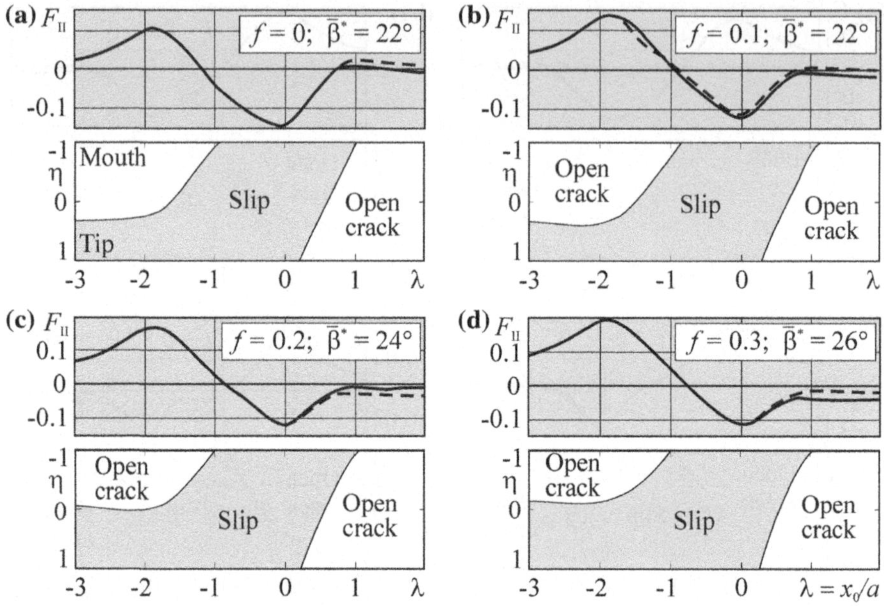

Fig. 4.10 Maps of the crack faces contact and the SIF F_{II} depending on the location of the counterbody for different values of the friction coefficient f between the contacting bodies; $f_c = 0.1$; $\varepsilon = 1.0$; the values of the corresponding characteristic angles $\bar{\beta}^*$ are taken from Table 4.2b

dashed lines in Figs. 4.9 and 4.10) is insignificant in all analyzed cases and can be observed mainly for large values of the friction coefficient f_c when the counterbody is located to the right of the crack mouth. In this case, the extreme values of the SIF ΔF_{II} are not affected.

It is worth noting that, in analyzing the kinetics of contact of the faces of shear edge macrocracks presented in Figs. 4.9 and 4.10, we can conclude that, for the indicated direction of motion of the counterbody, the crack always starts to close from the mouth (the only exception is the case of smooth contact ($f_c = 0$) in which the crack is simultaneously closes in the mouth and also, for a while, at the tip). Moreover, the higher the friction coefficient f_c between the crack faces, the faster the process of closing. However, the behavior of the friction coefficient f exerts almost no influence on the rate of the crack closure. After the full passage of the counterbody over the crack mouth (i.e., for $\lambda < -1$), the crack immediately begins to open from the mouth.

Finally, it is now reasonable to recall that the maps of contact of the crack faces in Figs. 4.9 and 4.10 are plotted not for the same angle β but for different angles $\beta = \bar{\beta}^*$ (Table 4.2). Hence, these maps illustrate the kinetics of opening displacements of the crack faces just in these cases depending on the friction coefficients f and f_c.

We now consider a more general case of the behavior of shear edge cracks where the following contact conditions on the crack faces are realized: normal opening

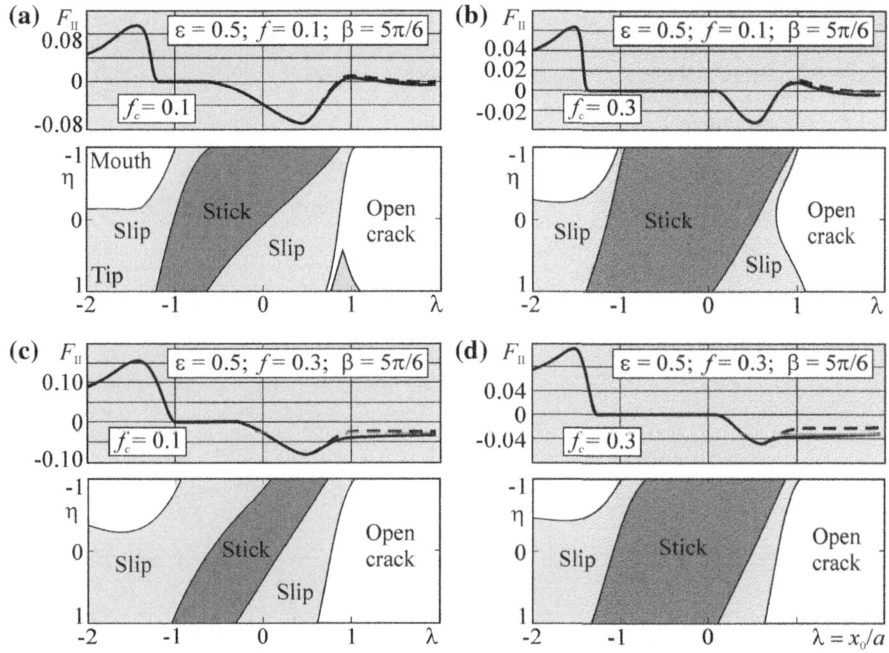

Fig. 4.11 Maps of the crack faces contact and the SIF $F_{\rm II}$ depending on the location of the counterbody

displacement of the faces, slipping with friction, and sticking. For this purpose, we plot the corresponding contact maps. The obtained dependences give more possibilities to analyze the behavior of shear edge cracks in a single cycle of contact of the bodies and enable us to estimate the relationship between $F_{\rm II}(\lambda)$ and the conditions of opening displacement, contact, and sticking of the faces.

The plots of the functions $F_{\rm II}(\lambda)$ for various values of the friction coefficients f and f_c and the corresponding contact maps for the inclination angle $\beta \approx \beta^* = 5\pi/6$ are presented in Fig. 4.11. As follows from the results presented in Fig. 4.11a, c that, for $f_c = 0.1$, the threefold increase in the friction coefficient f causes a significant reduction of the sections sticking in the crack mouth and has almost no influence on the same sections at the crack tip. For the friction coefficient $f_c = 0.3$, no phenomena of this kind are observed: a similar increase in the coefficient of friction between the contacting bodies almost does not affect the sections of sticking (Fig. 4.11b, d). It is also worth noting that, in the case of motion of the contact load from the right to the left, the contact and then the sticking of crack faces in the analyzed cases is always originated at the mouth for the friction coefficient $f = 0.3$. For $f = 0.1$, the crack closure starts from the tip and the first section of sticking appears in the mouth.

In conclusion, we add that the partial opening of the crack faces leads to an insignificant shift of the curve $F_{\rm II}(\lambda)$ to the side of lower values as compared with the case of full contact (the dashed lines in Fig. 4.11) but only in the case where

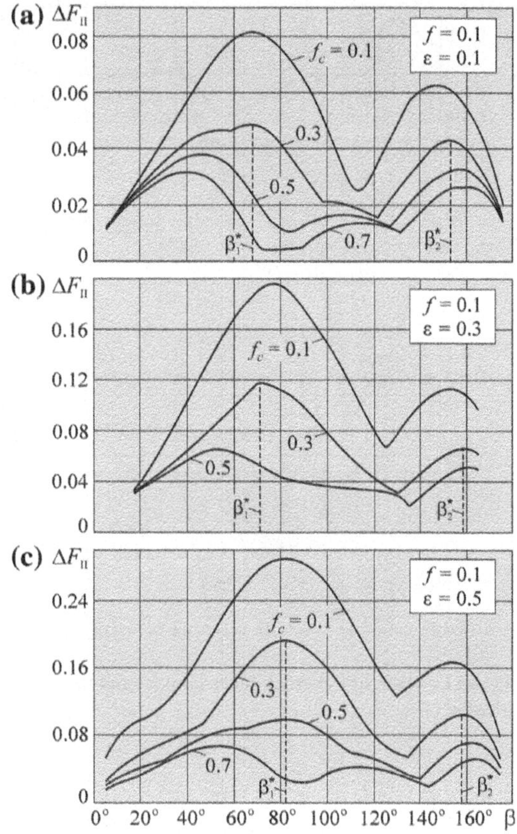

Fig. 4.12 Dependences of the normalized SIF range $\Delta F_{\mathrm{II}} = \Delta K_{\mathrm{II}}/(p_0\sqrt{\pi a})$ on the angle of crack orientation β for $f = 0.1$ and different relative crack lengths $\varepsilon = l_0/a$: $\varepsilon = 0.1$ (**a**); $\varepsilon = 0.3$ (**b**); $\varepsilon = 0.5$ (**c**)

the counterbody is located to the right of the crack mouth and does not reach the locations for which the minimum of the SIF F_{II} is attained. Thus, values of the SIF range ΔF_{II} determined as a result of the solution of a simpler problem for the case of full contact of the crack faces can be used for the investigation of the processes of contact fatigue fracture.

Special angle [33]. Up to now, a small range of the angles of crack orientation ($\beta = 140°\ldots170°$) was studied in order to establish the angles of orientation of the edge crack most favorable for its propagation in rolling bodies by the shear mechanism. In what follows, we present the dependences of the SIF range ΔK_{II} in a contact cycle (that controls the process of crack growth in rolling bodies by the mode II mechanism) on the angle β of orientation of the edge crack with the regard for the friction coefficient f between the wheel and the rail, the friction coefficient f_c between the crack faces, and the relative crack length $\varepsilon = l_0/a$ (see Figs. 4.12 and 4.13). The numerical analyses were carried out for the arbitrarily oriented edge cracks with relative lengths $\varepsilon = 0.1$, 0.3 and 0.5 and the orientation angles β varying from 5° to 175°. To determine (estimate) the nonnormalized values of the SIFs K_{I}, K_{II} and $K_{\mathrm{I}\theta}$, we used the following values of the operating parameters of the wheel-rail systems: the maximum Hertzian pressure $p_0 = 1100$ MPa [93], the length of the

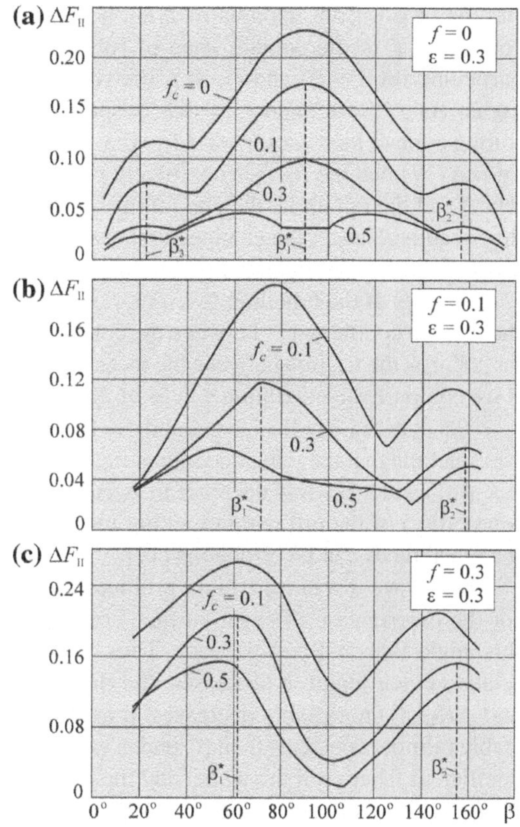

Fig. 4.13 Dependences of the normalized SIF range $\Delta F_{\mathrm{II}} = \Delta K_{\mathrm{II}}/(p_0\sqrt{\pi a})$ on the angle β of crack orientation for different friction coefficients f between the wheel and the rail: $f = 0$ (**a**); $f = 0.1$ (**b**); $f = 0.3$ (**c**); $\varepsilon = l_0/a = 0.3$

contact zone (section) $2a = 14$ mm, the coefficients of slipping with friction in the rolling contact $f = 0, 0.1, 0.3$ and 0.5, and the friction coefficients between the crack faces in the rail $f_c = 0, 0.1, 0.3, 0.5$ and 0.7.

The investigation of the SIF range ΔK_{II} was carried out for the fixed friction coefficient $f = 0.1$ between the wheel and the rail, for different relative lengths of the crack $\varepsilon = l_0/a$ (Fig. 4.12), for a fixed relative crack length $\varepsilon = 0.3$, and for different friction coefficients f between the wheel and the rail (Fig. 4.13).

The common feature of the curves presented in Figs. 4.12 and 4.13 is the presence of two ΔK_{II} maxima at the angles $\beta = \beta_1^*$ and $\beta = \beta_2^*$ corresponding to the most favorable orientations for the propagation of shear cracks inclined either in the direction of the wheel motion (if the angle $\beta = \beta_2^* > \pi/2$ is shallow) or in the direction close to the perpendicular (the angle $\beta = \beta_1^* \leq \pi/2$). As follows from Fig. 4.12, the values of both maxima of ΔK_{II} increase with the relative crack length ε ($0 < \varepsilon \leq 0.5$) increase. It is clear that, if the ε value increases further, then, starting from a certain time, both $|K_{\mathrm{II}}|$ and ΔK_{II} maxima begin to decrease as the crack tip moves away from the boundary of the half-plane subjected to the action of the

Hertzian load. Both maxima of ΔK_{II} become much higher increase as the friction coefficient f in the contact zone of rolling bodies increases (Fig. 4.13). Thus, in particular, for $\beta = \beta_2^*$ and $\varepsilon = 0.3$, the value of $\Delta K_{\mathrm{II}}(\beta_2^*)$ becomes 2.5 and 2.6 times higher for $f_c = 0.3$ and $f_c = 0.5$, respectively, as the friction coefficient f in the rolling contact increases from 0.1 to 0.3. This fact is of high importance because it is known [17] that the friction coefficient $f = 0.1$ corresponds the friction between the wheel and the rail under the conditions of wet weather, while $f = 0.3$ corresponds to dry weather. Hence, the changes in the weather conditions noticeably affect the SIF range ΔK_{II} for the surface cracks in the rails.

The plots of the functions $\Delta K_{\mathrm{II}}\,(\beta, f, f_{c,}\ \varepsilon)$ in Figs. 4.12 and 4.13 demonstrate that the friction coefficient f_c between the contacting crack faces exerts a strong influence on ΔK_{II}: as the friction between the faces increases, ΔK_{II} noticeably decreases. Thus, if we compare the maximum values of ΔK_{II} for the friction coefficients $f_c = 0.1$ and $f_c = 0.5$ (see Fig. 4.13b; $f = 0.1$ and $\varepsilon = 0.3$) for both angles $\beta = \beta_1^*$ and $\beta = \beta_2^*$, then we conclude that the indicated intensification of friction causes a 2.7-fold decrease in $\Delta K_{\mathrm{II}}(\beta_1^*)$ and a 2.2-fold decrease in $\Delta K_{\mathrm{II}}(\beta_2^*)$. Thus, the prediction of the service life (durability) of the rail without taking into account the presence of friction between the crack faces can be erroneous [17].

Finally, we discuss the main problem of evaluation of the angle most favorable for the propagation of shear surface cracks in the railhead. It is not clear whether this angle is $\beta = \beta_1^*$ or $\beta = \beta_2^*$. Thus, as friction coefficient f_s increases and the relative crack length ε decreases, the difference between the maxima of $\Delta K_{\mathrm{II}}(\beta_1^*)$ and $\Delta K_{\mathrm{II}}(\beta_2^*)$ noticeably decreases. It is also worth noting that the angle β_2^* is more stable (almost constant) than β_1^* under variations of the crack length and the friction coefficient f between the wheel and the rail.

It follows from Figs. 4.12 and 4.13 that the angle $\beta = \beta_1^*$ is more favorable for the propagation of shear cracks. However, the results of additional investigations of the quantity max $K_{\mathrm{I}\theta}$ (β, ε, $p_{0.}$ f, f_c), which controls the process of crack growth by the normal opening (mode I) mechanism, revealed an important specific feature (see Table 4.3, $\varepsilon = 0.5, f = 0.3$, and $f = 0.5$). In this case, the values of max ΔK_{II} corresponding to the directions most favorable for crack propagation according to the mechanism of shear (angles β_1^* and β_2^*) are compared with the threshold values of the characteristics cyclic crack growth resistance in transverse shear $\Delta K_{\mathrm{II},th} =$ 13.01 MPa·(m)$^{1/2}$ [9], while the values of max $K_{\mathrm{I}\theta}$ are compared with the tensile threshold value $K_{\mathrm{I},th} = 4.3$ MPa·(m)$^{1/2}$ [94] for rail steels. For the angle $\beta = \beta_1^*$, the values of max ΔK_{II} and max $K_{\mathrm{I}\theta}$ exceed the corresponding threshold values for $\Delta K_{\mathrm{II},th}$ and $K_{\mathrm{I},th}$. For the angle $\beta = \beta_2^*$, this is observed only for max ΔK_{II}. This means that the cracks close to vertical (oriented at the angle $\beta = \beta_1^*$) may grow not only by the mechanism of shear but also by the opening (mode I) mechanism, unlike the shallow cracks oriented at the angle $\beta = \beta_2^*$ for given values of the parameters ε, f_c, f, and p_0. Hence, the cracks oriented at the angle $\beta = \beta_1^*$ may either branch or deviate from the initial direction $\beta = \beta_1^*$. Thus, the angle $\beta = \beta_2^* = 150°\ldots160°$ is more favorable for the growth of shear cracks than the angle $\beta = \beta_1^*$.

Table 4.3 Comparison of two mechanisms of fracture for the critical angles β_k^* ($k = 1, 2$) of orientation of the edge crack; $\varepsilon = 0.5$

f_c	k	β_k^*	max ΔK_{II}	max $K_{\mathrm{I}\theta}$	β_k^*	max ΔK_{II}	max $K_{\mathrm{I}\theta}$
		$f = 0.3$			$f = 0.5$		
0.1	1	68°	55.55	9.22	55°	69.92	24.16
	2	152°	39.35	8.91	149°	52.52	19.86
0.3	1	66°	41.87	9.27	52°	59.61	24.35
	2	160°	27.05	8.33	158°	42.00	16.99
0.5	1	64°	28.62	9.33	50°	49.98	24.49
	2	162°	21.98	8.03	160°	34.76	16.18

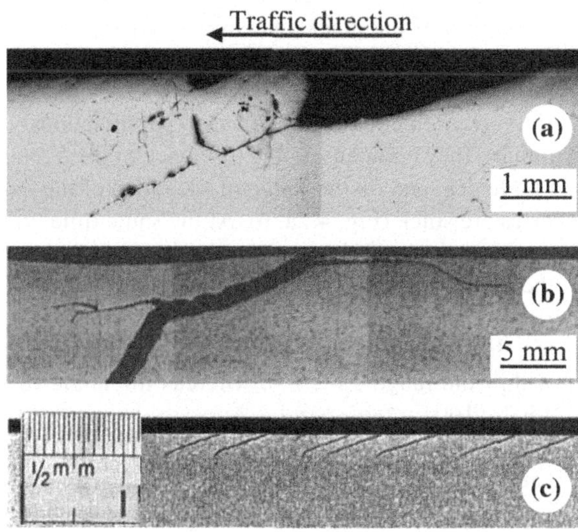

Fig. 4.14 Typical contact fatigue surface defects in the railhead: pitting [79] (**a**), squat ("dark spot") [79] (**b**), checks [76] (**c**)

Hence, in view of the accumulated numerical data, the angle $\beta = \beta_2^*$ can be regarded as the characteristic angle, which can be found from the following conditions:

$$\frac{\partial}{\partial\beta}\left[\Delta K_{\mathrm{II}}(\beta, \varepsilon, p_0, f, f_c)\right]_{\beta=\beta_k^*} = 0, \quad k = 1, 2; \tag{4.5}$$

$$\begin{gathered} \Delta K_{\mathrm{II}}\left(\beta_2^*, \varepsilon, p_0, f, f_c\right) > \Delta K_{\mathrm{II,th}}, \\ \max K_{\mathrm{I}\theta}\left(\beta_2^*, \varepsilon, p_0, f, f_c\right) < \max K_{\mathrm{I}\theta}\left(\beta_1^*, \varepsilon, p_0, f, f_c\right), \\ \max K_{\mathrm{I}\theta}\left(\beta_1^*, \varepsilon, p_0, f, f_c\right) > \Delta K_{\mathrm{I,th}}, \end{gathered} \tag{4.6}$$

where ε, p_0, f, and f_c are fixed parameters. The analysis of these contact fatigue defects, such as pitting (Fig. 4.14a), the left branch of a squat ("dark spot") (Fig. 4.14b), and checks (Fig. 4.14c) demonstrates that these defects originate, in fact, from the edge crack inclined at the characteristic angle.

Note that, in a special case where the friction coefficient in the contact between the wheel and the rail $f = 0$, the curve ΔK_{II} (β) has an additional maximum for

$\beta = \beta_3^* = 15°...20°$ (Fig. 4.13a). In this case, it is possible to assume that, for the conditions required for the formation of the right branch of a squat-type ("dark spot") damage are created for $f = 0$ (e.g., if the car wheels move by inertia), (Fig. 4.14b).

Consequences

1. The increase in the friction coefficient f_c between the crack faces causes both a significant decrease in maximum values of the SIFs $|F_{II}|$ and ΔF_{II} in a contact cycle and the enlargement of the zones of sticking of the crack faces zones (Fig. 4.2).
2. The range of the locations of the counterbody characterized by the sticking of crack faces (in the analyzed cases) enlarges in the case where the horizontal projection of the crack increases due to the growth of crack length and as the angle of crack orientation increases (i.e., the complementary angle ($\bar{\beta} = \pi - \beta$) decreases).
3. If the friction coefficients f and f_c simultaneously increase within the analyzed ranges, they compensate the influence of each other both on the SIF range ΔK_{II} (Table 4.1) and on the range of locations of the counterbody in which the crack faces are stuck (Fig. 4.2a, b). At the same time, the simultaneous increase in the coefficient f_c and decrease in the coefficient f cause an significant decrease in the SIF range ΔK_{II}.
4. For the analyzed ranges of operating parameters, the relative length $\varepsilon = \varepsilon^*$ of the shear crack corresponding to the maximum SIF range ΔF_{II}, as a rule, lies within the range 2.0...2.6, whereas the angle of shear crack orientation $\beta^* = 150°...160°$.
5. The increase in the coefficient of sliding friction f between the rolling bodies causes a significant increase in the SIF $|K_{II}|$ and a decrease in the zone of sticking between the crack faces, especially if the counterbody is located over the crack mouth ($\beta = 160°$) or has already passed through the mouth (Fig. 4.2b).
6. The contact and subsequent sticking of the crack faces in the analyzed cases (for small $\bar{\beta} = \pi - \beta$ and the counterbody moving from the right to the left) always originates from the crack mouth for the friction coefficient $f = 0.3$. At the same time, for $f = 0.1$, the crack starts to close from the tip, while sticking is first observed in the mouth.
7. The existence of the characteristic angle $\beta = \beta^*$ of propagation of edge (surface) cracks in rail heads is theoretically established and the criteria for its determination are proposed (Eqs. (4.5) and (4.6)). The values of the angle β^* obtained by using the characteristics of cyclic crack resistance of rail steels under the conditions of transverse shear are in good agreement with numerous available engineering data. It can be assumed that this angle is basic for the formation of typical surface contact fatigue defects, such as pits, checks, squat ("dark spots)," etc. in rolling bodies.
8. A hypothesis about the mechanism of formation of widespread contact fatigue defects called squats ("dark spots") in the rails is proposed.

4.1.2 Edge Crack Under Normal Opening Conditions. Pitting Formation

We consider the case of cyclic contact of two bodies (cyclic rolling) realized under the conditions of boundary lubrication (see the beginning of Chap. 4). Under these conditions, certain amounts of incompressible liquids (water or lubricants) penetrate into the crack mouth in the process of rolling and create a pressure between the crack faces distributed in a certain way. We model the action of this pressure (Fig. 4.15) by a uniformly distributed normal pressure given by the formula

$$p_1 = r \cdot p_0 \cdot \sqrt{1-\lambda^2}; \qquad \lambda = x_0/a, \tag{4.7}$$

or by a pressure varying along the crack faces according to a linear law

$$p_1(t) = r \cdot p_0 \cdot (1 - t/l) \cdot \sqrt{1-\lambda^2}, \qquad 0 \le t \le l. \tag{4.8}$$

In these formulas, l is the running crack length and the parameter r varies from 0 up to 1 and determines the level of pressure on the crack faces relative to the pressure over its mouth. We also assume that the pressure $p_1(\mathrm{t})$ acts only in the case where the contact load covers the crack mouth. In the other cases, we have $p_1 = 0$ and $p_1(\mathrm{t}) = 0$.

Influence of pressure acting along the crack faces on the stress intensity factors. We consider loading schemes depicted in Fig. 4.15a, b. First, we determine the stress intensity factors at the tip of an initially rectilinear inclined edge crack in the elastic half-plane. The model contact load (4.1) moves along its boundary. In the presence of lubricants (lubricating liquids) between the bodies in cyclic contact, the crack faces suffer the action of a uniformly distributed normal pressure with intensity p_1 (4.7) or a linearly distributed pressure with intensity $p_1(t)$ (4.8) in the case where the load covers the crack mouth. It is assumed that the crack faces are not in contact due to the presence of lubricants. Thus, the problem of determination of the SIF is reduced to the solution of the corresponding Eq. (3.155) (with regard for condition (3.160)) of the first basic problem of the elasticity theory for a half-plane weakened by an edge crack. The corresponding loading schemes are shown in Fig. 4.15. The

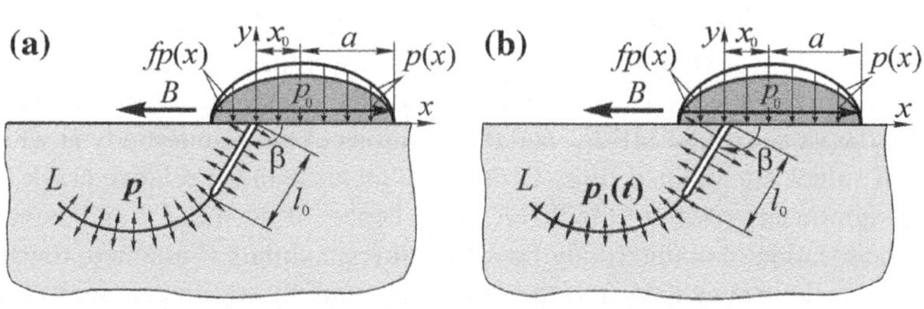

Fig. 4.15 Computational schemes in the following cases: **a** uniform pressure on the crack faces; **b** linearly distributed pressure; **B** is the direction of movement of the counterbody

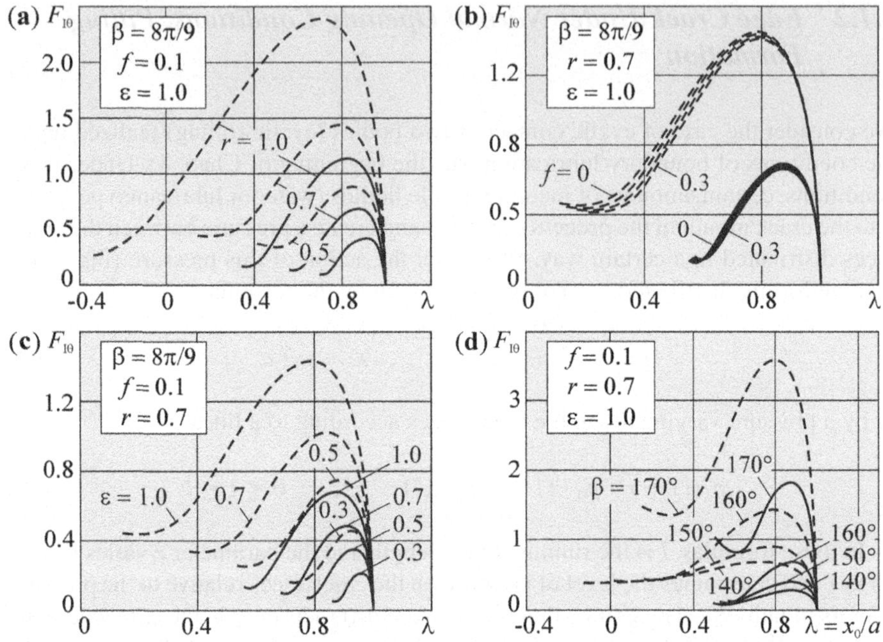

Fig. 4.16 Dependences of the normalized SIF $F_{I\theta}$ on the location of a contact load over the mouth of the edge crack in the cases of pressure uniformly distributed on the crack faces (dashed lines) and linearly distributed pressure (solid lines)

right-hand side $P(\eta)$ of this equation is determined by relation (3.174). The numerical solution of Eq. (3.155) is found by the method of mechanical quadratures.

We now estimate the influence of the basic operating parameters of rolling couples (f, β, r and ε) and the type of loading on the crack faces, on the values of the normalized mixed-mode SIF $F_{I\theta} = K_{I\theta}/\left(p_0\sqrt{\pi a}\right)$ in a contact cycle. These dependences are presented in Fig. 4.16.

In analyzing the results presented in Fig. 4.16, we can make the following conclusions: the values of the SIF $F_{I\theta}$ noticeably increase with the intensity of pressure acting upon the crack faces. Assume that the counterbody moves from the right to the left. In this case, the hazardous locations of the counterbody for which $\max F_{I\theta}(\lambda)$ is attained are realized as soon as this body begins to cover the crack mouth (Fig. 4.16a). The increase in the friction coefficient f between the contacting bodies almost does not affect the values of the SIF $F_{I\theta}$ and the locations of the counterbody at which its maximum values are attained (Fig. 4.16b). The increase in the relative crack length causes a significant growth of the SIF $F_{I\theta}$ and, hence, of the $\max F_{I\theta}$ and, hence, the shift of the location of counterbody for which this maximum is attained, toward the side of the crack tip (Fig. 4.16c). If the slope of an initially rectilinear crack changes, then we observe a clear trend of decrease in the values of the SIF $F_{I\theta}$ as the angle β decreases. The increase in $\max F_{I\theta}$ with the angle β can be explained by the fact that

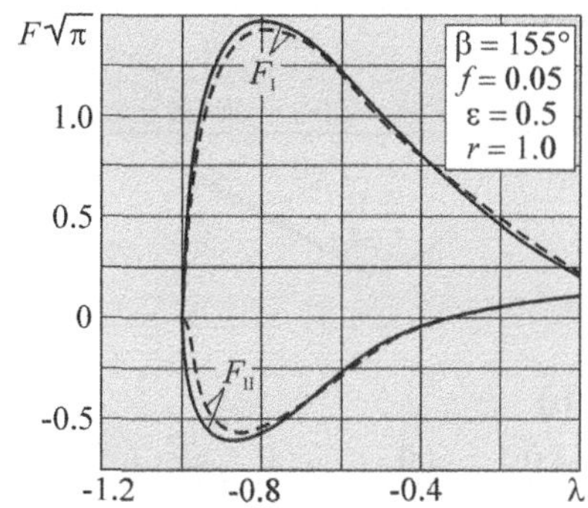

Fig. 4.17 Comparison of the obtained dependences $F(\lambda)$ (solid line) with A. F. Bower's results [10] (dashed line)

the crack approaches the boundary of the body (Fig. 4.16d). The transition from the uniform pressure acting on the crack faces to a pressure distributed according to a linear law causes a significant decrease in the SIF $F_{I\theta}$ in all analyzed cases. However, all described above are preserved for the linear distribution of loading over the crack faces.

To substantiate the constructed dependences, we perform the comparative analysis of our results with the results obtained by Bower [10] for the case of pressure uniformly distributed over the crack faces (Fig. 4.17). From this figure, we see that the curves $F_I(\lambda)$ and $F_{II}(\lambda)$ obtained under the simultaneous action of the contact load and wedging pressure on the crack faces are in good agreement with the data published by Bower. We observe an insignificant difference within the range $\lambda = -0.9\ldots-0.7$ for F_I and $\lambda = -1.0\ldots-0.8$ for F_{II}, i.e., for the locations of the counterbody corresponding to the maxima of the corresponding SIFs.

Paths of propagation of the edge crack under the action of moving load on the boundary of the body and the pressure of lubricant on the crack faces. Assume that the presence of liquid (water or lubricant) in the contact zone (in the process of rolling) leads to the wedging of the crack faces. In turn, this implies that crack propagates according to the mode I mechanism. Hence, the amplitude of mixed-type SIF $\Delta K_{I\theta} = \max K_{I\theta}$ is a parameter controlling fracture. In order to construct the trajectories (paths) of development of an edge crack (Fig. 4.15), we use a step-by-step algorithm similar to the algorithm described in Chap. 2 (see Sect. 2.7). Here, in addition to the increment of path h, we introduce a step $\Delta\lambda$ related to the motion of the counterbody in a contact cycle [14, 88]. In each stage of construction of the path, the increment of crack path h is measured from the crack tip in the direction specified by the angle $\theta = \theta^{**}$ (Fig. 2.21). The auxiliary step $\Delta\lambda$ is used to seek the extrema of amplitude of the parameter $\Delta K_{I\theta}$ ($\max K_{I\theta}$) in a contact cycle . Suppose that, in each cycle, the crack grows only for the locations of the counterbody ($\lambda = \lambda^* = x_0^*/a$) guaranteeing the maximum of SIF $K_{I\theta}$, i.e., that $\max K_{I\theta}(\lambda) = K_{I\theta}(\lambda^*)$. In each step of construction of the trajectory (for the corresponding number of contact cycles), the quantities λ^*, θ^{**}, and ΔK are regarded as constant. The SIFs K_I, K_{II} and $K_{I\theta}$ are determined in each step of construction from the solution of SIE (3.155) of the problem of elasticity theory for the half-plane weakened by an edge curvilinear crack

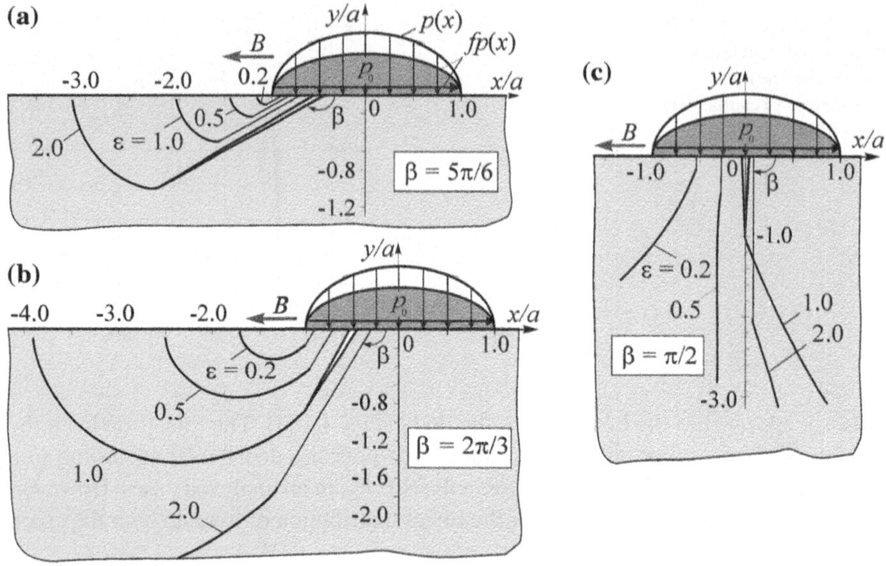

Fig. 4.18 Paths of a crack propagation depending on its initial length $\varepsilon = l_0/a$ for different inclination angles β to the boundary ($f = 0.1$; $r = 1.0$): **a** $\beta = 5\pi/6$; **b** $\beta = 2\pi/3$; **c** $\beta = \pi/2$

of the corresponding configuration. Every increment of the trajectory is approximated by a polynomial of the third power (see Sect. 2.7).

We now construct the paths of propagation of an edge initially rectilinear crack inclined to the boundary of the rolling body under the conditions of boundary lubrication by using the model schemes depicted in Fig. 4.15. In this case, the lubricating liquid exerts uniform pressure upon the crack faces (relation (4.7)) or creates a linearly distributed pressure vanishing at the crack tip (relation (4.8)). In Figs. 4.18, 4.19 and 4.20, we present the trajectories obtained in the case of uniform pressure acting upon the crack faces, while the case of linear distributed pressure is illustrated in Fig. 4.21. The paths were constructed as functions of the angle β of inclination of the crack to the edge of the half-plane, relative initial crack length $\varepsilon = l_0/a$, the coefficient of sliding friction f between the rolling bodies, and the parameter r characterizing the intensity of pressure exerted by liquid upon the crack faces. In our calculations, we use the following ranges of the parameters: $\varepsilon = 0.2\ldots2.0$. According to the conditions of boundary lubrication, $f = 0.05\ldots0.15$ and the relative pressure of a lubricant $r = 0.1\ldots1.0$. It is known from the experimental data that, in the case of unidirectional rolling, the initial macrocracks in a driven body are most often initiated at the angles $\bar{\beta} = 180°–\beta = 15°\ldots45°$ to its boundary (the direction of motion of the counterbody). Hence, in our calculations, we prefer the angles $\beta = 8\pi/9$ and $\beta = 5\pi/6$. However, some other angles were also considered: $\beta = \pi/6$; $\pi/2$; $2\pi/3$; and $3\pi/4$.

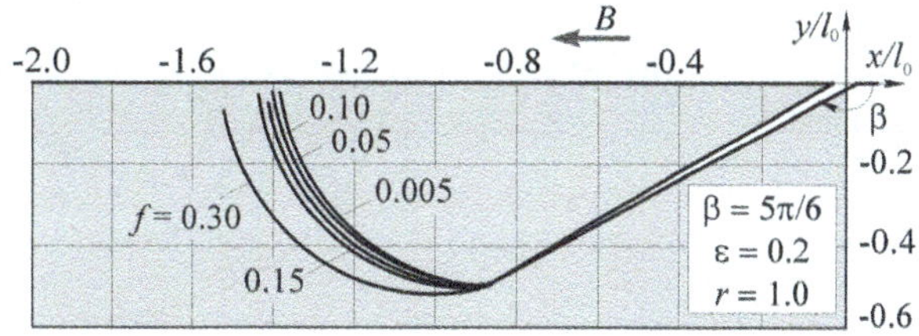

Fig. 4.19 Paths of crack propagation as functions of the friction coefficient f between the rolling bodies

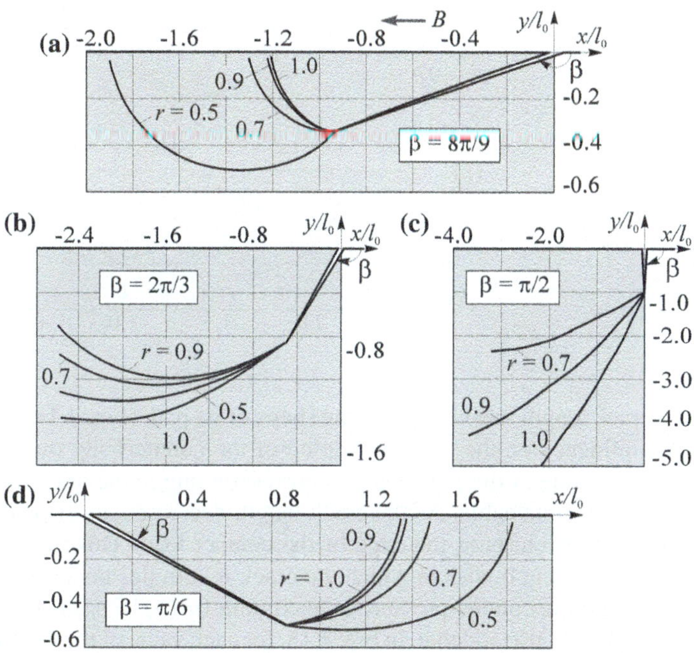

Fig. 4.20 Paths of crack propagation as functions of the intensity r of pressure of the lubricant on its faces for different initial slopes of the crack to the boundary of the body; $f = 0.1$; $\varepsilon = l_0/a = 0.2$

In Fig. 4.18, we show the paths of an edge crack depending on its initial length and orientation for a fixed value of the friction coefficient in contact ($f = 0.1$) and the same intensity of pressure of the lubricant upon the crack faces ($r = 1.0$).

The analysis of the paths and numerical data obtained in the course of their construction shows that the slope β of the initial crack to the boundary of the body strongly affects the shape of the crack path (Fig. 4.18a–c). The more acute the initial orientation angle (β > π/2), the higher the rate of propagation of the crack to the boundary (Fig. 4.18a, b). The stress intensity factors along the paths increase, especially as the crack tip approaches the boundary. The cracks initially oriented at the angles β ≈ π/2 penetrate deep into the bulk of the material (Fig. 4.18c).

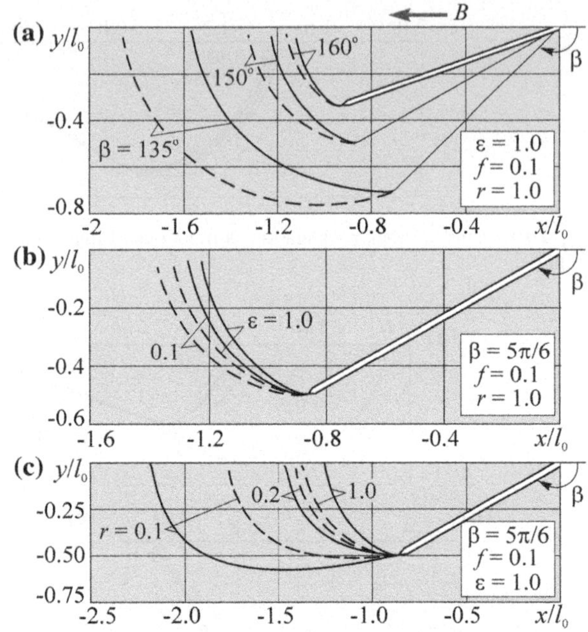

Fig. 4.21 Paths of the edge macrocrack propagation by the mode I mechanism under the action of a moving contact load on the boundary of the half-plane and the pressure of lubricant on the crack faces (the dashed curves—the uniform distribution of pressure and the solid curves—the linear distribution)

The initial crack length weakly affects the shape of its path (Fig. 4.18a, b). At the same time, its influence on the critical location of the counterbody over the crack mouth, i.e., on the value of the parameter λ^* corresponding to the maximum of $K_{I\theta}$ in a contact cycle, is noticeable. As the crack length decreases, its critical location under the counterbody shifts to the edge of the contact load. This means that the mutual location of rolling bodies in which the crack mouth lies under the center of the contact load and the contact pressure is maximum ($p_1(\lambda) = p_0$) is not the most hazardous location. We also add that, in Fig. 4.18, for each value of the relative length ε, the location of the initial crack is specified according to its value of λ^*. It is worth noting that the values of λ^* remain almost constant for the entire period for crack growth in all analyzed cases.

The paths of an edge crack depicted in Fig. 4.19 show that the decrease in the contact friction between the rolling bodies somewhat accelerates the process of crack propagation toward the boundary of the body (the path becomes steeper) but exerts no essential influence on the shape of the path.

As can be expected, a significant influence on the growth paths of edge cracks is exerted by the pressure of lubricants in the crack (Fig. 4.20). We now consider the cracks initially inclined to the edge of the half-plane at acute angles ($\beta = 8\pi/9$, Fig. 4.20a, and $\beta = \pi/6$, Fig. 4.20d). The increase in the pressure of liquid in the crack (in the parameter r) strongly promotes a more rapid motion of the crack to the edge of the half-plane.[1] For the values of the parameter r lower than the indicated values, the crack does not propagate ($K_I < 0$). This result supports the hypothesis about the key influence of lubricants on the development of the pitting. However, for the cracks initially inclined to the edge at large angles ($\beta \approx \pi/2$), the increase in the level of pressure in the crack leads to their deepening into the material (Fig. 4.20d). The transition from one trend to another one can be observed for the paths corresponding to the angle $\beta = 2\pi/3$, as the parameter r changes from 0.9 to 1.0 (Fig. 4.20b).

The results of the numerical analyses of the crack propagation paths in the case of linear dist-ributions of the liquid (lubricant) pressure over the crack faces in the rolling process under the boun-dary lubrication conditions are shown in Fig. 4.21.

The changes in the distribution of pressure of the lubricant on the crack faces have ambiguous consequences for the development of the crack (Fig. 4.21c). Thus, for the maximum value of the parameter r ($r = 1.0$), the transition from the uniform distribution to the distribution according to a linear law causes an insignificant decrease in the path length. For low values of r ($r = 0.1$), the same transition leads to its significant elongation. This fact can be explained by a general trend to increase in the length of the crack growth path observed as the pressure on its faces decreases, which happens if the uniform distribution is replaced by the linear distribution. In general, the constructed paths correspond to the stable motion of the crack toward the edge of the body for both analyzed types of loading [19, 40, 41]. This confirms the well-known hypothesis by Way [99] about the decisive influence of lubricants on the formation of pits.

Note that the tenfold increase in the relative length of the initial crack ($\varepsilon = 0.1 \ldots 1.0$) weakly affects the shape of the path (Fig. 4.21b). This reveals an insignificant decrease in its length observed as ε increases and the crack propagates to the surface of contacting body and forms a pit.

Finally, we note that the numerical solutions of the SIE were obtained by the method of mechanical quadratures for the order of the systems of the corresponding complex algebraic equations $N = 60$. The step of motion of the load was $\Delta\lambda = \Delta x_0/a = 0.1$, while the step of increment of the path $H = 0.05l_0$ where l_0 is the initial crack length.

Consequences

1. The values of SIF $F_{I\theta}$ strongly increase with the parameter r of the intensity of pressure on the crack faces. In this case, the most hazardous locations of the counterbody corresponding to the attainment of $\max F_{I\theta}$ are realized as soon the counterbody begins to cover the crack mouth (Fig. 4.16a) in its motion from the right to the left.
2. The increase in the coefficient of friction f between the contacting bodies has almost no influence both on the values of the SIF $F_{I\theta}$ and on the locations of

[1] However, for a given direction of motion of the counterbody from the right to the left (Fig. 4.15), the attainment of high pressures (large r) of the lubricant in cracks inclined at small angles β ($\beta \approx \pi/6$) is unlikely.

the counterbody (the values of λ) at which its maximum values are attained (Fig. 4.16b).

3. The increase in the relative crack length causes a significant increase in SIF $F_{I\theta}$ and, hence, in max$F_{I\theta}$. In this case, the location of the counterbody for which this maximum is realized shifts to the side of the macrocrack tip (Fig. 4.16c).
4. In the case where the slope of an initially rectilinear crack changes, we observe a clear trend to decrease in the values of the SIF $F_{I\theta}$ as the angle β decreases. The increase in max$F_{I\theta}$ with the angle β can be explained by the approach of the crack to the boundary of the body (Fig. 4.16d).
5. The transition from the uniform pressure on the crack faces to the pressure distributed according to a linear law leads to a significant decrease in the SIF $F_{I\theta}$ in all analyzed cases. Moreover, all trends described above are preserved for this type of loading on the crack faces.
6. Under the simultaneous action of the contact load and wedging of the crack faces, the SIF F_I and F_{II} practically coincide with the results obtained by Bower [10].
7. The penetration of large amounts of liquid in the crack mouth leads to the formation of high pressure on the faces ($r = 1.0$) and causes a decrease in the path length in the case of replacement of the uniform distribution by a distribution according to a linear law. In the presence of insignificant amounts of liquid in the crack mouth ($r = 0.1$), we observe the opposite picture.
8. A decrease in the angle of crack orientation β causes a significant increase in the lengths of crack paths.
9. The changes in the relative length of the initial crack ($\varepsilon = 0.1\ldots1.0$) weakly affect the shape of the path.
10. The constructed paths reveal the stable appearance of the crack on the edge of the body for both analyzed types of loading, which theoretically confirms the well-known hypothesis by Way [99] about the decisive influence of lubricants on the formation of pitting.

4.1.3 Residual Contact Lifetime of Rolling Pair Elements by Criterion of Pitting Formation with Consider the Cyclic Crack Growth Resistance Characteristics of Materials

The residual lifetime is evaluated according to the criterion of formation of pitting, i.e., according to the number of rolling cycles prior to its formation by using the procedure described in Chap. 2 (Sect. 2.3). It is assumed that if the sites of pitting appear on the surfaces of the rolling bodies, then the bodies lose theirs operating characteristics and require either replacement or repair.

To compute the lifetime according to the known solutions of the Hertz problem [62], we establish the maximum levels of contact pressure p_0 and the length of the contact zone $2a$ for the given rolling couple as follows:

$$p_0 = \sqrt{\frac{P}{\pi\eta b} \cdot \frac{R_1 + R_2}{R_1 R_2}}; \quad 2a = \frac{4}{\sqrt{\pi}}\sqrt{\eta\frac{P}{b} \cdot \frac{R_1 R_2}{R_1 + R_2}}; \quad \eta = \frac{1 - \nu_1^2}{E_1} + \frac{1 - \nu_2^2}{E_2}, \tag{4.9}$$

where ν_1 and ν_2 are Poisson's ratios, E_1 and E_2 are Young's moduli, R_1 and R_2 are the radii of the rolling bodies, b is the width of rolling bodies, and P is the total normal load acting upon the rolling couple.

The values of p_0 and $2a$ obtained in this way are introduced in the computational scheme as operating parameters. The application of these parameters in the evaluation of the residual lifetime of elements of a specific rolling couple allows us to consider not only the characteristics of cyclic crack growth resistance of materials of the elements but also their physical sizes and the level of applied loading in the course of operation.

Since our calculations are performed within the framework of linear fracture mechanics, we start to measure the residual lifetime from the time when the crack reaches the macrolevel and begins to propagate by the shear mechanism. (Thus, the period of macrocrack initiation is not considered in this case.) This approach has a reasonable substantiation, namely, the experimental data on the contact fatigue of cyclically contacting bodies demonstrate [50] that the number of rolling cycles required for a crack to pass from the germinal state to the level of macrocrack is much smaller (by orders of magnitude) than the number of cycles leading to the appearance of a surface defect (pit).

The residual lifetime is first calculated for the shear stage of crack propagation of when it develops rectilinearly by elongating with the help of the values of the SIF range K_{II} in each step of calculations. These steps do not change for the entire period of realization of the numerical algorithm. However, if the crack starts to propagate by the mode I mechanism, then its lifetime is determined along the curvilinear sections of the path with variable increments by using maximum values of the mixed-type SIF $K_{I\theta}$.

In Table 4.4, we present the characteristics of cyclic crack growth resistance of steels for the elements of wheel-rail systems and backup rolls of the rolling mills whose lifetimes are investigated in this section.

Lifetime of the backup rolls of rolling mills estimated by the formation of pitting. The rolls of rolling mills are high-cost elements of the rolling installation due to their unhandiness and complexity of their production. Therefore, the determination of the causes of decrease in the lifetime of rolls and conditions under which the lifetime may increase is an important scientific and engineering problem.

It is known that the main causes of the loss of serviceability by the surfaces of the rolls are the appearance of spalling and pitting. These defects are removed by using the technology of regrinding, which enables us to get pure surfaces of the

Table 4.4 Characteristics of cyclic crack growth resistance of some steels

Steel	ΔK_{th}	ΔK_{1-2}	ΔK_{2-3}	ΔK_{fc}	ΔK^{*}	n	v_0	q
	MPa·(m)$^{1/2}$						m/cycles	
Roll steel								
SKH (shear) [79]	18.0	20.4	40.0	47.8	–	–	2.56×10^{-10}	0.99
9KhF (9XΦ) (opening) [2]	10.9	14.1	35.9	53.0	–	–	6.5×10^{-8}	1.06
Rail steel								
RSB12 (shear) [9]	13.01	16.5	73.5	92.4	–	–	3.84×10^{-7}	1.41
75KhGST (75XΓCT) (opening) [94]	4.3	5.2	29.1	37.0	19.7	3.48	1.06×10^{-7}	1.26
Wheel medium-strength steel								
65G (65Γ) (shear) [70]	–	17.0	50.0	–	33.4	1.15	–	–
65G (65Γ) (opening) [82]	–	11.7	34.5	–	29.8	3.20	–	–

rolls by removing the defective layer. However, on the other hand, this decreases the diameter of the rolls. A single operation of regrinding removes a layer with a thickness of 3…10 mm from the surface [77]. As a rule, after 5…15 regrindings, the rolls are finally removed from the rolling mill. From the viewpoint of fracture mechanics, it is necessary to study how the main operation factors (environment and load) affect the crack propagation and the appearance of pits or spalls in the surface layer of the roll, thus decreasing its contact lifetime.

The contact lifetime estimated by the development of pitting was determined for a backup roll of the rolling mill under the conditions of boundary lubrication in contact between the backup and working rolls (this lubrication is determined by the technical process of rolling). Note that the characteristics of cyclic crack resistance of SKH and 9KhF roll steels are available in the literature only for one of the mechanisms of crack propagation. Therefore, in view of the closeness of the chemical compositions of roll steels (Table 4.4), we assume that the behavior of roll steels is similar to the behavior of SKH steel in the shear stage and to the behavior of 9KhF steel in the stage of opening displacement [23, 46, 54].

For both stages of crack propagation, we computed the lifetime for various values of the friction coefficients f_c between the crack faces and the initial angles β of inclination of the crack to the edge of the body (Table 4.5). The lower block of data in Table 4.5 characterizes the dependence of lifetime on the condition of transition from the shear mechanism to the opening mechanism chosen in our calculations (see relation (2.29)). In this relation, the parameters of pressure on the crack faces r_1, r_2 and r_3 correspond to the conditions $K_{I\theta}(r) \geq K_{I,1-2}$, $K_{I\theta}(r) \geq 3/2 \cdot K_{I,1-2}$, and

Table 4.5 Residual contact durability of the backup roll evaluated according to the development of pitting in SKH and 9KhF (9XΦ) steels with regard for the stages of shear or normal opening in the propagation of the edge crack, respectively ($a = 6$ mm, $f = 0.1$)

	r	$l_{0\tau}$, mm	$l_{0\sigma}$, mm	$N_{g\tau} \cdot 10^{-6}$, cycles	$N_{g\sigma} \cdot 10^{-6}$, cycles	$N_g \cdot 10^{-6}$, cycles
f_c	$\beta = 160°, r = r_3, p_0 = 1500$ MPa					
0.0	0.1350	1.15	3.21	5.87	0.05	5.93
0.1	0.1035	1.66	4.72	8.77	0.08	8.85
β	$f_c = 0.1, r = r_3, p_0 = 1500$ MPa					
150°	0.1370	1.43	4.69	9.35	0.13	9.48
160°	0.1035	1.66	4.72	8.77	0.08	8.84
r	$f_c = 0.1, \beta = 160°, p_0 = 1500$ MPa					
r_1	0.1035	1.66	4.72	8.77	0.08	8.85
r_2	0.1365	1.66	4.72	8.77	0.02	8.79
r_3	0.1660	1.66	4.72	8.77	0.01	8.78

$K_{\mathrm{I}\theta}(r) \geq 2 \cdot K_{\mathrm{I},1\text{–}2}$, respectively. Note that the choice of the condition of start of the opening displacement of a crack branch (transition to the mode I mechanism) strongly affects the lifetime $N_{g\sigma}$ in the stage of crack propagation by the mode I mechanism. Thus, the transition from the condition r_1 to the condition r_3 causes a decrease in $N_{g\sigma}$ by about an order of magnitude. Note that, in our calculations, the crack faces suffer the action of a uniformly distributed pressure of lubricant in the stage of opening displacement.

The analysis of the numerical results presented in Table 4.5 indicate that the lifetime $N_{g\tau}$ determined in the stage of shear exceeds the lifetime $N_{g\sigma}$ in the stage of opening displacements by two orders of magnitude and is determining. As compared with the case of smooth contact ($f_c = 0$), the presence of friction between the crack faces with a coefficient $f_c = 0.1$ increases the residual lifetime N_g by 50%. The decrease in the slope of the initial macrocrack causes a significant increase in the residual lifetime both for the stage of shear and for the stage of normal opening displacement (mechanism).

Lifetime of elements of the wheel-rail couple computed with regard for the formation of pitting in RSB12 and 75KhGST (75XГCT) rail steels and in 65G (65Г) medium-strength wheel steel. It is known that the railway transport is very important for the national economy. The problem of guaranteeing of the regularity of railway transportations is a strategic task for our country. The solution of this problem requires the creation of the criteria of evaluation of the serviceability and lifetime of the critical objects of the railway transport. The wheel-rail systems are regarded as an important class of objects of this kind (Fig. 4.22).

We compute the lifetimes of these objects in the cases of shear and opening mechanisms of propagation of edge macrocracks in elements of the wheel-rail systems [18, 35, 52, 85] for various characteristics of cyclic crack growth resistance of steels, different conditions of transition between the indicated mechanisms [85], and various

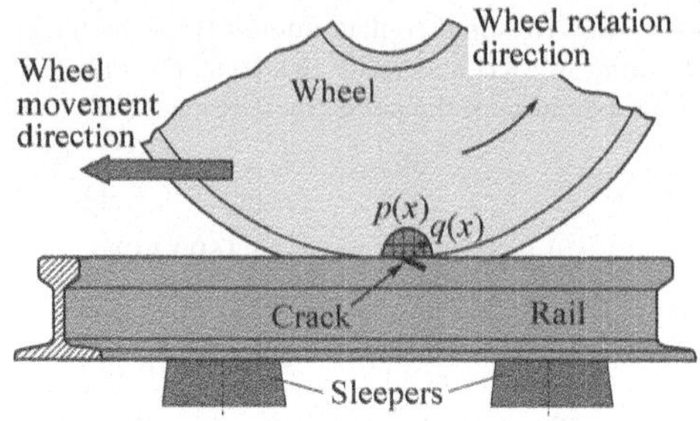

Fig. 4.22 Scheme of contact interaction in the wheel-rail system

distributions and levels of pressure of the lubricant on the crack faces. By using the obtained results, in view of the above-mentioned and other factors, we can optimize the choice of materials and operating parameters for the wheel-rail systems.

Note that the diagrams of fatigue fracture of materials according to the mode II mechanism, including the case of rail steels, are practically absent in the available literature. Hence, in our calculations, we use the fatigue fracture diagrams (FFD) and the characteristics of cyclic crack growth resistance in normal opening (Table 4.3) for 75KhGST rail steel with the structure of lamellar pearlite. In the case of shear, we use the FFD plotted for RSB12 steel whose chemical composition and structure are similar to the chemical composition and structure 75KhGST steel.

Prior to the analysis of the numerical results, we present some experimental and in-service data on the contact lifetime of rail steels. On the basis of the engineering data presented in [102], it was indicated that the total lifetime of the wheel-rail system in rolling constitutes $N = 10^6 \ldots 10^7$ cycles. In [47], it was experimentally established that $N = 1.5\times10^6$ (900A steel for $p_0 = 1100$ MPa). In [12], for rail pearlitic steels (STD, HH, HH-1, BS11.0 and CrMo), it was demonstrated that $N \approx (1\ldots2)\times10^6$ cycles if the maximum level of contact pressure varies within the range $p_0 = 1000\ldots1300$ MPa. It was also assumed [9, 10, 79] that the shear stage is longer than the normal opening stage.

In Table 4.6, we present the numerical data for three types of conditions for the transitions from the stage of initiation to the stage of shear and from the stage of shear to stage of normal opening displacement (see relations (2.23), (2.26), (2.28), and (2.29)): For the first type of conditions, the characteristics $K_{\mathrm{II},1\text{-}2}$ and $K_{\mathrm{I},1\text{-}2}$, respectively, are the key parameters in the first and second transitions; for the second type, (marked by *), the key characteristics are $K_{\mathrm{II}}(v_{th})$ and $K_{\mathrm{I},1\text{-}2}$ in the first and second transitions, respectively, and for the third type (marked by **), the role of these parameters is played by $K_{\mathrm{II}}(v_{th})$ and $K_{\mathrm{I}}(v_{th})$ for the first and second transitions, respectively. The analysis of the numerical results obtained for the indicated types of conditions enables us to determine the conditions describing the process of pitting formation most accurately. The data presented in Table 4.6 were obtained for the uniform distribution of pressure on the crack faces.

The analysis of the numerical data yields the following conclusions: for the first type, we get somewhat underestimated values of the lifetime N_g, for the second type, these values are higher, and for the third type, we get the highest values closest to

Table 4.6 Residual lifetime of RSB12 and 75KhGST (75XΓCT) rail steels ($a = 7$ mm)

	$l_{0\tau}$, mm	$l_{0\sigma}$, mm	$N_{g\tau} \cdot 10^{-6}$, cycles	$N_{g\sigma} \cdot 10^{-6}$, cycles	$N_g \cdot 10^{-6}$, cycles
f_c	$\beta = 150°$; $r = 0.1$; $f = 0.1$; $p_0 = 1100$ MPa				
0.05	1.01	5.04	0.11	3.04	3.15
0.05*	**0.76**	**5.04**	**0.39**	**3.04**	**3.44**
0.05**	0.76	4.01	0.39	6,59	6,98
0.10	1.01	2.71	0.09	0.43	0.51
0.10*	**0.76**	**2.71**	**0.38**	**0.43**	**0.80**
0.10**	0.76	2.21	0.37	1.95	2.32
0.15	1.01	1.70	0.06	0.22	0.28
0.15*	**0.76**	**1.70**	**0.35**	**0.22**	**0.57**
0.15**	0.76	1.38	0.33	0.89	1.22
0.20	1.01	1.16	0.02	0.15	0.17
0.20*	**0.76**	**1.16**	**0.31**	**0.15**	**0.46**
0.20**	0.76	0.93	0.27	0.64	0.91
β°	$r = 0.1$; $f = 0.1$; $f_c = 0$; $p_0 = 1100$ MPa				
145°	1.02	2.46	0.08	0.67	0.75
145°*	**0.72**	**2.46**	**0.39**	**0.67**	**1.06**
145°**	0.72	2.00	0.38	4.66	5.05
150°	1.01	2.71	0.09	0.43	0.52
150°*	**0.76**	**2.71**	**0.38**	**0.43**	**0.80**
150°**	0.76	2.21	0.37	1.95	2.32
155°	1.07	3.74	0.10	0.53	0.63
155°*	**0.82**	**3.74**	**0.39**	**0.53**	**0.92**
155°**	0.82	3.10	0.38	2.44	2.82
160°	1.20	2.13	0.08	0.16	0.24
160°*	**0.93**	**2.13**	**0.40**	**0.16**	**0.56**
160°**	0.93	1.77	0.38	0.82	1.20
f_c	$\beta = 150°$; $r = 0.1$; $f = 0.1$; $p_0 = 1100$ MPa				
0.00	1.01	2.71	0.09	0.43	0.52
0.00*	**0.76**	**2.71**	**0.38**	**0.43**	**0.80**
0.00**	0.76	2.21	0.37	1.95	2.32
0.05	1.33	2.71	0.08	0.43	0.51
0.05*	**0.83**	**2.71**	**0.51**	**0.43**	**0.94**
0.05**	0.83	2.21	0.50	1.95	2.45
0.10	1.55	2.71	0.10	0.43	0.53
0.10*	**0.97**	**2.71**	**0.82**	**0.43**	**1.25**
0.10**	0.97	2.21	0.79	1.95	2.74
0.15	2.08	2.71	0.09	0.43	0.51
0.15*	**1.28**	**2.71**	**0.90**	**0.43**	**1.33**

(continued)

Table 4.6 (continued)

	$l_{0\tau}$, mm	$l_{0\sigma}$, mm	$N_{g\tau} \cdot 10^{-6}$, cycles	$N_{g\sigma} \cdot 10^{-6}$, cycles	$N_g \cdot 10^{-6}$, cycles
f_c	$\beta = 150°; r = 0.1; f = 0.1; p_0 = 1100$ MPa				
0.15**	1.28	2.21	0.84	1.95	2.80
0.20	2.59	2.71	0.03	0.43	0.46
0.20*	**1.79**	**2.71**	**0.83**	**0.43**	**1.26**
0.20**	1.79	2.21	0.72	1.95	2.67

the experimental data. However, in the third case, the lifetime in shear $N_{g\tau}$ is much lower than the lifetime in normal opening $N_{g\sigma}$, which contradicts the experimental data [3]. Hence, the data on $N_{g\tau}$, $N_{g\sigma}$, and N_g obtained in the second case prove to be most suitable. The reasoning presented above implies that the second type of conditions is optimal for the numerical analysis. The conclusions formulated in what follows, in the first turn, are mainly base on the corresponding results (marked by * in Table 4.6).

The data presented in Table 4.6 also implies that the decrease in the angle β by only 15° promotes a more than twofold increase in the total lifetime N_g. In this case, the variations of lifetime in the stage of opening displacement is more pronounced, which is explained by the elongation of the path observed as the angle β decreases (Fig. 4.18). The pressure of lubricants also strongly affects the lifetime of rails. The elevation of pressure on the crack faces decreases the residual lifetime, accelerates the transition to the stage of opening displacement (shortening the length of the shear crack), and accelerates the process of crack growth in the stage of opening displacement. The increase in friction between the crack faces causes a significant increase in the lifetime in the stage of shear, which leads, in turn, to a considerable increase in the residual lifetime.

In general, the results of numerical calculations presented in Table 4.6 confirm a natural hypothesis that the increase in friction between the faces of the edge crack strongly inhibits its growth and leads to an increase in the lifetime of rails (in the shear stage of crack growth). At the same time, the elevation of pressure of liquid upon the crack faces accelerates its growth and decreases lifetime (in the opening stage). The angle of initial orientation of the crack also strongly affects the residual lifetime. Thus, the lifetime significantly increases as the indicated angle decreases both in the stages of shear and normal opening displacement.

Theoretical curves of contact fatigue for rail and wheel steels. Sizes of crumbling (pitting) particles. In practice, there are numerous attempts to construct the dependences of durability (lifetime) of the elements of rolling couples on the contact load, i.e., the so-called contact fatigue curves. It is clear that the presence of these dependences in the engineering calculations would be helpful to give adequate predictions the service life of designed elements of a couple according to the a priori known conditions of operation of these elements. However, due to the complexity of realization of these experiments and their high cost, the researchers mostly restrict themselves to several points on the coordinate system p_0–N_g. The required curves

Table 4.7 Residual lifetimes of RSB12 and 75KhGST (75XГCT) rail steels as functions of the contact load and the distribution of pressure of the lubricant over the crack faces

p_0, MPa	$l_{0\tau}$, mm	$l_{0\sigma}$, mm	$l_{c\sigma}$, mm	$N_{g\tau} \cdot 10^{-6}$, cycles	$N_{g\sigma} \cdot 10^{-6}$, cycles	$N_g \cdot 10^{-6}$, cycles
$f = 0.1, f_c = 0.1, r = 0.1, \beta = 150°, a = 7$ mm						
Linear distribution						
600	3.038	11.07	16.84	2.3913	1.6265	4.0178
800	1.831	7.07	11.11	1.6161	1.4780	3.0942
1000	1.279	5.20	8.17	0.8098	1.4103	2.2201
1200	0.824	3.98	6.69	0.6250	1.1963	1.8212
1400	0.627	3.27	5.39	0.4485	1.0403	1.4888
1600	0.522	2.80	4.80	0.3003	0.9290	1.2293
1800	0.441	2.47	4.38	0.2894	0.8292	1.1186
Uniform distribution						
600	3.038	5.629	9.18	1.7282	0.7583	2.4865
800	1.831	3.931	6.33	1.4772	0.5771	2.0542
1000	1.279	3.092	5.14	0.7536	0.4837	1.2373
1200	0.824	2.424	4.18	0.5881	0.3897	0.9778
1400	0.627	2.002	3.51	0.3802	0.3381	0.7183
1600	0.522	1.709	3.14	0.3065	0.3048	0.6113
1800	0.441	1.489	2.93	0.2742	0.2832	0.5575

are obtained on the basis of these data by means of interpolation (and sometimes of extrapolation). The drawbacks of this method are obvious.

In this chapter, on the basis of the dependences of the residual lifetime of rail and wheel steels [31, 40, 42, 85, 89, 90] on the maximum contact pressure p_0 obtained by using numerical methods, we make an attempt to construct the theoretical contact fatigue curves for elements of the wheel-rail system. In order to substantiate the obtained results, the theoretical dependences were supplemented with the available experimental data.

The investigated model studies two types of distributions of pressure of the lubricants on the crack faces. Therefore, in constructing contact fatigue curves, it is important to determine how the variations of these distributions affect the shape and mutual arrangement of the obtained curves. For this purpose, in Tables 4.7 and 4.8, we present the dependences of the residual lifetime on the maximum contact pressure p_0 for both types of action of the lubricant upon the crack faces in the opening stage of crack propagation.

The results of calculations presented in Tables 4.7 and 4.8 imply that the variations of the maximum level of contact pressure in the wheel-rail system strongly affects the residual lifetime in both stages of propagation of edge cracks. This influence is especially pronounced in the stage of crack growth by the shear mechanism in which

Table 4.8 Residual lifetime of 65G (65Г) wheel steel as a function of the contact load and the distribution of pressure of the lubricant over the crack faces

p_0, MPa	$l_{0\tau}$, mm	$l_{0\sigma}$, mm	$l_{c\sigma}$, mm	$N_{g\tau} \cdot 10^{-6}$, cycles	$N_{g\sigma} \cdot 10^{-6}$, cycles	$N_g \cdot 10^{-6}$, cycles
$f = 0.1, f_c = 0.1, r = 0.1, \beta = 150°, a = 7$ mm						
Uniform distribution						
600	4.44	15.37	21.94	0.1749	0.4837	0.6586
800	2.79	10.75	15.32	0.1050	0.3638	0.4689
1000	1.95	8.12	11.90	0.0755	0.2976	0.3731
1200	1.44	6.44	9.46	0.0593	0.2481	0.3074
1400	1.12	5.36	8.05	0.0487	0.2178	0.2666
1600	0.86	4.51	6.59	0.0414	0.1879	0.2293
1800	0.73	3.91	5.81	0.0350	0.1713	0.2063
Linear distribution						
600	4.44	36.44	49.06	0.4290	0.9018	1.3308
800	2.79	24.57	34.46	0.2193	0.6980	0.9173
1000	1.95	17.64	24.52	0.1190	0.5781	0.6971
1200	1.44	13.31	19.29	0.0741	0.5038	0.5779
1400	1.12	10.37	12.75	0.0563	0.4313	0.4877
1600	0.86	8.41	10.12	0.0460	0.1526	0.1986
1800	0.73	7.01	8.41	0.0374	0.0954	0.1329

the lifetime varies almost by an order of magnitude within the accepted range of values of p_0. It is also worth noting that, as the load increases, the lifetime in the stage of normal opening approaches the lifetime observed in the shear stage. In this case, for the linear distribution of pressure on the crack faces, the stage of opening displacement becomes longer than the shear stage even for $p_0 = 1000$ MPa. This effect is explained by the decrease in the initial values of $l_{0\sigma}$. Note that, within the framework of the formulated model, these values also determine the final length of the shear crack.

Note that the transition from the uniform distribution of pressure on the crack faces to the linear distribution causes a noticeable increase in the lifetime for both stages of crack propagation and an increase in initial crack lengths ($l_{0\sigma}$). The last observation implies that, for the linear distribution, the rectilinear section of the path is longer and the pit is deeper than for the uniform distribution.

By using the values of the residual lifetime for rail and wheel steels given in Tables 4.7 and 4.8, we can plot the theoretical contact fatigue curves (Fig. 4.23).

To complete the obtained picture, we use the initial and final values of crack length in the stage of opening displacement to construct the table of characteristic sizes of pitting (Table 4.9) formed for different values of the contact pressure for both types of distributions of pressure of the lubricant over the crack faces.

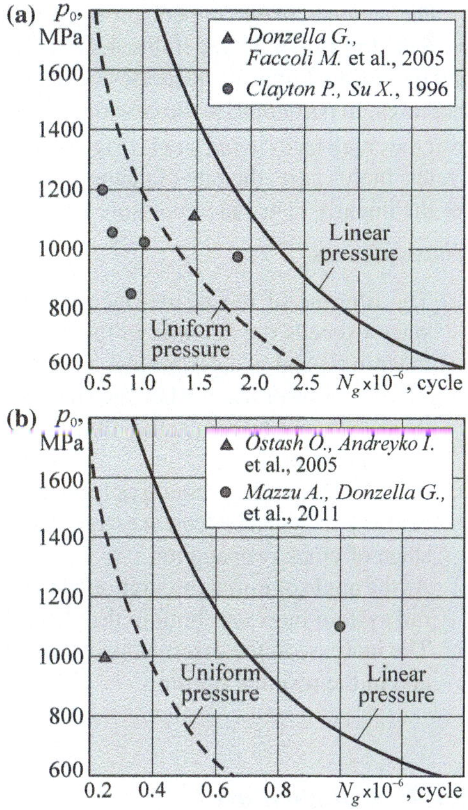

Fig. 4.23 Theoretical contact fatigue curves for: **a** RSB12 and 75KhGST (75XΓCT) rail steels; **b** 65G (65Γ) medium-strength wheel steel; $f = 0.1$; $f_c = 0.1$; $r = 0.1$; $\beta = 150°$; $a = 7$ mm

Table 4.9 Sizes of defects and the number of rolling cycles required for their initiation in RSB12 and 75KhGST (75XΓCT) rail steels

p_0, MPa	Pit depth, mm	Pit length, mm	$N_g \cdot 10^{-6}$, cycles	Pit depth, mm	Pit length, mm	$N_g \cdot 10^{-6}$, cycles
Uniform distribution				*Linear distribution*		
600	2.81	7,57	2.4865	5.53	13.57	4.0178
800	1.96	5.29	2.0542	3.53	8.88	3.0942
1000	1.54	4.34	1.2373	2.60	6.53	2.2201
1200	1.21	3.40	0.9778	1.99	5.23	1.8212
1400	1.00	2.93	0.7183	1.63	4.40	1.4888
1600	0.85	2.65	0.6113	1.40	3.93	1.2293
1800	0.74	2.49	0.5575	1.23	3.62	1.1186

The analysis of the dependences computed for the corresponding loads p_0 and presented in Fig. 4.23 and Table 4.9 enables us to conclude that the sizes of defects and the residual lifetime decrease according to a logarithmic law as the external load increases. In particular, in the case of uniform distribution of the pressure of lubricant over the crack faces in rail steels (Fig. 4.23*a*), we get $N_g = \ln[(p_0 - 500)/2500]^{-3/4}$ cycles. In this case, the rate of changes in the lifetime and its value are always higher for the linearly distributed pressure than for the uniform distribution.

Consequences

1. The lifetime of the nearsurface layer of a backup roll computed in the shear stage exceeds the lifetime in the normal opening stage by about two orders of magnitude and is determining.
2. The elevation of friction between the faces of an edge crack noticeably decelerates its growth and increases the lifetime of the wheel-rail system in the shear stage of crack propagation.
3. The increase in the pressure of lubricant on the crack faces accelerates the process of crack growth and decreases the lifetime of the wheel-rail system in the opening stage of crack propagation.
4. As the angle of initial orientation of the crack decreases, the lifetime of the wheel-rail system increases both in the shear and opening stages of crack propagation.
5. The increase in the external load causes a significant decrease in the sizes of pits and in the residual lifetime.

4.1.4 Conclusions

- The increase in the friction coefficient f_c between the crack faces causes an noticeable decrease in the extreme values of the SIF K_{II} and an increase in the zones of sticking of the crack faces in a rolling cycle (Fig. 4.2a).
- In the analyzed cases, the interval of locations of the counterbody for which the crack faces are stuck increases with the horizontal projection the crack due to the increase in its length and in the angle of orientation of the crack (i.e., as the supplementary angle decreases).
- For the considered ranges of operating parameters, the relative length ($\varepsilon = l_0/a$) of the shear crack, corresponding to the maximum SIF range K_{II} is, as a rule, within the limits 2.0…2.6 and the angle of orientation of the shear crack belongs to the interval $\beta^* = 150°\ldots160°$.
- The friction coefficient f between the rolling bodies strongly affects the SIF K_{II} and ΔK_{II}, and the formation of sticking of the faces of a gently sloping ($\beta \approx \beta^*$) crack only if the counterbody is located over the crack. However, this influence is very weak prior to the indicated time (Fig. 4.2b) in the case where the counterbody moves from the right to the left.
- The values of the mixed-type SIF $K_{I\theta}$ significantly increase with the parameter r of intensity of pressure on the crack faces. In this case, the hazardous locations of the

counterbody for which max$F_{I\theta}$ is attained are realized as soon as the counterbody begins to cover the crack mouth (Fig. 4.16a).

- The increase in the relative crack length leads to a significant growth of the SIF $K_{I\theta}$ and, hence, of max$K_{I\theta}$; moreover, the location of the counterbody for which this maximum is realized shifts to the side of the crack tip (Fig. 4.16c).
- The transition from the uniform pressure on the crack faces to the pressure distributed according to the linear law causes a significant decrease in the SIF $K_{I\theta}$ in all analyzed cases. Moreover, all trends discussed above are preserved for the indicated type of loading of the crack faces.
- The constructed trajectories of propagation of edge cracks reveal the stable appearance of the cracks on the edge of the body for both considered types of distributions of pressure over the crack faces, which confirms the well-known hypothesis by Way [99] concerning the decisive influence of lubricants on the formation of pits.
- The lifetime of the nearsurface layer of a backup roll determined for the shear stage is determining and exceeds the lifetime obtained for the opening stage by about two orders of magnitude.
- The elevation of friction between the faces of edge cracks strongly decelerate the process of crack growth and causes an increase in the lifetime of the wheel-rail system in the shear stage of crack propagation.
- The increase in the external load (in the maximum levels of contact pressure p_0) causes a significant decrease in the sizes of pits and in the residual lifetime.

4.2 Spalling

The phenomenon of spalling (see Figs. 4.24 and 4.25) of a part of the contact surfaces of cyclically contacting solid bodies is a typical contact fatigue defect for a number of tribojoints (bearings, wheel-rail systems, cylinders of rolling mills, etc.). It is clear that the appearance of these defects in elements of rolling couples negatively affects their serviceability and lifetime. Therefore, the lifetime of these tribojoints (rolling couples) can be also estimated by the criterion of spalling formation.

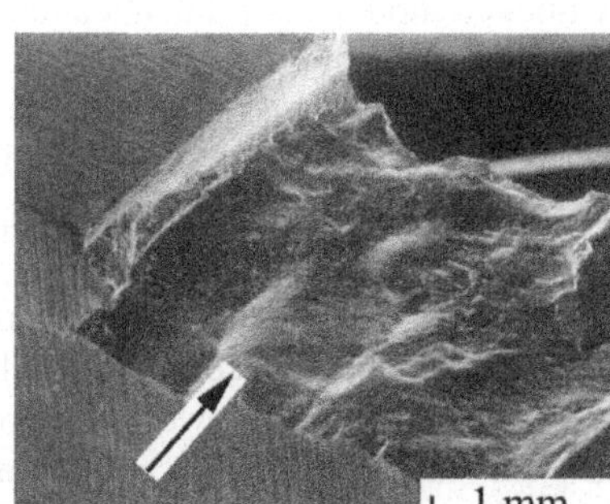

Fig. 4.24 Spalling in bearing steel under contact fatigue test [83]

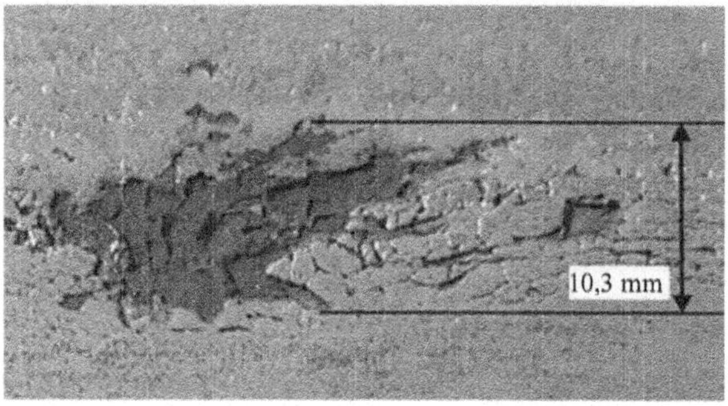

Fig. 4.25 Spalling from the rail surface

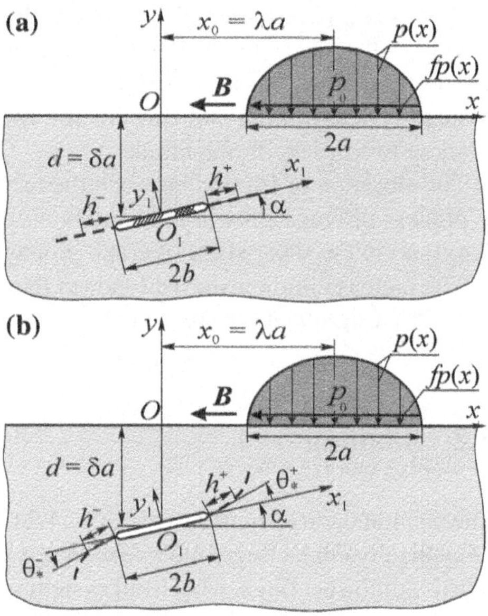

Fig. 4.26 Computational schemes of the problem **a** shear stage; **b** opening stage; ***B*** is the direction of motion of the counterbody

In view of the above reasoning, we now consider the kinetics of crack propagation under the contact surface of one body in a couple within the framework of the formulated model. Assume that a microcrack appears in the bulk of the material and is located under the contact surface at a depth $d = \delta a$ (δ is the relative depth of location of the crack). In the process of cyclic loading of the body, this microcrack is transformed into a macrocrack, which first grows rectilinearly under the action of shear stresses and then begins to propagate curvilinearly by the mode I mechanism (Fig. 4.26).

In the contact zone, under the action of a contact load moving along the boundary of the half-plane, the faces of internal crack are either open, or slide with friction, or are stuck (Fig. 4.26a). However, in order to simplify the problem, we assume that, in the shear stage, the faces of macrocrack are in contact along the entire length under the conditions of sliding with friction. If the crack begins to develop curvilinearly by the mode I mechanism, then we assume that the crack faces are not in contact (open) along the entire crack length (Fig. 4.26b). In finding the trajectories of propagation of the subsurface crack, we trace its growth from both tips in the shear and opening stages.

Note that the direction of tangential forces in the wheel-rail system is chosen depending on the type of railway wheels. Thus, the direction of action of the friction forces from the (driven) wheels of cars onto the surface of the rail coincides with the direction of motion of the car. For the locomotive (driving) wheels, the friction forces act in the opposite direction if the train moves in the traction mode. If we consider the action of friction forces on the surface of a railway wheel, then these directions are diametrically opposite. Thus, the directions of motion of the counterbody (B) and the tangential forces $fp(x)$ in Fig. 4.26 correspond to the action of car wheels on the rail or the action of the rail on a locomotive in the course of motions of the train in the traction mode.

4.2.1 Subsurface Crack in Contact Zone Under Transverse Shear Conditions

The stage of transverse shear for a subsurface macrocrack has a significant difference from the shear stage for the edge macrocrack due to its development at a certain distance from the contact surface. The indicated difference is reduced, in the first turn, to a significant duration of the crack propagation explained by possible large distances from the edge of the half-plane, where the acting stresses are much lower than in the surface layer. This effect becomes stronger for the horizontal subsurface cracks because both crack tips may develop for infinitely long time without attaining the values of $\Delta K_{\mathrm{II}fc}$ critical for shear or the values of $K_{\mathrm{I}th}$, which enable us to pass to the mode I mechanism of crack propagation. The presented reasoning implies that the lifetime computed in this stage by the criterion of spalling formation can be sometimes much larger than the lifetime obtained for edge cracks. Moreover, the contact durability of the elements of tribojoints is found just in the shear stage of propagation of subsurface macrocracks.

Influence of the operating parameters of rolling couple on the stress intensity factors K_{II} at the tips of subsurface crack in a single contact cycle. We now establish (within the framework of the accepted computational model) the dependences of the normalized SIF $F_{\mathrm{II}} = K_{\mathrm{II}}/\left(p_0\sqrt{\pi a}\right)$ on the location of the counterbody on the edge of the half-plane containing a horizontal subsurface crack for various values of the friction coefficient f_c between the faces. These dependences demonstrate, to the full extent, the dynamics of changes in the stress-strain state at the crack tips in a single contact cycle (Fig. 4.27).

The analysis of the plots indicates that if the contact load moves from the right to the left, then we have $\min F_{\mathrm{II}}^{-} = -0.095$ and $\max F_{\mathrm{II}}^{-} = 0.232$ for $f_c = 0$ at the left crack tip and $\min F_{\mathrm{II}}^{-} = -0.072$ and $\max F_{\mathrm{II}}^{-} = 0.151$ for $f_c = 0.4$. Thus, the increase

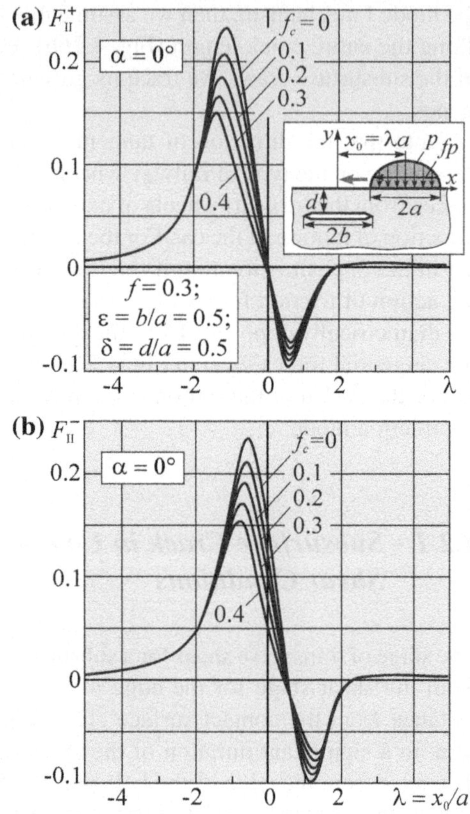

Fig. 4.27 Dependences of the normalized SIFs at (**a**) the right (F^+_{II}) and (**b**) left (F^-_{II}) tips of a horizontal crack on the location of the counterbody for various values of the friction coefficient f_c between the crack faces

in the friction coefficient causes a decrease in the minimum and maximum absolute values of the SIF by 25%, and 35%, respectively. Similarly, for the right tip, we get $\min F^+_{II} = -0.099$ and $\max F^+_{II} = 0.234$ for $f_c = 0$ and $\min F^+_{II} = -0.072$, $\max F^+_{II} = 0.155$ for $f_c = 0.4$. Hence, as f_c increases at the right tip, the minimum and maximum absolute values of the SIF decrease by 27% and 34%, respectively. For the gradual increase in the friction coefficient f_c by 0.1, we conclude that it leads to a decrease in $\min\left|F^-_{II}\right|$ and $\max F^-_{II}$ by 6.8% and 10.2%, respectively, on the average, whereas $\min\left|F^+_{II}\right|$ and $\max F^+_{II}$ decrease, on the average, by 7.6% and 9.8%.

Hence, for a horizontal subsurface macrocrack, the changes in the friction coefficient between the faces affect (almost identically and significantly) the maximum and minimum normalized SIF F_{II} both at the right and left crack tips. In the case of motion of the counterbody from the right to the left, the times when the maxima of F_{II} are attained at the left and right tips are different. Indeed, the SIF first attains its maximum value at the right crack tip for $\lambda = -0.8\ldots-0.5$ and then at the left tip for $\lambda = -1.6\ldots-1.1$.

Consider the case of a subsurface shear macrocrack inclined at an angle $\alpha = 45°$ to the horizontal (Fig. 4.28). The main distinction from the case of horizontal crack

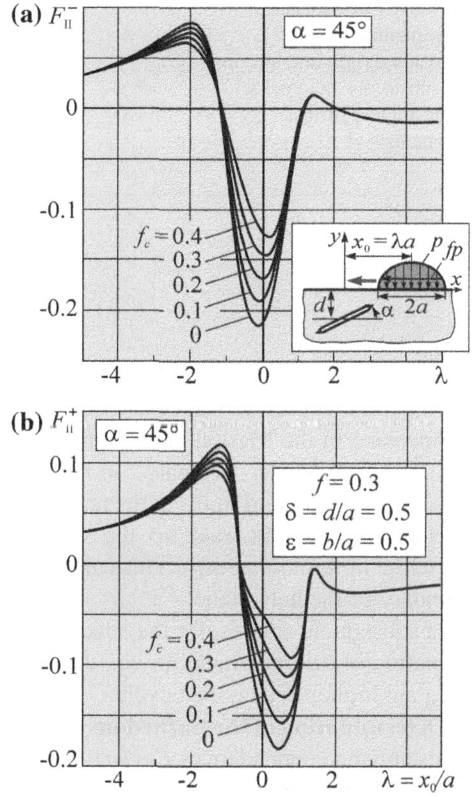

Fig. 4.28 Dependences of the normalized SIFs (**a**) F_{II}^- and (**b**) F_{II}^+ in the inclined subsurface macrocrack on the location of the counterbody for different values of the friction coefficient f_c between the crack faces

is that the maximum and minimum absolute values of the SIF F_{II} at the right tip are higher than the corresponding values at the left tip. This is explained, first of all, by the difference between the distances from these tips to the contact surface. Moreover, if we pass from a horizontal crack to an inclined crack, then we observe a deceleration of growth of the maximum values of the SIF F_{II} and a more intense increase in the minimum absolute values of the SIF F_{II} as the friction coefficient f_c decreases. For the inclined crack, the SIF K_{II} attains the extrema at both tips when the load moves over the crack center. For the horizontal crack, this happens when the load has already passed the crack (Figs. 4.27 and 4.28). In general, the trend to decrease in the SIF range ΔF_{II} remains invariable as f_c increases, as in the case of a horizontal crack.

It follows from Fig. 4.28 that, for the crack inclined at the angle $\alpha = 45°$, we get $\min F_{II}^- = -0.213$ and $\max F_{II}^- = 0.084$ for $f_c = 0$ at the left tip and, in addition, $\min F_{II}^- = -0.126$ and $\max F_{II}^- = 0.066$ for $f_c = 0.4$. Thus, the increase in the friction coefficient f_c causes a decrease in the minimum and maximum absolute values of the SIF F_{II} by 41% and 22%, respectively. Similarly at the right tip, we find $\min F_{II}^+ = -0.183$ and $\max F_{II}^+ = 0.117$ for $f_c = 0$ and $\min F_{II}^+ = -0.093$ and

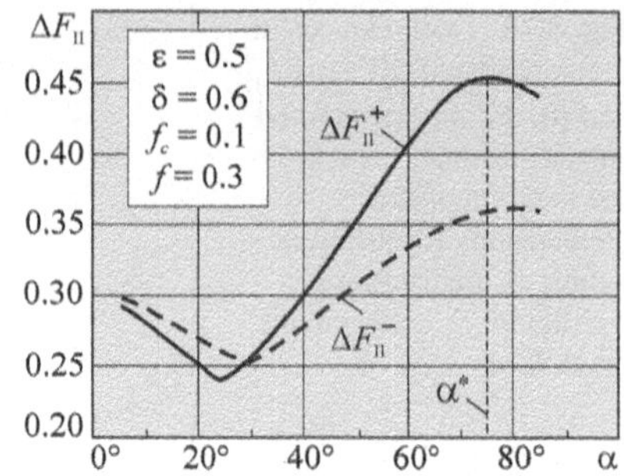

Fig. 4.29 Dependence of the normalized SIF ΔF_{II} range at the right and left tips of a subsurface macrocrack on the angle α of its orientation

$\max F_{\text{II}}^+ = 0.091$ for $f_c = 0.4$. Thus, as f_c increases at the right tip, the minimum and maximum absolute values decrease by 49% and 22%, respectively. For the gradual increase in the friction coefficient f_c by 0.1, the numerical results indicate that $\min|F_{\text{II}}^-|$ and $\max F_{\text{II}}^-$ decrease, on the average, by 12.3% and 6.1%, respectively, whereas $\min|F_{\text{II}}^+|$ and $\max F_{\text{II}}^+$ decrease, on the average, by 15.5% and 6.1%.

Hence, at the right crack tip, the SIF F_{II} changes faster than at the left tip (solely due to the minimum values). This can be explained by the closeness of this tip to the boundary of the half-plane.

In subsequent studies, for the adequate evaluation of the residual lifetime, we use the notion of driving crack tip, i.e., the tip at which the values of max SIF $|K_{\text{II}}|$ and ΔK_{II} are higher in a contact cycle.

Determination of the hazardous orientation of the subsurface crack. By using the established dependences $F_{\text{II}}(\lambda)$ for the right and left crack tips and assuming that the value of the SIF range ΔF_{II} at the driving crack tip is responsible for the fracture, we plot the function $\Delta F_{\text{II}}(\alpha)$ at the right and left tips of the subsurface crack. Then we determine the angle of orientation α guaranteeing the attainment of max $\Delta F_{\text{II}}(\alpha)$, i.e., the angle guaranteeing the validity of the necessary condition of extremum:

$$\left.\frac{\partial[\Delta F_{\text{II}}(\alpha)]}{\partial\alpha}\right|_{\alpha=\alpha^*} = 0. \tag{4.10}$$

In Fig. 4.29, we present the plots of the dependence of ΔF_{II} on the inclination angle of the subsurface macrocrack. Note that, as a result of variation of the angle α, one crack tip moves toward the edge of the half-plane, whereas the other tip moves away from the edge. Hence, the values of the angle α guaranteeing the extremum of ΔF_{II} are different for different crack tips. It follows from the dependences presented in Fig. 4.29 that, as the slope of the rectilinear subsurface shear macrocrack increases, the SIF range ΔF_{II} attains its minimum value and then its maximum value first at the right crack tip (this is the driving tip as follows from the values of ΔF_{II} at the extreme points) and then at the left tip.

Summarizing the arguments presented above consideration, we conclude that the higher the values of the SIF range ΔF_{II} in the initial stage of crack propagation, the higher the probability for this crack to begin the process of its propagation. Hence,

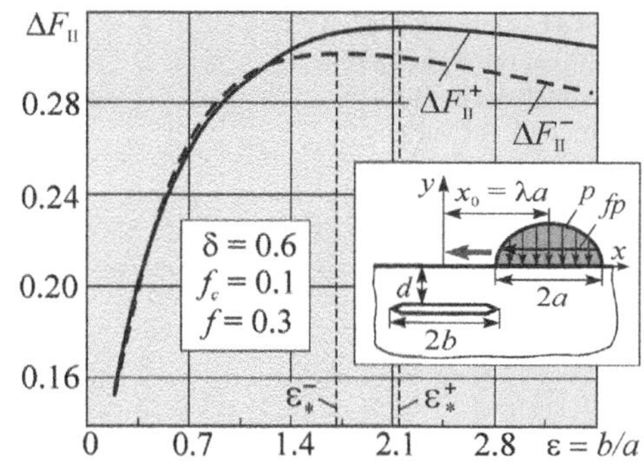

Fig. 4.30 Dependences of the normalized SIF ΔF_{II} range at the left and right tips of a horizontal crack on the relative length ε

we can assume that the most and least hazardous slopes of the subsurface crack are $\alpha = \alpha^* = 75°$ and $\alpha = 25°$, respectively (we make choice among the values of ΔF_{II} for the right tip because this is the driving tip in the analyzed case).

Critical length of the horizontal subsurface crack. As a result of calculations similar to those performed above for the SIF range ΔF_{II} at the driving crack tip as a function of the relative length ε, we get the range of values of ε, for which ΔF_{II} continuously increase and, hence, guarantee the permanent growth of the crack (provided that $\Delta F_{II}(\varepsilon) \geq \Delta K_{IIth}$). In particular, the dependences constructed in this way also indicate the values of $\varepsilon = b/a$ guaranteeing max $\Delta F_{II}(\varepsilon)$. Thus, we write the condition similar to (4.10) in the following form:

$$\left.\frac{\partial[\Delta F_{II}(\varepsilon)]}{\partial\varepsilon}\right|_{\varepsilon=\varepsilon*} = 0. \tag{4.11}$$

It is worth noting that the presence of the maximum values of ΔF_{II} at the left and right tips of a horizontal subsurface macrocrack imposes restrictions on the rectilinear growth of the crack by the shear mechanism because ΔF_{II} increases only in a certain range of values of the relative crack length ε. Hence, the larger the interval, the larger (more dangerous) the crack size.

It follows from Fig. 4.30 that, for the right branch of the horizontal crack, the range of values of ε in which the SIF range ΔF_{II} increases, is much larger, than for the left branch. Hence, when the left crack tip stops, the right tip continues to propagate for a certain time. Moreover, the SIF range ΔF_{II} at the left crack tip after the attainment of its maximum begins to decrease faster than at the right tip, which is also an argument for the validity of the observations made above about the possibility longer growth of the right branch of the crack. At the right and left tips, $\Delta F_{II}(\varepsilon)$ attains the maximum values for $\varepsilon_*^+ = 2.2$ and $\varepsilon_*^- = 1.7$, respectively.

We also note that, despite the horizontal location of the crack in the body, the values of the SIF range ΔF_{II} at the right tip exceed the values of ΔF_{II} at the left tip by 5%.

Characteristic depths of location of horizontal subsurface crack. The depths of location subsurface macrocrack noticeably affect both the stress-strain state at the tip and its subsequent propagation. It is known that the horizontal subsurface cracks

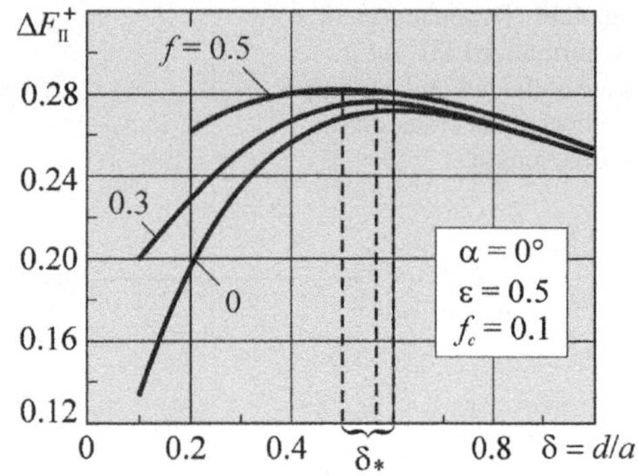

Fig. 4.31 Dependences of $\Delta F_{\text{II}}(\delta)$ at the right tip of a horizontal subsurface macrocrack for various values of the friction coefficient f

initiated and propagating in the material form a certain strip of their accumulation (the experimental data give the optimal relative depth of their location of about $\delta = d/a \approx 0.7$), where the indicated development is most probable for these cracks. Therefore, the value of depth is, as a rule, chosen in the vicinity of the indicated value.

In the present work, the depth of location horizontal shear macrocrack is determined by analogy with the hazardous orientation (angle α^*) and the interval of critical lengths. For this purpose, we determine the presence of maximum at the right crack tip (which is driving for the given directions of motion of the counterbody and friction forces) in the plot of the function $\Delta F_{\text{II}}^{*}(\delta)$. It is clear that the problem posed in this way possesses a solution only with a single extremum in the indicated dependence, and this extremum determines the required depth δ. The corresponding necessary condition can be represented in the following form:

$$\left.\frac{\partial\left[\Delta F_{\text{II}}^{+}(\delta)\right]}{\partial \delta}\right|_{\delta=\delta*} = 0. \tag{4.12}$$

The results of calculations are presented in Fig. 4.31, where each plotted dependence has a point of maximum. Hence, most likely, the cracks are mainly concentrated at depths of about $\delta = \delta_*$, thus guaranteeing the highest value of ΔF_{II}^{+}.

By analyzing the influence of the friction coefficient f on the value of δ_*, we conclude that the zone (depth) of favorable propagation shifts toward the rolling surface as the indicated coefficient increases. Thus, for $f = 0$ and $f = 0.5$, the hazardous depths of location are $\delta_* = 0.6$ and $\delta_* = 0.5$, respectively. Recalculating to the known sizes of the contact section (e.g., for a wheel-rail couple with $2a = 14$ mm), we get the depth range $d^* = 3.5\ldots4.2$ mm.

The accumulated results are in good agreement with the results obtained in [59].

Consequences

1. The SIF $|K_{II}|$ and the maximum value of the SIF range ΔK_{II} strongly decrease (Figs. 4.27 and 4.28), as the friction coefficient f_c between the faces of horizontal and inclined cracks increases, as in the case of edge cracks (Sect 4.1.1).
2. The SIF K_{II} at both tips of the inclined cracks attain the maximum absolute values when the contact load moves over the crack center (Fig. 4.28); at the same time, for horizontal cracks, these values are attained if the load has already passed over the crack (Fig. 4.27).
3. The most hazardous slope of the subsurface crack is realized for the angle $\alpha^* = 75°$, while the safest angle is $\alpha = 25°$ (for a fixed configuration of the parameters f, f_c, ε and δ; see Fig. 4.29).
4. For the right tip of the horizontal shear subsurface crack, the interval of relative crack lengths ε for which the SIF range ΔK_{II} increases is much larger than for the left tip. Hence, after the stop of the left crack tip, the right tip continues to propagate for a certain time.
5. The increase in the friction coefficient f between rolling bodies shifts the zone of favorable development of horizontal subsurface cracks toward the rolling surface (Fig. 4.31).

4.2.2 Subsurface Crack in Contact Zone Under Normal Opening Conditions

The stress-strain state in the zone of contact rolling is characterized by a high gradient and rapidly varies due to the relative motion of the rolling bodies [92]. Hence, the tips of the subsurface crack readily pass from the zone of compression into the zone of tension. Thus, according to the computational model (Chap. 2), the process of crack growth is controlled by the maximum values of the parameter $K_{I\theta}$ and estimated by using the σ_θ-criterion. Under these conditions, there exists a possibility of curvilinear crack growth, including the possibility of crack propagation toward the edge of the body and, hence, the formation of spalling-type defects. In addition, the mode I stage of crack propagation is of interest due to the possibility of tracing the specific features of propagation of both tips of the curvilinear crack in the process of cyclic contact of rolling bodies.

Influence of the operating parameters of rolling couple on the mixed-type stress intensity factors $K_{I\theta}$ at the tips of subsurface crack under the conditions cyclic contact. First, we consider the general case of subsurface crack oriented at any angle to the edge of the half-plane. We performed calculations for various angles of inclination α of the initially rectilinear crack to the boundary of the half-plane ($\alpha = 0$; $\pm\pi/12$; $\pm\pi/6$; $\pm\pi/4$; $\pm\pi/3$; $\pi/2$).

In Fig. 4.32, we show the plots of the stress intensity factors for an arbitrarily oriented subsurface crack depending on the location of the contact load on the edge of the half-plane. It follows from the presented results that if the contact load moves

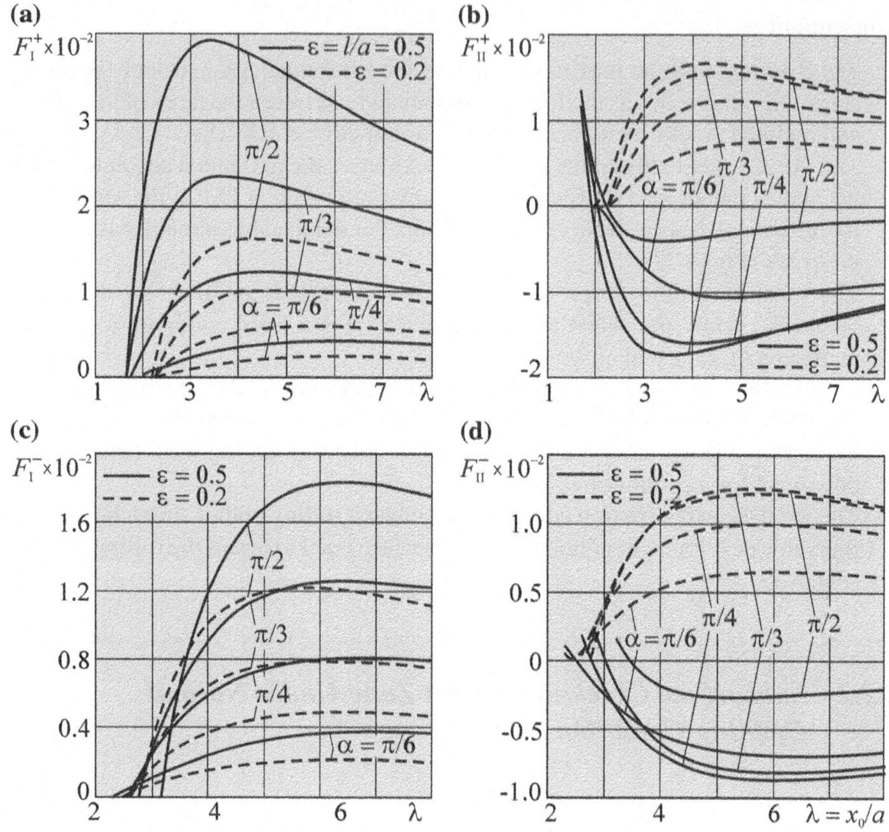

Fig. 4.32 Normalized SIFs $F_{I,II}^{\pm}(\lambda)$ depending on the location of contact load and the angle α of crack orientation; $f = 0.3$; $\delta = d/a = 0.7$

along the edge of the half-plane from the right to then left, then an arbitrarily oriented crack is open ($K_I^{\pm} > 0$ and $\delta(t) = v_n^+(t) - v_n^-(t) > 0$; see relation (3.218)) if the contact load is located at a sufficiently large distance to the right from the right crack tip ($\lambda = x_0/a > 2.0$). This range of λ is large because the values of SIFs $F_I^{\pm}(\lambda)$ slowly decrease to zero. The larger the slope of the subsurface crack to the horizontal, the higher the maximum values of the SIFs $F_I^{\pm}(\lambda)$. The maximum values of $F_I^{\pm}(\lambda)$ increase with the relative crack length. At the right crack tip, the stress intensity factors $F_I(\lambda)$ are higher than at the left tip (the same trend is also observed for the absolute value of the coefficient $|F_{II}(\lambda)|$).

Horizontal crack. To study the influence of the friction coefficient f on the values of the normalized SIFs F_I and F_{II} in a contact cycle, we now consider a horizontal subsurface crack located at a relative depth $\delta = d/a = 0.7$ and has the relative length $\varepsilon = b/a = 0.5$.

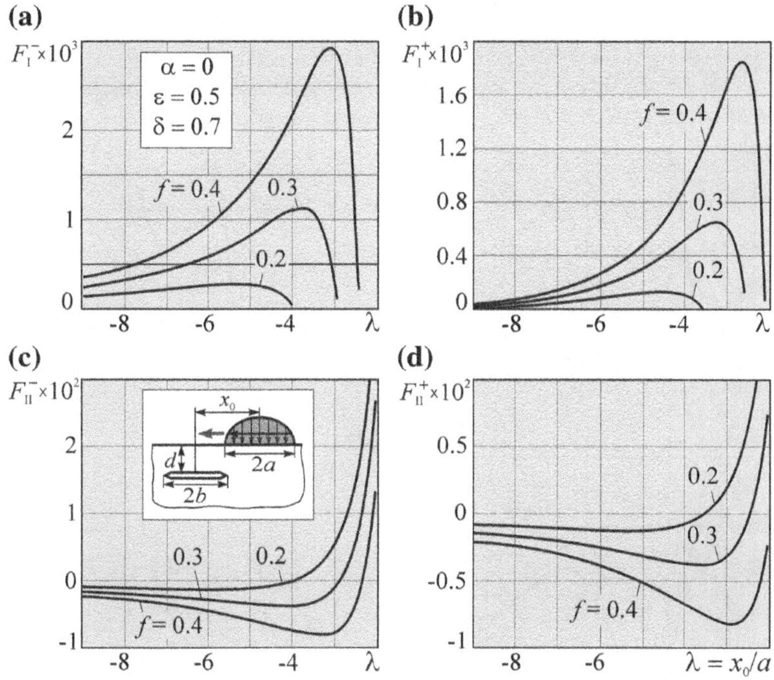

Fig. 4.33 Normalized SIFs F_{I} and F_{II} at the right and left tips of a horizontal subsurface macrocrack depending on the location (λ) of the counterbody for different values of the friction coefficient f

Since the maximum value of contact pressure p_0, which has no influence on the normalized SIFs F_{I} and F_{II}, and the friction coefficient f remain basic operating parameters in the opening stage of crack propagation, we study the dependences of the SIF $F_{\text{I}}^{\pm}$ and $F_{\text{II}}^{\pm}$ on the friction coefficient f for a single rolling cycle. These dependences (Fig. 4.33) enable us to determine the driving tip of the crack and indicate thee locations of the counterbody guaranteeing the absence of contact of the crack faces (in this case, the necessary condition $K_{\text{I}} > 0$ is satisfied).

The presented plots show that the increase in the friction coefficient f in contact between the rolling bodies leads to a significant increase in the SIF F_{I} at the right and left crack tips and shifts the value of λ corresponding to the maximum of F_{I} to the right. Further, since the positive sign of F_{I} shows that the crack is open at the corresponding crack tip. Moreover, an increase in its values directly affects the range of λ for which the crack remains open. Hence, for high values of the friction coefficient f, this range of λ is always larger than for low values.

We also note that the values of F_{II} in the analyzed interval are, as a rule, negative. Hence, the increase in the friction coefficient f causes an increase in $|F_{\text{II}}|$ at the right and left crack tips.

Further, we study the influence of the relative crack length ε and the depth of its location under the surface on the SIF at the right and left tips. We also computed the

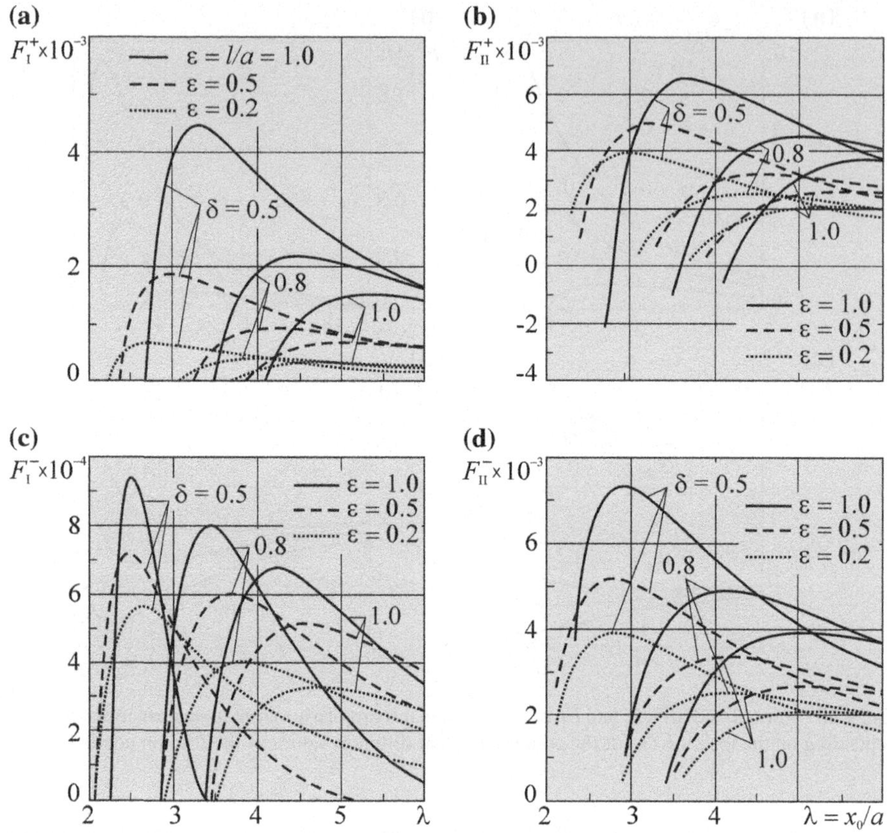

Fig. 4.34 Normalized SIFs $F_{I,II}^{\pm}(\lambda)$ for the horizontal crack depending on the location (λ) of the contact load; $\delta = d/a; f = 0.3; \alpha = 0°$

stress intensity factors $F_I^{\pm}(\lambda)$ and $F_{II}^{\pm}(\lambda)$ in the course of motion of the counterbody along the edge of the half-plane. The results of calculations are presented in Fig. 4.34.

As a result of the analysis of the SIFs depicted in Fig. 4.34, we can make the following conclusions:

- Under the conditions of the investigated problem and, in particular, for a given direction of contact tangential forces, the range of locations of the load (the parameter $\lambda = x_0/a$) under which the horizontal crack remains open is located to the right of the crack and varies within the range $2.0 < \lambda < 9.0$. The higher values of λ are not analyzed due to the extremely small corresponding values of the stress intensity factors in the vicinities of the crack tips. For these values, the crack practically does not interact with the contact load.
- If the contact load moves along the edge of the half-plane, then the stress intensity factors $F_I^{\pm}(\lambda)$ and $F_{II}^{\pm}(\lambda)$ attain their maxima at certain locations. The closer the crack to the half-plane boundary, the higher the maximum values.
- The maximum values of the SIFs $F_I^{\pm}(\lambda)$ and $F_{II}^{\pm}(\lambda)$ increase with the relative length of the horizontal crack.

Note that the plots of $F^{+}_{\mathrm{I,II}}(\lambda)$ are in good agreement with the plots presented in [68] and the plots for the SIF $F_{\mathrm{II}}(\lambda)$ in Fig. 4.34 are presented solely for the values of λ such that $K_{\mathrm{I}}(\lambda) > 0$.

In order to determine the configurations of values of the operating parameters, orientations, lengths, and the depths of location of the crack, as well as the characteristics of cyclic crack growth resistance of the material for which we observe the transition from the stage of crack propagation by the shear mechanism to the stage of propagation by the opening mechanism, we plotted the dependences of $F_{\mathrm{I}\theta}$ on these parameters. On the basis of these dependences, we established the values of the angle α and length ε such that, after their attainment, the macrocrack propagates by the opening mechanism (i.e., the condition $\max K^{\pm}_{\mathrm{I}\theta} > \Delta K_{\mathrm{I}th}$ is satisfied) for high- and medium-strength 65G wheel steels. The maximum levels of pressure p_0, the length of the contact section $2a$, and the coefficient of sliding friction f between the rolling bodies were chosen within the ranges of the operating parameters for the wheel-rail systems: $p_0 = 1500$ MPa and 2000 MPa, $2a = 14$ mm, and $f = 0.2$; 0.3; 0.4.

The characteristics of crack growth resistance presented in Table 4.10 and the indicated operating parameters enable us to determine the normalized values of $F_{\mathrm{I}th}$ corresponding to $\Delta K_{\mathrm{I}th}$ for high- and medium-strength steels. By using the formula $F = K/\left(p_0\sqrt{\pi a}\right)$ for the loads $p_0 = 1500$ MPa and $p_0 = 2000$ MPa, we get $F^{1\epsilon}_{\mathrm{I}th}(1500) = 2.97\times10^{-2}$, $F^{2\epsilon}_{\mathrm{I}th}(2000) = 2.22\times10^{-2}$, $F^{1c}_{\mathrm{I}th}(1500) = 3.15\times10^{-2}$, and $F^{2c}_{\mathrm{I}th}(2000) = 2.36\times10^{-2}$. In Fig. 4.35a, b, we present the corresponding dependences. They reveal an insignificant predominance of the left tip for the horizontal crack ($\alpha = 0$). However, as the inclination angle α increases, the right tip gradually turns into the leading tip [i.e., $F^{+}_{\mathrm{I}\theta}(\alpha) > F^{-}_{\mathrm{I}\theta}(\alpha)$], which can be explained by the fact that it approaches the boundary of the half-plane. By using the established values of $F^{1\epsilon}_{\mathrm{I}th}$, $F^{2\epsilon}_{\mathrm{I}th}$, $F^{1c}_{\mathrm{I}th}$, and $F^{2c}_{\mathrm{I}th}$ and the fact that the obtained curves are monotonically increasing (at least up to the angle $\alpha = 56°$), we can find the slopes of the crack starting from which it may propagate (by the opening mechanism). For $f = 0.4$ and the left crack tip, the load $p_0 = 1500$ MPa (i.e., the values of $F^{1\epsilon}_{\mathrm{I}th}$ and $F^{1c}_{\mathrm{I}th}$) corresponds to the angles $\alpha = 33°$ for high-strength steel and to $\alpha = 40°$ for medium-strength steel. At the same time, for the load $p_0 = 2000$ MPa (i.e., for $F^{2\epsilon}_{\mathrm{I}th}$ and $F^{2c}_{\mathrm{I}th}$), we get the angles $\alpha = 17°$ and $\alpha = 19°$ for the high- and medium-strength steels, respectively. For the right tip, the load $p_0 = 1500$ MPa corresponds to the angles $\alpha = 20°$ and 22° for high- and medium-strength steels, respectively. Moreover, for $p_0 = 2000$ MPa, we get $\alpha = 13°$ and 14° for high- and medium-strength steels, respectively. In addition, it follows from Fig. 4.35b that the threshold values of $F^{1}_{\mathrm{I}th}$ and $F^{2}_{\mathrm{I}th}$ can be attained for the friction coefficient $f = 0.3$ at the right crack tip. In this case, for $p_0 = 1500$ and 2000 MPa, we, respectively, obtain $\alpha = 44°$ and 30° for the high-strength steel and $\alpha = 48°$ and 32° for the medium-strength steel. As follows from Fig. 4.35a, b, for the friction coefficient $f = 0.2$, the threshold values of $F^{1}_{\mathrm{I}th}$ and $F^{2}_{\mathrm{I}th}$ under the same conditions are not attained at the tips for any orientation.

Table 4.10 Characteristics of cyclic crack growth resistance of wheel steels [82]

Steel	ΔK_{Ith}	ΔK_I^*	ΔK_{Ifc}	n
	MPa·(m)$^{1/2}$			
High-strength wheel steel	6.6	27.5	65	3.1
Medium-strength wheel steel	7.0	29.8	100	3.2

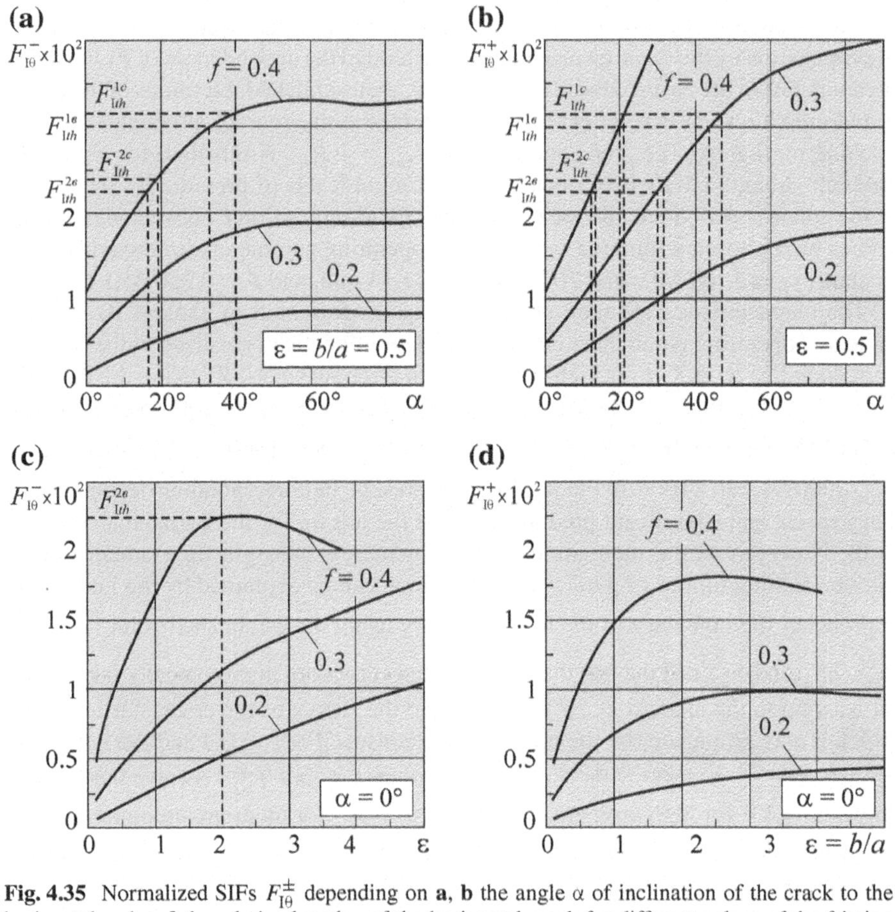

Fig. 4.35 Normalized SIFs $F_{I\theta}^{\pm}$ depending on **a**, **b** the angle α of inclination of the crack to the horizontal and **c**, **d** the relative length ε of the horizontal crack for different values of the friction coefficient in the case of contact of rolling bodies; $\delta = d/a = 0.7$; $\alpha = 0$

The analysis of the dependences of $F_{I\theta}$ on the relative length ε of the horizontal crack given in Fig. 4.35c, d shows that the threshold values of $F_{Ith}^{1\theta}$ and F_{Ith}^{1c} under the load $p_0 = 1500$ MPa are not attained in the investigated ranges of the parameters ε and *f*. The crack may start to propagate by the opening mechanism only from the left tip under the load $p_0 = 2000$ MPa for the friction coefficient $f = 0.4$ and the relative length $\varepsilon = 2.0$. In view of the chosen parameters of loading, we can evaluate the probable length of spalling as follows: $2b = 2\varepsilon a = 28$ mm. The depth of this defect can be found by using the parameter δ equal to 0.7. Hence, $d = \delta a = 4.9$ mm.

It is worth noting that the angle θ^{*-} specifying the direction of propagation of the horizontal crack from the left tip strongly depends both on the relative crack length ε and on the friction coefficient *f*. As a result, the left tip may either penetrate deeper into the material (for small relative crack lengths) or move to the surface of the body. This is not true for the angle θ^{*+} at the right tip. A horizontal crack always propagates from the right tip to the boundary of the half-plane. For $f = 0.2$, the transition from the propagation of the left crack tip into the bulk of the material to its development toward the boundary starts for $\varepsilon \geq 1.6$. For $f = 0.3$ and $f = 0.4$, this happens for $\varepsilon \geq 2.2$ and $\varepsilon \geq 3.0$, respectively.

In general, the analysis of the results presented in Fig. 4.35 **demonstrates that the increase of the friction coefficient in contact between the bodies, the slope α of the initial crack to the horizontal, the intensity of contact pressure (the parameter p_0) and the relative crack length ε (provided that the crack is open) promotes the onset of its propagation by the mode I mechanism** $(K_{I\theta}^{\pm} \geq \Delta K_{Ith})$. We also note that the difference between the characteristics of crack growth resistance of high- and medium-strength wheel steels (Table 4.10) weakly affects the parameters α, *f*, and ε (but not p_0) for which we observe the transition from the shear mechanism to the opening mechanism. The obtained results almost do not reveal the configurations of parameters for which the horizontal crack in wheel steels passes to the propagation by the opening mechanism in the course of rolling under the conditions of dry friction. It is clear that the decrease in the crack resistance of the material and, in particular, in its threshold characteristic ΔK_{Ith}, the increase in the level of loading and friction (p_0, *f*), and the deviation of the crack from its horizontal orientation promote the transition from the shear mechanism of crack propagation to the opening mechanism and to the formation of spalling.

Conditions of full opening of the horizontal crack. Within the framework of the posed problem, we assume that the development of horizontal subsurface macrocracks is possible only in the case of complete absence of contact of the crack faces. Hence, the difference between the normal displacements for the crack faces (relation (3.128)) must be always positive, namely,

$$\Delta v_0(t) = \left(\frac{2G}{1+æ}\right)\left(v_0^+(t) - v_0^-(t)\right) = \mathrm{Re}\int_{-1}^{\eta} \frac{u(\xi)d\xi}{\sqrt{1-\xi^2}} \frac{|\omega'(\eta)|}{\omega'(\eta)} =$$

Table 4.11 Maximum values of the normalized mixed-type SIFs $F_{I\theta}^{-}$ and $F_{I\theta}^{+}$ for the right and left tips of the horizontal macrocrack and the corresponding values of the angles of deviation θ^{*-} and θ^{*+}

$\varepsilon = b/a$	$f = 0.3$				$f = 0.4$			
	$\max F_{I\theta}^{-}$	θ^{*-}	$\max F_{I\theta}^{+}$	θ^{*+}	$\max F_{I\theta}^{-}$	θ^{*-}	$\max F_{I\theta}^{+}$	θ^{*+}
$\delta = d/a = 0.5$								
0.5	0.006	−1.190	0.007	−1.110	0.014	−1.174	0.015	−1.092
1.0	0.009	−1.202	0.010	−0.999	0.018	−1.193	0.021	−1.002
2.0	0.011	−1.218	0.009	1.180	0.019	−1.223	0.010	0.035
3.0	Contact of the faces				Contact of the faces			
$\delta = 1.0$								
0.5	0.003	−1.171	0.003	−1.150	0.008	−1.152	0.008	−1.130
1.0	0.005	−1.184	0.005	−1.105	0.011	−1.162	0.012	−1.082
2.0	0.007	−1.190	0.008	−1.005	0.015	−1.175	0.017	−0.990
3.0	0.008	−1.185	0.008	1.215	0.016	−1.186	0.020	−0.861
4.0	0.009	−1.196	0.011	1.198	0.016	−1.198	0.018	−0.616
5.0	0.009	−1.204	0.014	1.210	0.016	−1.210	0.020	1.190
6.0	0.009	−1.212	0.016	1.203	0.015	−1.218	0.023	1.202
7.0	0.008	−1.219	0.018	1.206	0.015	−1.225	0.025	1.225
8.0	0.008	−1.224	0.020	1.216	Contact of the faces			
9.0	0.008	−1.228	0.022	1.230	Contact of the faces			
10.0	Contact of the faces				Contact of the faces			

$$= -\frac{2}{N}\mathrm{Re}\sum_{n=1}^{N} u(s_n)\sum_{r=1}^{N-1}\frac{T_r(s_n)U_r(\eta)}{r}\,\frac{|\omega'(\eta)|}{\omega'(\eta)} > 0, \qquad |\eta| < 1, \tag{4.13}$$

Therefore, it is necessary to establish the values of the operating factors and the parameters of crack for which condition (4.13) is satisfied.

For this purpose, we first construct the table of maximum values of the SIF $F_{I\theta}$ at the right and left tips of an open subsurface crack for different values of the relative depth δ, relative length ε, and friction coefficient f. These results are shown in Table 4.11.

It follows from Table 4.11 that, for large depths ($\delta = 1.0$), the horizontal subsurface crack is open even for large relative lengths $\varepsilon \approx 7.0$ (which, in turn, indicates the possibility of long-term stage of crack growth by the mechanism of shear). At the same time, as the crack approaches the contact surface ($\delta = 0.5$), the admissible, from the viewpoint of openness, relative crack length decreases down to the value $\varepsilon = 2.0$. The increase in the friction coefficient f also has a similar influence on the interval of lengths for which the subsurface crack remains completely open. Thus,

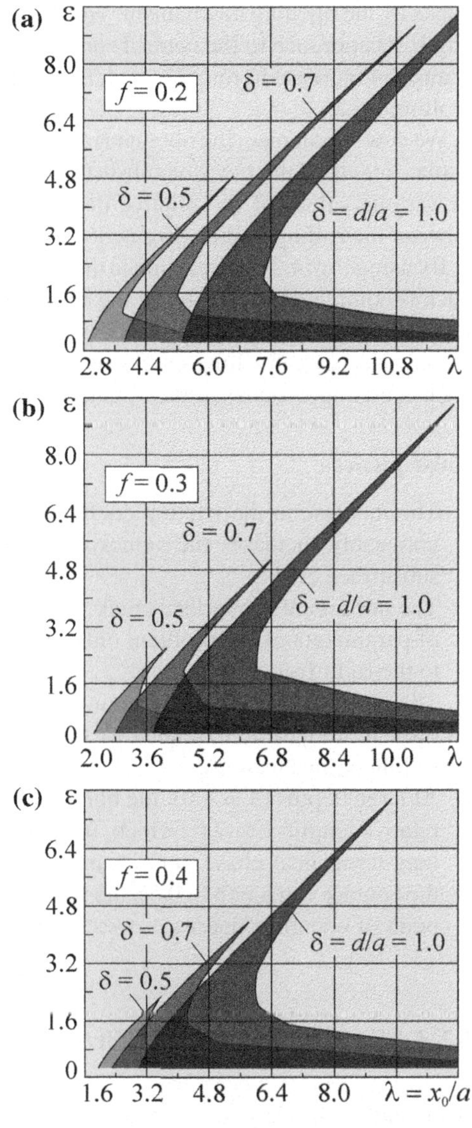

Fig. 4.36 Diagrams of full (along the entire crack length) normal opening of the horizontal subsurface crack

for $\delta = 1.0$, the transition from $f = 0.3$ to $f = 0.4$ decreases the maximum admissible length to $\varepsilon = 2.0$.

The described phenomenon of partial or complete crack closure (contact of its faces) is explained either by the approaching of the contact load (decrease in the relative depth δ) or by the increase in the tangential component of the load caused by the increase in the friction coefficient f. This yields the following important conclusion: the high-intensity external load (the distance to the counterbody is compensated by the level of loading) suppresses the development of long horizontal subsurface cracks by the opening mechanism. We also note that the increase in the relative crack length, its approach to the contact surface, and the increase in the friction coefficient f cause an increase in max $F_{I\theta}$ and, hence, the growth of the SSS at the crack tip in a rolling cycle.

We now supplement the obtained results by plotting the diagram of normal opening displacements of the faces of subsurface macrocrack in the coordinate system (λ–ε) for various values of the relative depth δ and the friction coefficient f in contact between the rolling bodies (Fig. 4.36).

By using Fig. 4.36, we can make the following prediction: A horizontal subsurface crack is completely open only if the counterbody is located to the right of its center. The ranges of λ in which the crack is completely open enlarge for short cracks and strongly decrease for long cracks as the friction coefficient f increases and as the crack center approaches to the boundary of the body.

Consequences

1. The increase in the friction coefficient f in contact between the rolling bodies noticeably increases the values of the SIF F_I at the right and left tips of the subsurface crack.
2. The horizontal subsurface crack is completely open within the analyzed ranges of parameters of the problem only in the case where the counterbody is located to the right from its center.
3. The range of locations of the counterbody for which the crack remains completely open is noticeably extended as the coefficient of sliding friction between the rolling bodies increases.
4. At large depths ($\delta = 1.0$), the horizontal subsurface crack is open even for large relative lengths $\varepsilon \approx 7.0$ (which, in turn, reveals the possibility of realization of a long-term shear phase of crack propagation). However, as the crack approaches the contact surface ($\delta = 0.5$), the relative crack length admissible from the viewpoint of openness decreases to $\varepsilon = 2.0$.

4.2.3 Subsurface Cracks Growth Paths

The results of investigations of subsurface cracks presented above reveal numerous crucial conditions, which should be satisfied in order to guarantee that the crack propagating by the mode I mechanism would form a spalling-type defect. Unlike the case of the edge crack, for this purpose, it is now insufficient to guarantee that the mixed-type SIF $K_{I\theta}$ is greater than K_{Ith} (the threshold of crack growth by the opening mechanism). Since the liquid does not penetrate into the subsurface crack and, hence, does not wedge out this crack, it is necessary to perform permanent control of the absence of contact of the crack faces under condition (4.13) in the process of growth of this crack. However, this is also insufficient to initiate spalling. In this case, it is necessary to guarantee that one branch of a subsurface crack starts its development in the direction of the rolling surface and reaches this surface prior to the attainment

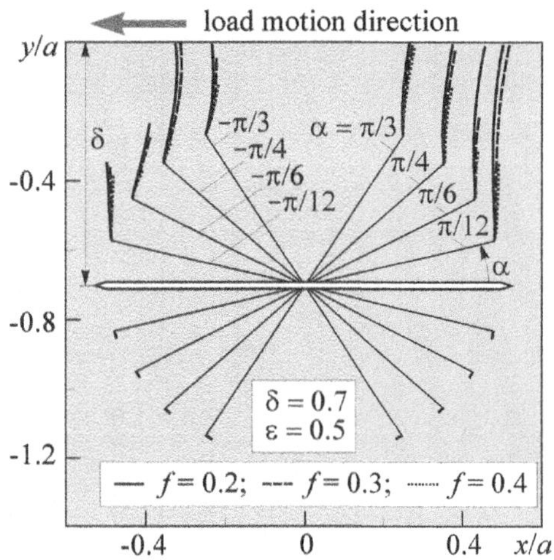

Fig. 4.37 Paths of the subsurface crack propagation depending on the angle α of its inclination for different values of the friction coefficient f (in 75KhGST (75XΓCT) steel)

the critical value of the SIF $K_{I\theta}$ by the opposite branch (if they do not propagate in the same direction). This reasoning enables us to conclude that the mechanism of spalling is quite complicated.

Development of an arbitrarily oriented crack by the mode I mechanism. In the first stage of investigation, we focus our attention on arbitrarily oriented subsurface cracks in 75KhGST rail steel [20, 22, 53]. We now assume that the relative depth of location of the center of subsurface crack is equal to $\delta = 0.7$ and set the relative length $\varepsilon = 0.5$. This enables us to freely orient a crack in the damaged body.

The analysis of the growth paths of arbitrarily oriented subsurface cracks depicted in Fig. 4.37 enables us to make the following conclusions: Within the ranges of the analyzed parameters characterizing the mutual friction of contacting bodies, as well as the length and location of the initial crack in the rolling body, the process of crack growth always occurs in the opposite directions from the left and right tips. The development of the crack in the direction of the edge of the half-plane occurs only from the tip nearest to this edge (in this case, this is the driving tip), and the branch of the crack path from the other tip slowly grows into the bulk of the material. The length of this branch of the crack path at the tip remote from the edge of the half-plane is always much smaller than for the tip close to the edge. This can be explained by the higher values of the SIF $K_{I\theta}$ at the tips closer to the surface, which, in turn, leads to higher crack growth rates. Hence, the branches of the path of the inclined subsurface macrocrack propagating into the bulk of the material are not seriously dangerous due to their small lengths and growth rates. The decrease in the friction coefficient between contacting rolling bodies causes an insignificant deviation of the paths toward the direction of motion of the contact load for the branches of the crack approaching the boundary of the half-plane. The behavior of the friction coefficient f does not exert any influence on the paths propagating from the crack tips remote from the half-plane boundary.

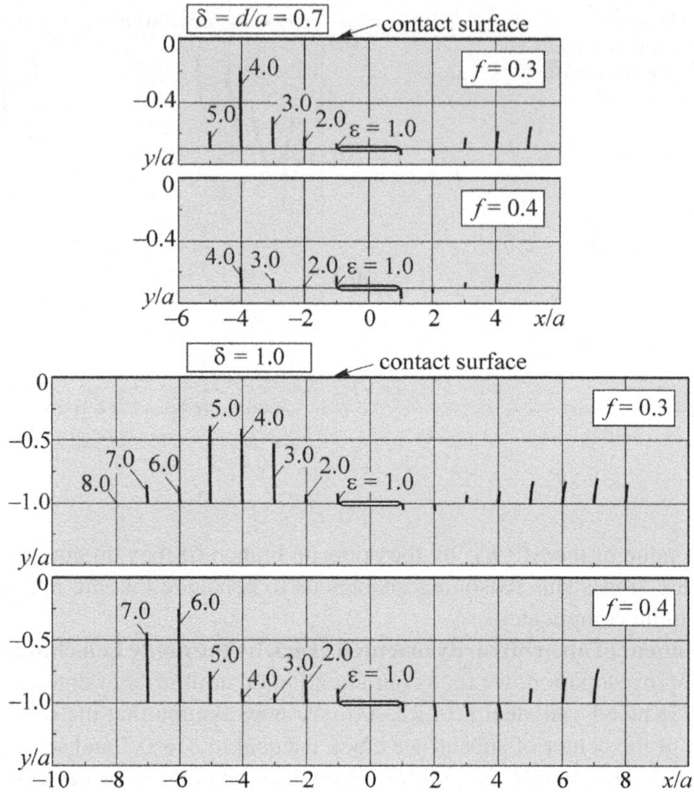

Fig. 4.38 Dependences of the paths of a horizontal crack propagation on the relative depth of its location $\delta = d/a$ and the relative length $\varepsilon = b/a$ (75KhGST (75ХГСТ) rail steel)

Note that the characteristics of crack growth resistance of 75KhGST steel (Table 4.4) are required for the construction of the paths by using relation (2.32).

Horizontal crack. Formation of spalling. We now study specific features of the development of subsurface cracks in the zone of rolling contact in RSB12 and 75KhGST (75ХГСТ) rail steels.

First, we consider a simple case in which the process of growth of a horizontal subsurface crack in the stage of shear is not studied and the crack length is postulated and guaranteed by the condition of normal opening displacement of the crack over the entire its length and the condition of crack growth by the normal opening mechanism, i.e., by the condition $\max K_{I\theta} > \Delta K_{Ith}$ at both tips. In this way, we construct the paths (trajectories) of growth of the subsurface horizontal crack by the mode I mechanism for 75KhGST rail steel depending on the depth of location of the crack and the friction coefficient between the rolling bodies (between the wheel and the rail under the conditions of dry friction). The chosen values of the relative crack length are shown in Fig. 4.38 (starting from $\varepsilon = 1.0$).

For the long horizontal subsurface cracks (Fig. 4.38) located at small relative depths ($\delta = 0.7$ and 1.0), we see that the variations of the relative crack length strongly affect the initial directions of its propagation from the tips and may lead to their sharp changes. For the relative crack length $\varepsilon < 2.0$, the crack develops into the bulk of the material from the right tip and to the boundary of the body from the left tip (a similar trend of growth of a horizontal crack was also discovered in [69]). For longer cracks with $\varepsilon \geq 2.0$, their development from both tips is realized toward the boundary of the body. In this case, all crack growth paths are almost perpendicular to the boundary. After an insignificant period of growth, the faces of the horizontal crack come in contact (at that time, the calculations are terminated) and the crack growth is decelerated or completely stopped (according to the σ_θ-criterion). Note that, for a small depth of location of the subsurface crack ($\delta = 0.2$), the crack faces are initially in contact prior to its growth for all analyzed values of the friction coefficient ($f = 0.2\ldots0.4$) and the relative crack length ($\varepsilon = 0.5\ldots10$). Hence, it is quite difficult for a horizontal crack to appear on the surface of the body by the mode I mechanism.

Further, we consider the case where the subsurface crack first propagates by the shear (mode II) mechanism and then, in the stage of normal opening (mode I) mechanism, it is described by the procedure proposed in Sect. 2.4. In the calculations performed in the stage of shear, we used the characteristics of cyclic crack growth resistance of RSB12 rail steel (Table 4.4).

The growth paths of the horizontal subsurface crack presented in Fig. 4.39 demonstrate the influence of the relative depth δ, the maximum contact load p_0, and the friction coefficient f on the size and shape of a defect.

As follows from Fig. 4.39a, the crack located at the relative depth $\delta = d/a = 0.1$ grows to the boundary of the body from both tips. The boundary is first approached by the right branch of the crack with formation (initiation) of a spalling-type defect. For an insignificant increase in the depth of location of the crack, the paths originating from the left tip reach the contact surface, whereas the paths originating from the right tip propagate into the bulk of the material. The same picture is observed as the maximum value of contact pressure p_0 increases up to 1700 MPa (Fig. 4.39b). In all other considered cases, including the entire analyzed range of the friction coefficient f (Fig. 4.39c), the crack penetrates deeper into the material and, therefore, creates a potentially hazardous defect, which can hardly be detected. Note that the onset of spontaneous crack propagation at the corresponding tip is marked by*.

It should also be mentioned that the ability of subsurface cracks in rolling bodies to form branches spontaneously propagating into the bulk of the material was indicated in the book [98]. The authors of this book treat these defects as inadmissible.

For a horizontal crack relatively close to the contact surface ($\delta = 0.3$ and $\delta = 0.1$), the specific features of its propagation are better illustrated by Fig. 4.40. In the cases where one (driving) tip moves to the rolling surface and serves as a cause of spontaneous fracture, which results in spalling, the other (driven) tip moves, as a rule, very slowly from or to the boundary. In Fig. 4.41, we present a schematic diagram of some trends of growth of the subsurface crack and the formation of contact fatigue damage, i.e., spalling.

Thus, the increase in the distance of between crack and the contact surface, the decrease in the maximum contact pressure, or the decrease in the friction coefficient always lead the increase in the maximum length of subsurface shear crack.

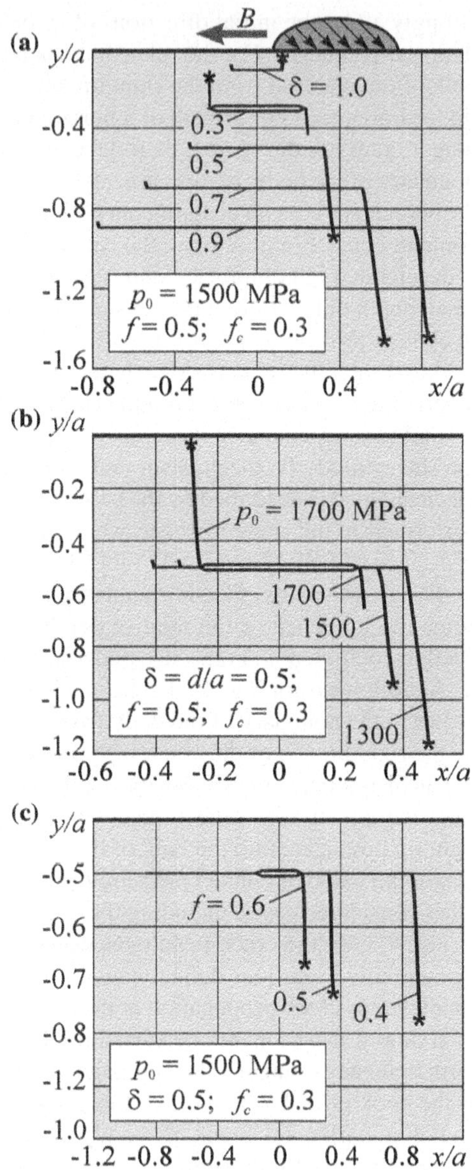

Fig. 4.39 Paths of a horizontal subsurface macrocrack propagation for different values of **a** the relative depth of location δ; **b** the maximum contact pressure p_0; **c** the friction coefficient f; (for RSB12 and 75KhGST (75ХГСТ) rail steels)

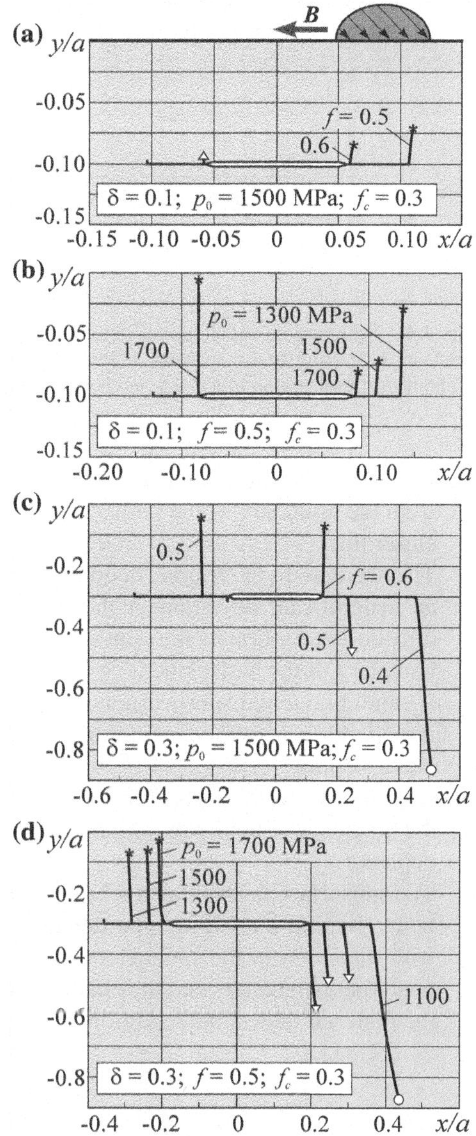

Fig. 4.40 Paths of the horizontal subsurface macrocrack propagation in the rail steels (RSB12, 75KhGST (75XΓCT)) depending on the friction coefficient f in contact between the wheel and the rail and on maximum value of contact pressure p_0 for the relative depth of the crack location $\delta = d/a = 0.1$ (**a**, **b**) and $\delta = 0.3$ (**c**, **d**); *—condition of spontaneous fracture $\Delta K_{I\theta} > \Delta K_{Ifc}$ is fulfilled; o—crack penetrated into material and stopped ($\Delta K_{I\theta}^{+} < \Delta K_{Ith}$); ∇—calculation is stopped because at the other crack tip spontaneous fracture started

Consequences

1. Within the ranges of the considered parameters characterizing the mutual sliding friction of rolling bodies, as well as the length and location of the crack in the body, the process of growth of a crack inclined at a certain angle always occurs in the opposite directions for left and right tips.

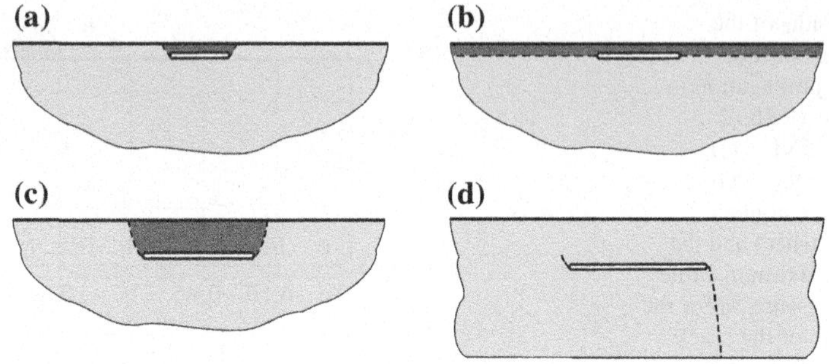

Fig. 4.41 Schemes of contact fatigue defects caused by the propagation of a subsurface horizontal crack in the rolling body: spalling of finite length (**a**), spalling of infinite length (**b**), gap (**c**), spontaneous propagation of a crack and possible fracture of the product (**d**)

2. The length of the growth path (trajectory) of an inclined crack from the tip remote from the boundary of the body (half-plane) is always much smaller than for the closer tip.
3. The decrease in the friction coefficient between contacting rolling bodies causes an insignificant deviation of the paths of an arbitrarily oriented crack in the direction of motion of the contact load for the branches that propagating to the boundary of the half-plane.
4. A small horizontal subsurface crack ($\varepsilon < 2.0$) creates a spalling-type defect only if its depth is insignificant and in the presence of high external loads (p_0) in its close vicinity.
5. Long horizontal subsurface cracks ($\varepsilon \geq 2.0$) always develop from both tips toward the boundary of the body, which, in turn, leads to the formation of spalling-type defects.
6. The subsurface horizontal cracks may develop either to the surface of the rolling body creating a spalling-type defect or a gap or into the bulk of the material with transition to the spontaneous development for some configurations of the operating parameters and the characteristics of cyclic crack growth resistance of the material under the conditions of transverse shear and normal opening [21, 24, 27, 41].

4.2.4 *Residual Contact Lifetime of Rolling Pair Elements by Spalling Formation Criterion*

We computed the residual lifetime by using the two-stage model [31, 40, 42] described in Chap. 2 and applied in the present section to subsurface cracks. Since the crack growth rate at both tips in each stage of crack propagation is not connected with the value of ΔK_{Ifc}, the critical crack growth is not simultaneously realized for the left and right branches. Therefore, in order to determine lifetime, we choose a number of rolling cycles guaranteeing the validity of the condition $\Delta K_{I\theta} > \Delta K_{Ifc}$ at one tip [21].

In what follows, we present the results of evaluation of the residual contact lifetime of rail steels damaged by subsurface cracks according to the criterion of spalling formation. Note that, in the stage of crack growth by the shear mechanism, it is necessary to know the final length of the shear crack regarded, as indicated in Chap. 2, as the initial length for the stage of normal opening. Therefore, we start the numerical analysis of the residual lifetime from the second (opening) stage of crack propagation.

Lifetime of the elements of wheel-rail couples. Influence of the operating parameters. The problem of durability of the railroad rails and car and locomotive wheel couples remains quite urgent. Therefore, in view of the abundance of subsurface cracks, their fracturing ability, and the complexity of their examination, we study and predict the residual lifetime according to the criterion of spalling formation in the nearsurface layer of rails made of RSB12 or 75KhGST (75ХГСТ) steel.

To find the crack growth rates in rail steels, we use the Yarema–Mykytyshyn approximation formula (2.14) [100] and the characteristics of cyclic crack growth resistance of rail steels presented in Table 4.4. We perform calculations to evaluate the influence both of the parameters of location of the subsurface crack and of the operating factors, such as the level of contact load (p_0), the friction force in contact between the wheel and the rail, and the level of friction between the crack faces on the residual lifetime.

Arbitrarily oriented subsurface crack. Consider an arbitrarily oriented subsurface crack located in the nearsurface contact zone for the ranges of operating parameters typical of the wheel-rail system.

The results of calculations (Table 4.12) show that:

- the increase in the friction coefficient f_c between the crack faces causes a significant increase both in the initial crack length $l_{0\tau}$ in the stage of shear and in the lifetime in this stage;
- the increase in the coefficient of sliding friction f between the wheel and the rail decreases the initial lengths $l_{0\tau}$ and $l_{0\sigma}$ and the lifetime in both stages of macrocrack propagation;
- the increase in the angle of inclination α of the crack causes an insignificant increase in its initial length $l_{0\tau}$ in the stage of growth by shear and a significant decrease in the initial length $l_{0\sigma}$ in the opening stage; the lifetime increases in both stages;
- the gradual deepening of the crack into the material (the increase in δ) causes an insignificant increase in the initial length in the shear stage and a noticeable increase in the initial length in the opening stage; the lifetime increases in both stages;
- a regularity similar to the previous regularity can be also obtained if the maximum value of contact pressure p_0 decreases.

Horizontal crack. For the subsurface horizontal cracks ($\alpha = 0$), in Tables 4.13 and 4.14, we present the numerical data the lifetimes in the stages of propagation of subsurface cracks by the mode II ($N_{g\tau}$) and mode I ($N_{g\sigma}$) mechanisms, as well as the corresponding cracks lengths.

Table 4.12 Dependences of the residual lifetime of the nearsurface layer in RSB12 and 75KhGST (75XΓCT) rail steels damaged by subsurface cracks initially inclined at the angle α on the main parameters of the problem p_0, f, f_c, and δ

	$l_{0\tau}$, mm	$l_{0\sigma}$, mm	$N_{g\tau}\cdot 10^{-5}$ cycles	$N_{g\sigma}\cdot 10^{-5}$ cycles	$N_g\cdot 10^{-5}$ cycles
f_c	$p_0 = 1500$ MPa; $\delta = d/a = 0.5; f = 0.3; \alpha = \pi/6$				
0.1	0.164	3.000	2.7453	2.5792	5.3245
0.3	0.407	3.000	4.4758	2.5792	7.0550
0.5	1.090	3.000	9.9840	2.5792	12.563
f	$p_0 = 1500$ MPa; $\delta = 0.5; f_c = 0.3; \alpha = \pi/6$				
0.1	0.498	6.871	6.4993	–	–
0.3	0.407	3.000	4.4758	2.5792	7.0550
0.5	0.330	1.012	4.1331	2.2960	6.4291
α	$p_0 = 1500$ MPa; $\delta = 0.5; f = 0.3; f_c = 0.3$				
$\pi/6$	0.407	3.000	4.4758	2.5792	7.0550
$\pi/4$	0.737	1.579	7.6741	5.4396	13.114
$\pi/3$	0.750	1.130	6.5885	17.037	23.626
δ	$p_0 = 1500$ MPa; $f_c = 0.3; f = 0.3; \alpha = \pi/6$				
0.3	0.356	1.384	3.8261	2.4051	6.2313
0.5	0.407	3.000	4.4758	2.5792	7.0550
0.7	0.434	4.911	5.1272	2.8381	7.9653
p_0, MPa	$\delta = 0.5; f = 0.3; f_c = 0.3; \alpha = \pi/6$				
1500	0.407	3.000	4.4758	2.5792	7.0550
1300	0.540	3.458	5.8064	2.5627	8.3691
1100	0.752	3.994	8.1741	2.449	10.623

The analysis of the results presented in Tables 4.12, 4.13, and 4.14 enables us to make the following conclusions: the variations of the friction coefficients f and f_c affect the initial length $l_{0\tau}$ and the residual lifetime as in the case of arbitrarily oriented cracks in the same direction but not so rapidly. The variations of the relative depth of location δ of the subsurface horizontal macrocrack reveal a clear minimum at $\delta = 0.5$ both for the initial length and lifetime in the stage of shear. This confirms the results presented in Figs. 4.39 and 4.40. As could be expected, the increase in the maximum pressure p_0 in the contact zone of rolling bodies decreases the lifetime in both stages of crack growth. In this case, in the shear stage, the lifetime decreases much faster than in the opening stage.

As already indicated, the horizontal subsurface crack forms spalling-type defects not in all cases. Therefore, the parameters responsible for the formation of this kind of defects are presented in Table 4.13 in boldface. It is worth noting that all paths shown in Figs. 4.39 and 4.40 were constructed in the course of finding the lifetime for this table.

Table 4.13 Dependences of the residual lifetime of the nearsurface layer of RSB12 and 75KhGST (75XГCT) rail steels damaged by the horizontal ($\alpha = 0$) subsurface macrocrack on the main parameters of the problem: p_0, f, and f_c; $\delta = d/a = 0.5$

	$l_{0\tau}$, mm	$l_{0\sigma}$, mm	$N_{g\tau}\cdot 10^{-5}$ cycles	$N_{g\sigma}\cdot 10^{-5}$ cycles	$N_g\cdot 10^{-5}$ cycles
f_c	$p_0 = 1500$ MPa; $\delta = 0.5; f = 0.5$				
0.1	0.142	2.243	2.5334	2.3488	4.8822
0.3	0.287	2.243	3.6848	2.3488	6.0336
0.5	0.573	2.243	5.4969	2.3488	7.8457
f	$p_0 = 1500$ MPa; $\delta = 0.5; f_c = 0.3$				
0.4	0.287	6.079	3.7679	3.3821	7.1500
0.5	0.287	2.243	3.6848	2.3488	6.0336
0.6	0.287	1.026	3.5052	2.1001	5.6053
δ	$p_0 = 1500$ MPa; $f_c = 0.3; f = 0.5$				
0.3	**0.312**	**1.627**	**7.0772**	**2.0356**	**9.1128**
0.5	0.287	2.243	3.6848	2.3488	6.0336
0.7	0.307	3.601	3.9089	2.9852	6.8941
0.9	0.353	5.335	4.2383	3.4221	7.6604
1.0	0.382	6.346	4.8328	4.3622	9.1950
p_0, MPa	$\delta = 0.5; f = 0.5; f_c = 0.3$				
1700	**0.224**	**1.804**	**3.2737**	**2.1230**	**5.3968**
1500	0.287	2.243	3.6848	2.3488	6.0336
1300	0.381	2.871	4.3427	2.6206	6.9633
1100	0.531	3.851	5.4911	2.9226	8.4136
900	0.795	5.594	8.1438	3.1863	11.330

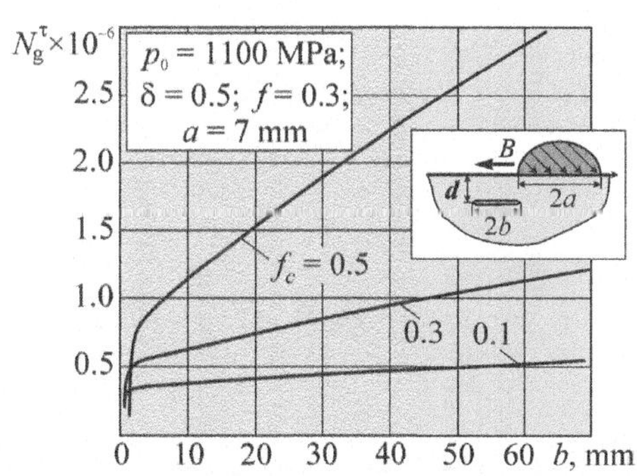

Fig. 4.42 Dependences of the lifetime $N_{g\tau}$ of RSB12 rail steel on the length of a subsurface horizontal shear crack $b = a/\varepsilon$

The data presented in Fig. 4.39 and 4.40 and in Tables 4.13 and 4.14 demonstrate that it is difficult for the subsurface horizontal crack to reach the surface, i.e., to satisfy the condition $\Delta K_{I\theta} > \Delta K_{Ith}$. This condition is realized under high contact pressures (p_0) for very high friction coefficients (f). Earlier, in [26], this was shown in the case of steels of the backup rolls of rolling mills. This probably explains the appearance of engirdling spalling in the rolls of rolling mills.

Table 4.14 Residual contact lifetime evaluated according to the criterion of spalling formation in the nearsurface zone of RSB12 and 75KhGST (75ХГСТ) rail steels; $\delta = 0.1; 0.3$

	$l_{0\tau}$, mm	$l_{0\sigma}$, mm	$l_{c\sigma}$, mm	$N_{g\tau} \cdot 10^{-5}$, cycles	$N_{g\sigma} \cdot 10^{-5}$, cycles	$N_g \cdot 10^{-5}$, cycles	Termination of calculations
p_0, MPa	$\delta = 0.1; f = 0.5; f_c = 0.3;$						
1100	1.62	Crack growth parallel to the boundary					$K_{II} > K_{IIth}$; $\max K_{I\theta} < K_{Ith}$
1300	1.22	1.84	2.35	5.5005	1.9984	7.4989	Spalling, $\max K_{I\theta}^{+} > K_{Ifc}$
1500	0.98	1.46	1.66	3.6545	2.0889	5.7434	Spalling, $\max K_{I\theta}^{+} > K_{Ifc}$
1700	0.76	1.16	2.01	2.7600	1.9644	4.7244	Spalling, $\max K_{I\theta}^{-} > K_{Ifc}$
f	$\delta = 0.1; p_0 = 1500$ MPa; $f_c = 0.3;$						
0.4	1.22	Crack growth parallel to the boundary					$K_{II} > K_{IIth}$; $\max K_{I\theta} < K_{Ith}$
0.5	0.98	1.46	1.66	3.6545	2.0889	5.7434	Spalling, $\max K_{I\theta}^{+} > K_{Ifc}$
0.6	0.78	0.79	0.93	1.3192	2.0555	3.3747	Spalling, $\max K_{I\theta}^{+} > K_{Ifc}$
p_0, MPa	$\delta = 0.3; f = 0.5; f_c = 0.3;$						
1100	1.18	5.00	9,10	5.9242	–	–	Stop of the crack, $K_I^{+} < 0$
1300	0.84	3.98	7.26	4.6881	2.4827	7.1708	Spalling, $\max K_{I\theta}^{-} > K_{Ifc}$
1500	0.62	3.25	6.61	4.0729	2.3696	6.4425	
1700	0.48	2.71	6.69	3.2990	2.0734	5.3724	
f	$\delta = 0.3; p_0 = 1500$ MPa; $f_c = 0.3;$						
0.4	0.62	6.36	10.40	4.2966	–	–	Stop of the crack, $K_I^{+} < 0$
0.5	0.62	3.25	6.61	4.0729	2.3696	6.4425	Spalling, $\max K_{I\theta}^{-} > K_{Ifc}$
0.6	0.56	2.13	4.23	3.2207	1.9750	5.1957	Spalling, $\max K_{I\theta}^{+} > K_{Ifc}$
f_c	$\delta = 0.3; p_0 = 1500$ MPa; $f = 0.5;$						

(continued)

Table 4.14 (continued)

	$l_{0\tau}$, mm	$l_{0\sigma}$, mm	$l_{c\sigma}$, mm	$N_{g\tau} \cdot 10^{-5}$, cycles	$N_{g\sigma} \cdot 10^{-5}$, cycles	$N_g \cdot 10^{-5}$, cycles	Termination of calculations
0.1	0.32	3.25	6.61	2.4988	2.3696	4.8684	Spalling, $\max K_{\mathrm{I}\theta}^{-} > K_{\mathrm{I}fc}$
0.3	0.62	3.25	6.61	4.0729	2.3696	6.4425	
0.5	1.12	3.25	6.61	5.3620	2.3696	7.7316	
0.7	1.78	3.25	6.61	8.4600	2.3696	10.8296	

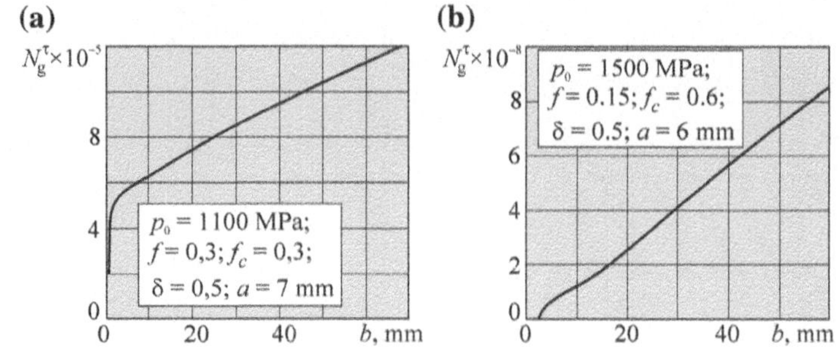

Fig. 4.43 Dependences of the lifetime for the shear stage of propagation of the horizontal crack on the admissible crack length for **a** RSB12 rail steel; **b** SKH roll steel

Contact lifetime according to the criterion of admissible length of subsurface crack. In some cases, neither the critical value of K_{IIfc} in the stage of shear, nor the initial value of K_{Ith} for the opening stage can be attained at any tip of the horizontal subsurface crack in the first (shear) stage of propagation if the crack grows along its continuation. Therefore, the crack length in the stage of shear may take large and hazardous sizes (Fig. 4.41). In this case, it is reasonable to introduce certain admissible (critical) lengths that should not be exceeded by the subsurface crack. In this connection, we computed the dependences of the residual lifetime on the admissible crack length in rail and roll steels in the process of propagation of subsurface cracks by the shear mechanism and presented these dependences in Figs. 4.42 and 4.43.

The characteristics of cyclic crack growth resistance for the SKH roll steel are taken from Table 4.11, while the crack growth rates in the material for both types of steel and the stages crack propagation were computed according to the Yarema–Mykytyshyn formula (2.14) [100].

As follows from the dependences $N_g^{\tau}(b)$ presented in Fig. 4.42, the lifetimes of the rail and roll steels differ by three orders of magnitude, which is explained, first of all, by different characteristics of crack growth resistance of these materials and, in particular, by the great difference between the orders of the characteristic v_0 (Table 4.4) and by different friction coefficients f_c of these steels. We also note that, with the exception of an insignificant initial section, the lifetimes of both types of steel in the shear stage of propagation of subsurface cracks obey a linear law. This enables us to use a linear extrapolation for the dependence $N_g^{\tau}(b)$ without cumbersome calculations.

Consequences

1. The increase in the friction coefficient f_c between the faces of subsurface horizontal crack causes a noticeable increase in the initial length $l_{0\tau}$ of the macrocrack and in the lifetime in the stage of crack propagation by the shear mechanism (Tables 4.12, 4.13 and 4.14).

2. The increase in the friction coefficient f decreases the initial lengths $l_{0\tau}$ and $l_{0\sigma}$ and lifetimes in both stages of propagation of the subsurface macrocrack.
3. The increase in the angle α of orientation of the crack causes an insignificant increase in the initial length $l_{0\tau}$ in the shear stage and a significant decrease in the initial length $l_{0\sigma}$ in the opening stage; the lifetimes of the contacting bodies increase in both stages.
4. The phenomenon of deepening of the crack into the bulk of the material (the increase in δ) causes an insignificant increase in its initial length in the shear stage and a significant increase in the initial length in the opening stage; the lifetimes of the contacting bodies increase in both stages.
5. The increase in the maximum pressure p_0 in the contact zone of rolling bodies causes a decrease in the lifetime in both stages of crack growth.
6. The lifetime in the shear stage of propagation of the subsurface horizontal crack increases as a linear function of crack length.

4.2.5 Conclusions

- The values of the SIF K_{II} at both tips of an inclined subsurface crack attain their maximum absolute values if the contact load moves over the crack center. For the horizontal crack, the maximum absolute values are attained when the load has already passed the crack (Figs. 4.27 and 4.28) in moving from the right to the left.
- The increase in the friction coefficient f in the contact zone between the rolling bodies strongly increases the SIF K_{I} at the right and left tips of the horizontal crack and extends the interval of locations of the counterbody for which the crack remains completely open in the contact cycle.
- The increase in the friction coefficient f between the rolling bodies shifts the depth of location of the horizontal crack, which is favorable for its propagation by shear, to the side of the rolling surface (Fig. 4.31) and decreases both the initial lengths $l_{0\tau}$ and $l_{0\sigma}$ and the residual lifetime in both stages of macrocrack propagation.
- The increase in the friction coefficient f_c between the crack faces causes a significant decrease in the SIF $|K_{\mathrm{II}}|$ and in the amplitude ΔK_{II} in a rolling cycle, as well as an increase in the initial length $l_{0\tau}$ of the macrocrack in the shear stage and a significant increase in lifetime in this stage (Tables 4.12, 4.13 and 4.14).
- Subsurface horizontal cracks may propagate either toward the rolling surface and form a spalling-type defect or a gap or into the bulk of the material with transition to spontaneous propagation for some configurations of the operating parameters and the characteristics of cyclic crack resistance of the material in the stages of transverse shear and normal opening.
- Within the ranges of the analyzed parameters, the horizontal cracks located at small distances from the rolling surface ($\delta = d/a = 0.1\ldots0.3$) may cause spalling for high levels of contact pressure (p_0); the cracks located deeper ($\delta = 0.3\ldots0.5$) exhibit a trend to long-term propagation under the surface, and the deep cracks ($\delta = 0.5\ldots1.0$ and more) under the conditions of high contact friction ($f = 0.5$)

and high levels of p_0 may form branches with their spontaneous propagation into the bulk of the material (Figs. 4.39 and 4.40).

4.3 Rolling Under Dry Friction Conditions or Contact Surface Wetting. Squat ("Dark Spot") Defects

4.3.1 Foreword Theses

The process of operation of railroad rails (under the conditions of rolling with sliding) is accompanied by the formation of defects of various types [60, 74, 102], including hazardous nearsurface squat defect (Fig. 4.44a) appearing on the running surface of the rail in the final stage of its development in the form of a "dark spot" (Fig. 4.44b). This defect can be encountered in backup rolls of the rolling mills [76, 77].

The squat ("dark spot") defect is characterized by the presence of two cracks growing from almost the same place on the working surface of the rail in the opposite directions. One of these cracks (principal) propagates at a small angle to the rolling surface in the direction of motion of the counterbody (train), whereas the other crack grows slower and at a smaller angle in the direction opposite to the motion of the counterbody. These cracks have branches growing upward to the surface and downward into the bulk of material. As a result, the rail surface settles at this site and becomes darker. Moreover, one of the lower branches of the principal crack may pass through the rail and cause its complete fracture. In [63, 64], it was also discovered that this hazardous type of the defects of rails is caused by the frequent alternation of dry and wet conditions in the course of motion of the transport. Sato, et al. [95] experimentally reproduced the alternation of the conditions of dry friction

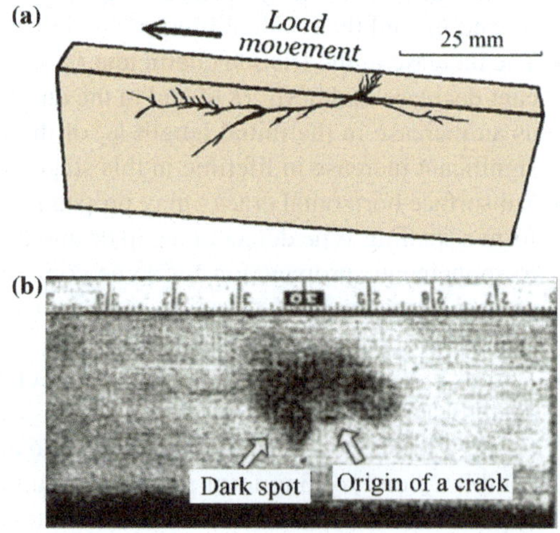

Fig. 4.44 Cross section of a squat ("dark spot") defect in the rail (**a**) [9]; defect of the "dark spot" on the runnimg surface of the rail (**b**) [61]

and wetting. Kaneta et al. [63, 64] obtained, for the first time under the laboratory conditions, defects of the "dark spot" quite similar to the observed on actual rails.

In the theoretical investigation of the process of propagation of surface cracks formed in the cause of rolling, the researchers mainly restricted themselves to the numerical analysis of the stress intensity factors and the initial directions of growth of rectilinear cracks. As model contact loads, the researchers mainly used Hertz contact forces, and the damaged rolling body was modeled by an elastic half-plane with edge crack.

Some works in this direction deal with the case of dry surface rolling contact studied with regard for the friction between the contacting crack faces. In some zones, the crack subjected to the action of moving external loads can be either closed (in this case, both slip and sticking of the crack faces are possible) or open. Problems of this kind were studied by Keer and Brayant [67], Bogdanski et al. [8], Fletcher and Beynon [49], Ringsberg and Bergkvist [93], and other researchers [4, 5, 101]. Bogdanski [7] additionally performed the numerical analysis of an important case of rectilinear cracks with one or two branches directed downward (the so-called "kinked" crack).

As shown in the previous section, the surface fracture of contacting bodies under the conditions of rolling is promoted by the presence of lubricants or water. Bower [10] studied three possible effects of the penetration of liquid into the crack: (a) the liquid lubricates the crack faces, (b) the liquid fills the crack and transfers the contact pressure upon its faces and (c) the liquid is partially captured by the crack and wedges it out. The effect of lubrication of the crack faces with liquid (a) important for the shear crack growth was also analyzed by the authors in [1, 49, 93]. The mechanisms of crack propagation caused by the pressure of liquid (b) and (c) were earlier considered by Keer and Brayant [67] and by Kaneta and Murakami [65]. The specific features of the initial stage of branching of the main branch of a squat-type defect in the backup rolls under the conditions of wetting were theoretically and experimentally studied by Frolish et al. [51].

The problems of cracks formed in rolling bodies were also studied in the three-dimensional statement and the plane surface semicircular and semielliptic cracks were analyzed in [66, 78].

In what follows, we perform the theoretical investigation of the formation of defects of the "dark spot"-type by using the two-dimensional model of fatigue fracture of rolling bodies formulated in Chap. 2 and also in the works [39, 87, 88].

It is known [6, 10, 67, 79] that, in rolling, the edge (surface) macrocrack propagates first at a small angle of $15^{\circ} \ldots 30^{\circ}$ to the surface in the direction of motion of the contact load by the shear mechanism. This stage of development of the crack is studied in Sect. 4.1.1. In what follows, we comprehensively study the stage of its subsequent development by the opening mechanism. The paths of crack propagation are studied in the case of unidirectional rolling depending on the friction coefficient between the rolling bodies and the initial length and orientation of the crack. The conditions of dry friction and wetting (lubrication) between the contacting bodies are modeled by large and small values of the coefficient of sliding friction, respectively. It is assumed that the liquid penetrating into the crack does not wedge out its faces and

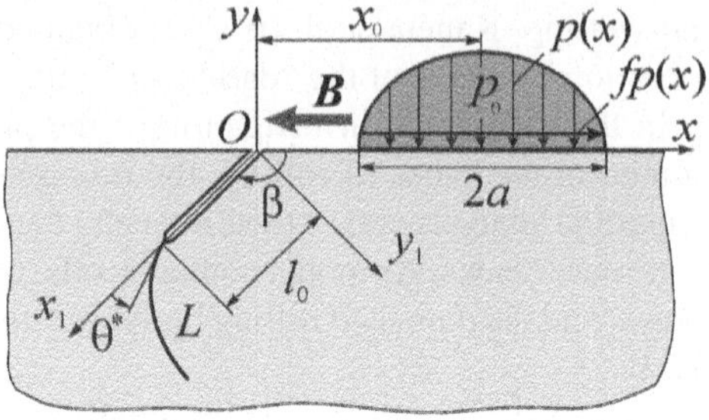

Fig. 4.45 Computational scheme of the model

note that the wedging influence of the liquid on the propagation of edge cracks and the formation of pitting is studied in Sect. 4.1.2 and earlier in [16, 38, 39, 43, 44]. In the numerical calculations, we use the criterion of generalized normal opening (σ_θ-criterion). The obtained numerical results are compared with the experimental data. We also consider the problem (very important for the engineering practice) of the mechanisms of formation of squats ("dark spots") in railroad rails.

In formulating the calculation problem of the squat defect, we assume that one (driven) body of a rolling couple is damaged by a surface macrocrack. A typical linear size of the contact section and the crack length are small as compared with the radius of curvature of the contact surface of the bodies. Therefore, instead of the actual body, in the two-dimensional statement of the problem, we consider the elastic half-plane weakened by an edge cut (crack). It is supposed that the rolling is cyclic and unidirectional. The contact influence of the second body (counterbody) upon the analyzed body is modeled by the repeated translational motion of the Hertz contact pressure $p(x)$ and tangential stresses $q(x)$ along the boundary of the half-plane in one direction (opposite to the Ox-axis, as shown in Fig. 4.45). This load is specified as follows:

$$p(x) = -\, p_o \sqrt{a^2 - (x - x_0)^2}/a\,, \quad q(x) = -fp(x), \quad |x - x_0| \le a, \quad y = 0, \tag{4.14}$$

where p_0 is the maximum pressure at the center of the contact section of length $2a$, f is the Amontons–Coulomb friction coefficient in contact. It is assumed that the initial crack is rectilinear, and the location of the contact section relative to the crack mouth is specified by the parameter x_0 (Fig. 4.45).

If, in the process of rolling, the bodies slip with sufficiently high friction ($f \ge 0.1$), this promotes the formation of zones of tension in the bodies, especially in the driven body and creates a possibility for the development of cracks in these bodies by the opening mechanism without contact of the crack faces. It is assumed that, under the conditions of wetting, the friction coefficient in the contacting rolling bodies decreases, and the amount of water penetrating into the crack is insufficient to cause its wedging it the process of rolling.

4.3.2 Shallow Edge Cracks Growth

Analysis of the SIF and the initial directions of crack growth. The stress intensity factors K_{I} and K_{II} for a curvilinear edge crack in the half-plane subjected to the action of a Hertz moving load (4.14) on its boundary and the values of these factors in each stage of construction of the crack growth path are determined as a result of the solution of the first basic problem of the theory of elasticity (for the half-plane with curvilinear edge cut; see Fig. 4.45) with boundary conditions imposed on the boundary of the half-plane:

$$\begin{aligned} &\sigma_y(x) - i\tau_{xy}(x) = -p(x) - iq(x), \qquad |x - x_0| \le a; \\ &\sigma_y(x) - i\tau_{xy}(x) = 0, \qquad |x - x_0| > a, \qquad y = 0, \end{aligned} \tag{4.15}$$

and on the crack faces

$$N^{\pm}(t) + iT^{\pm}(t) = 0, \qquad t \in L, \tag{4.16}$$

where L is the contour of a curvilinear crack.

This problem is reduced to the solution of the singular integral Eq. (3.155) with regard for condition (3.160) of boundedness of stresses on the edges of the crack mouth by the method of mechanical quadratures.

The calculations were performed for different friction coefficients between the contacting bodies $f = 0.1...0.4$ corresponding, in particular, to different weather conditions of operation of the engineering wheel-rail couples. As already indicated, the shear macrocrack is first formed in the driven body of the rolling couple. This crack propagates almost rectilinearly at a small angle of 15°…30° to the surface. Therefore, in our calculations, we mainly use the crack orientation angles $\bar{\beta} = 180°–\beta = \beta = 150°...165°$. The relative length of the initial rectilinear crack was chosen within the range $\varepsilon = l_0/a = 0.2...2.0$.

The analysis of the maximum values of the mixed-type SIF $K_{\mathrm{I}\theta}^*$ in a rolling cycle and the angles of deviation of the crack θ^* was carried out for the values of the friction coefficient $f = 0.1$ (wetting) and $f = 0.3$ (dry friction).

For a given direction of motion of the contact load (see Fig. 4.45), we show that the crack is open if the load is located mainly to the right of the crack mouth ($\lambda \ge 1$, $\lambda = x_0/a$). For the other λ, we observe a partial or full contact of the crack faces. Note that our results obtained for the SIFs K_{I} and K_{II} within the ranges of λ, where the crack is completely open, agree with the results of other authors and, in particular, with the data obtained by Ringsberg and Bergkvist [93] and Benuzzi et al. [5].

In the numerical analyses, the obtained maximum values of $K_{\mathrm{I}\theta}^*$ $\left(F_{\mathrm{I}\theta}^* = F_{\mathrm{I}\theta}(\lambda^*) = \max F_{\mathrm{I}\theta}(\lambda)\right)$ in a cycle are compared with the threshold value $K_{\mathrm{I}th} = 4.3$ MPa·(m)$^{1/2}$ for 75KhGST rail steel [94]. Thus, for the maximum pressure $p_0 = 1800$ MPa and a half length of the contact section $a = 7$ mm, its normalized value $F_{th} = K_{th}/(p_0\sqrt{\pi a}) = 0.016$. (Hence, for these values of the parameters a and

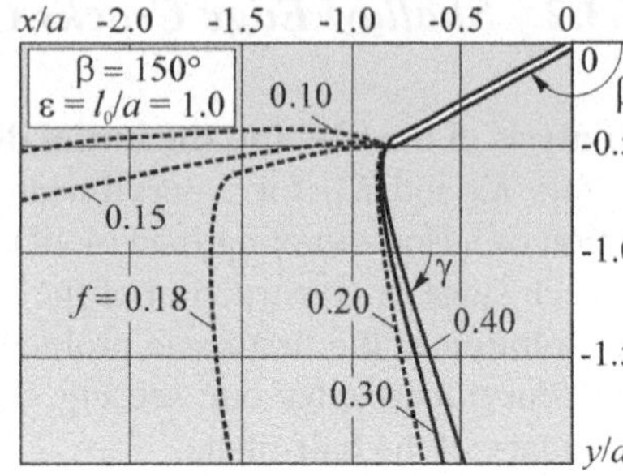

Fig. 4.46 Crack growth paths as functions of the friction coefficient f

p_0, the crack propagates for $F_{I\theta}^* > 0.016$). The results of calculations are presented in Table 4.15.

These results (Table 4.15) indicate that, for a low friction coefficient ($f = 0.1$), the values of $F_{I\theta}^* = K_{I\theta}^*/(p_0\sqrt{\pi a})$ are low: they vary within the range $F_{I\theta}^* = 0.008\ldots0.020$ (see Table 4.15a). Hence, only for some combinations of the parameters ε and β, the values of $F_{I\theta}^*$ exceed the threshold value and the crack begins to grow. For large angles of orientation of the crack $\beta = 165°$ and $160°$ (i.e., for very small angles of inclination to the boundary), the crack is immobile. At the same time, for $\beta = 155°$ and $150°$, the crack begins to propagate toward the surface (see Table 4.15b). Moreover, long ($\varepsilon \geq 1.0$) cracks start to grow for the angle $\beta = 155°$. At the same time, the process of propagation of quite short ($0.5 \leq \varepsilon \leq 1.0$) cracks starts for $\beta = 150°$.

For the friction coefficient $f = 0.3$, the values of $F_{I\theta}^*$ are much higher: they change within the range $F_{I\theta}^* = 0.030\ldots0.071$ (see Table 4.15c), and the crack of any length always starts to grow into the bulk of the material (see Table 4.15d).

We also studied the cracks making acute angles with the boundary ($\beta = 10°$; $20°$), which corresponds to the right branch of squat-type defects. For these cracks and $f = 0.1$, the values of the SIF are very low ($F_{I\theta}^* = 0.007\ldots0.011$) and, hence, for the chosen threshold value of F_{Ith}, the cracks do not start. For $f = 0.3$, all cracks begin to propagate; in this case, $F_{I\theta}^* = 0.042\ldots0.066$. In general, the cracks oriented at acute angles and starting, exhibit a trend to grow into the bulk of the material independently of their orientation and length.

Crack paths (trajectories). Here and in what follows, the crack growth paths are drawn in the figures by solid lines. The cracks propagate if $F_{I\theta}^*$ exceeds F_{Ith} for the chosen material and loading mode. However, the cracks may also propagate along the other presented paths. This is realized for a different choice of the operating parameters K_{Ith}, p_0, a and β. These cases are illustrated by the dotted lines in the figures.

It is shown that the crack growth paths strongly depend on the friction coefficient f between the rolling bodies: the higher the coefficient f, the stronger the trend to propagation into the material is exhibited by the crack. On the contrary, the lower the friction coefficient, the smaller the slope of the propagating crack to the contact surface (Fig. 4.46).

Table 4.15 Initial values of the normalized SIF $F^*_{I\theta}$ and the angles of propagation θ^* of the initial edge rectilinear crack for $f = 0.1$ (a, b) and $f = 0.3$ (c, d)

$f = 0.1$							$f = 0.3$					
β	160°	155°	150°	145°	20°	10°	160°	155°	150°	145°	20°	10°
ε	**(a)** $\max F_{I\theta}(\lambda) = F_{I\theta}(\lambda^*) = F^*_{I\theta}$						**(c)** $F^*_{I\theta}$					
0.2	0.0106	0.0107	0.0107	0.0110	0.0113	0.0107	0.0541	0.0602	0.0648	0.0679	0.0644	0.0504
0.3	0.0095	0.0094	0.0112	0.0153	0.0102	0.0105	0.0541	0.0590	0.0622	0.0638	0.0658	0.0546
0.5	0.0078	0.0112	0.0164	0.0166	0.0085	0.0096	0.0510	0.0539	0.0549	0.0554	0.0637	0.0581
0.7	0.0084	0.0139	0.0202	0.0131	0.0084	0.0087	0.0473	0.0487	0.0489	0.0489	0.0599	0.0585
1.0	0.0100	0.0167	0.0171	0.0106	0.0081	0.0105	0.0424	0.0424	0.0422	0.0420	0.0543	0.0571
1.2	0.0108	0.0180	0.0153	0.0098	0.0078	0.0110	0.0396	0.0392	0.0389	0.0387	0.0511	0.0556
1.4	0.0114	0.0189	0.0133	0.0090	0.0074	0.0114	0.0372	0.0366	0.0363	0.0360	0.0482	0.0540
1.6	0.0119	0.0195	0.0123	0.0084	0.0071	0.0113	0.0351	0.0345	0.0341	0.0338	0.0457	0.0524
1.8	0.0122	0.0200	0.0121	0.0079	0.0068	0.0112	0.0332	0.0326	0.0323	0.0319	0.0435	0.0508
2.0	0.0124	0.0203	0.0114	0.0075	0.0066	0.0111	0.0316	0.0310	0.0307	0.0304	0.0416	0.0492
	(b) θ^* [rad]						**(d)** θ^* [rad]					
0.2	0.787	0.753	0.726	*−0.948*	−0.754	−0.836	0.877	0.840	0.801	0.755	−0.797	−0.874
0.3	0.769	0.741	*−0.973*	*−1.152*	−0.733	−0.815	0.869	0.828	0.783	0.743	−0.771	−0.854
0.5	0.749	*−0.992*	*−1.165*	*−1.228*	−0.705	−0.780	0.855	0.802	0.765	0.743	−0.732	−0.826
0.7	*−0.829*	*−1.104*	*−1.218*	*−1.228*	*1.212*	−0.753	0.841	0.774	0.761	0.741	−0.703	−0.808
1.0	*−0.997*	*−1.164*	*−1.229*	*−1.228*	*1.220*	*1.215*	0.820	0.773	0.760	0.737	−0.674	−0.775
1.2	*−1.045*	*−1.183*	*−1.231*	*−1.230*	*1.227*	*1.219*	0.807	0.772	0.759	0.737	−0.661	−0.766
1.4	*−1.075*	*−1.194*	*−1.230*	*−1.230*	*1.226*	*1.230*	0.795	0.770	0.758	0.737	−0.653	−0.750
1.6	*−1.095*	*−1.201*	*−1.230*	*−1.230*	*1.229*	*1.226*	0.783	0.770	0.758	0.737	−0.646	−0.735
1.8	*−1.110*	*−1.206*	*−1.231*	*−1.230*	*1.225*	*1.227*	0.772	0.770	0.758	0.737	−0.643	−0.727
2.0	*−1.120*	*−1.210*	*−1.231*	*−1.231*	*1.230*	*1.230*	0.777	0.770	0.757	0.737	−0.636	−0.719

Comment: The values of θ^* for which the condition $\theta^* + \bar{\beta} < 0$ is satisfied are made italic (the crack starts to grow toward the surface)

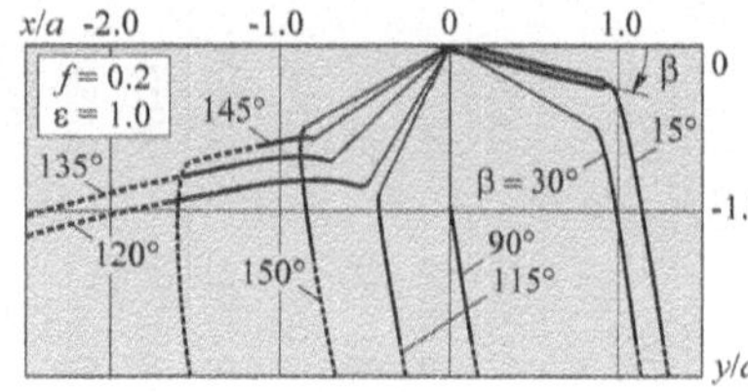

Fig. 4.47 Crack growth trajectories as functions of the angle β

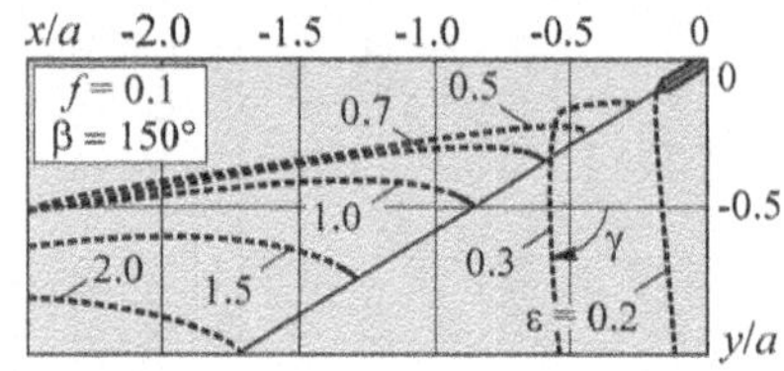

Fig. 4.48 Crack growth paths as functions of the initial relative crack length $\varepsilon = l_0/a$

For the low friction coefficient ($f = 0.1$), the behavior of the crack strongly depends on its orientation and length.

In Fig. 4.47, we show the crack growth paths for cracks with relative length $\varepsilon = 1.0$ depending on the angle of orientation β (for acute and obtuse angles). For the sake of comparison, we also present the paths for the angles β that are not typical of squat-type defects. There exists a range of angles $\beta = 130\ldots165°$ in which the crack begins to propagate toward the surface of the contacting body and then turns and grows along the surface with a certain deepening into the material. The cracks oriented at other angles immediately turn into the bulk of the material and then grow almost rectilinearly.

In Fig. 4.48, we illustrate the character of crack propagation with an angle of orientation $\beta = 150°$ depending on its initial length. It follows from this figure that fairly long initial cracks starting to propagate toward the boundary of the half-plane ($\varepsilon \geq 0.5$) do not reach the rolling surface and, hence, do not induce crumbling. They grow along the surface and somewhat deepen into the material. In this case, after the start of these cracks, the values of $F_{I\theta}^*$ along the paths decrease very rapidly and become lower than the threshold value F_{Ith}.

The cracks with smaller lengths (the curve $\varepsilon = 0.2$ in Fig. 4.48) turn downward immediately after the start and then grow rectilinearly.

Note that, in all similar cases, the final angle of crack propagation is practically independent of the length of the initial crack and its orientation (see Figs. 4.46, 4.47 and 4.48). The crack after an insignificant initial period rectilinearly grows into the bulk of the material at the same acute angle γ to boundary for given f to the side of the domain of action of tangential forces. The higher the level of contact friction, the stronger the boundary of the body "attracts" the crack (Fig. 4.46). For the friction coefficient $f = 0.1$ and $\varepsilon = 0.2$, this angle is equal to $\gamma \approx 85°$ (Fig. 4.48).

A crack with the relative length $\varepsilon = 0.3$ grows for a certain time parallel to the surface and then turns into the bulk of the material. As a result, its trajectory takes the form of a "knee". This behavior is intermediate between the cracks growing horizontally and the cracks growing into the material (the curves for $f = 0.18$, $\beta = 145°$, and $\varepsilon = 0.3$ are presented in Figs. 4.46, 4.47 and 4.48, respectively). We also note that this crack growth paths are in agreement with the actual defects formed in the rails: the presented photo taken from the work by Smith [97] (Fig. 4.49) demonstrates a crack in the form of a "knee".

For the high friction coefficient ($f = 0.3$), the crack begins to propagate into the material after its start and then grows rectilinearly at the angle $\gamma \approx 77°$ (see Figs. 4.46, 4.48 and 4.50). As already indicated, this angle is independent of the length and orientation of the crack.

It is worth noting that, for cracks growing into the bulk of the material, the values of $F^*_{I\theta}$ along the paths of their propagation decrease slower than for the cracks propagating parallel to the surface.

The analysis of crack paths performed under various conditions of rolling shows that the predictions of the behavior of propagating cracks solely on the basis of the initial values of θ^* made by some researchers can be erroneous. This becomes especially clear is we consider an example of the trajectory of propagation of a crack in the form of a "knee". Indeed, in order to determine the specific features of crack propagation under the conditions of a complex stressed state typical of contact rolling, it is necessary to construct the crack growth path.

It should be emphasized that the major part of calculations were performed for the number of complex linear algebraic equations (the discrete analog of the integral Eq. (3.155)) $N = 70$. In the construction of the path, the steps h (main step along the path) and $\Delta\lambda$ (auxiliary step corresponding to the motion of the counterbody) are

Fig. 4.49 Development of fatigue cracks in a rail [97]

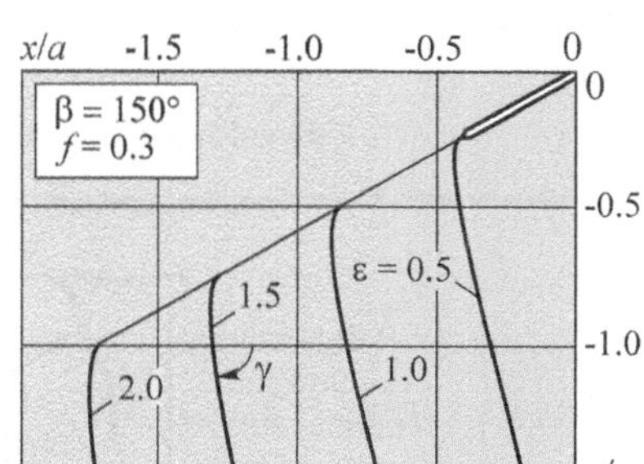

Fig. 4.50 Crack growth paths as functions of the initial relative crack length

set equal to $0.03a$ and $0.02a$, respectively. In this case, the evaluation of the initial values of $F^*_{I\theta}$ is performed, as a rule, with three or four correct significant digits.

Mechanisms of formation of squat-type defects. Kaneta et al. [63] experimentally reproduced the formation of a squat-type ("dark spot"-type) defect with the help of a two-roll machine for rolling and sliding contact. The schematic picture of the formation of defect of this type was proposed by Murakami et al. in [79] (see Fig. 4.51). It was established that the nuclei of macrocracks are initiated on the surface of the driven element of the rolling couple. First, they develop by the shear mechanism at a small angle to the rolling surface not only in the direction of motion of the load but also in the opposite direction.

The first (shear) stage of crack propagation in the case of rolling contact under the conditions of wetting has already been discussed above and in the work [32]. The accumulated numerical results show that the crack rectilinearly grows along its continuation by the shear mechanism. This is explained by the fact that the crack growth is realized if it is closed in the vicinity of the tip. Then, obviously, $K_I = 0$ and $\theta^* = 0$ according to the criterion of maximum tangential stresses.

The development of the left branch of the defect, i.e., in the direction of motion of the counterbody, unlike the behavior of the right branch, is accompanied, according to the experimental data from [63] and the scheme proposed by Murakami et al. [79], by the intense branching toward the surface of the body and from the surface. After a certain time, the crack separates into two main cracks. One of these cracks propagates parallel to the surface, while the second crack turns into the depth of the rail.

The results obtained in the present work enable us to make the following conclusions: The edge gently sloping crack formed in the nearsurface layer of the rolling body turns in the process of propagation from the surface into the depth of the material if the friction between the rail and the wheel is sufficiently high, i.e., under the conditions of dry friction (dry weather). This crack propagates into the bulk of the material by the mechanism of opening displacement and gradually transforms into a transverse (through) crack. This gives an answer (from the theoretical viewpoint) to the question posed in the work [79] about the causes of growth of the main branch of the crack from the boundary (see Fig. 4.51) and confirms the experimental data [92, 97] on the direction of its final growth (angle γ). Under the conditions of wetting, if the friction coefficient in contact between the rolling bodies becomes small, then

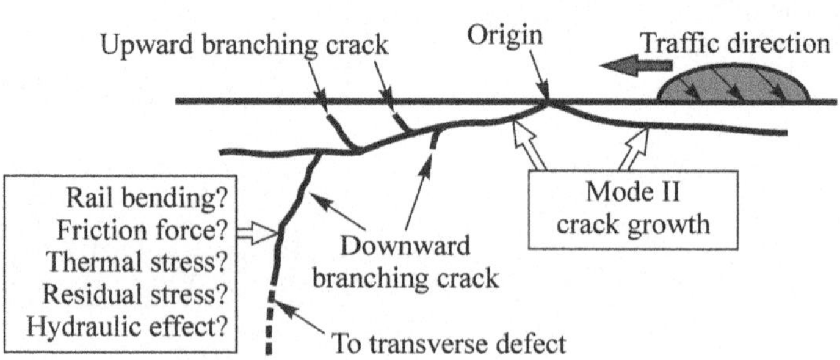

Fig. 4.51 Formation of a squat-type defect [79]

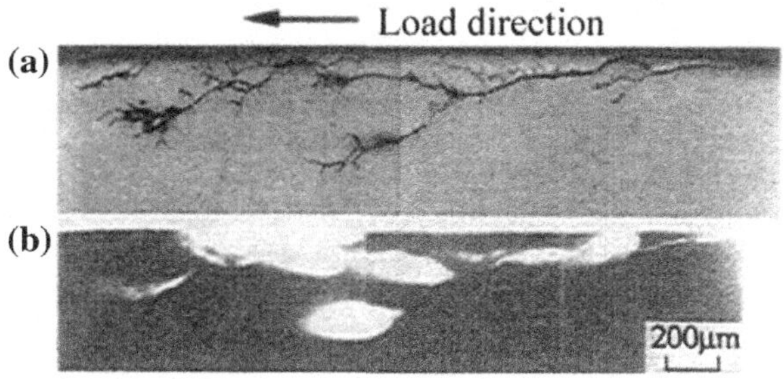

Fig. 4.52 Penetration of water into cracks under the conditions of rolling contact: **a** image made in the daylight; **b** fluorescent image; [57]

the crack does not propagate into the bulk of the material but may propagate by the opening mechanism in the direction parallel to the surface.

The right branch of a squat develops almost horizontally (parallel to the surface) (Fig. 4.51) and, most likely, does not propagate by the opening mechanism. In the analyzed cases, this crack may propagate by this mechanism under the conditions of dry friction with deepening into the material (Fig. 4.47) or, possibly, toward the surface under the conditions of wetting (Figs. 4.55, 4.56 and 4.57).

The nature (mechanism) of formation of the right branch of the squat can be, in our opinion, explained by using the results obtained in Sect. 4.1.1 ("Edge crack in the contact zone under the conditions of transverse shear"). In Fig. 4.13a, we present the plots of the function $\Delta F_{\text{II}}(\beta)$ for $f = 0$; $\varepsilon = 0.3$; $f_c = 0 \ldots 0.5$. For the indicated configuration of values of the parameters (in particular, $f = 0$), we revealed the third maximum of the function $\Delta F_{\text{II}}(\beta)$ realized for the angles of inclination of the crack $\beta = \beta_3^* \approx 15° \ldots 20°$, i.e., for the angles of propagation of the right branch of the squat. The condition of practical absence of sliding friction between the wheel and the rail ($f = 0$) is realized in the case of rolling of car wheels in the mode of motion of "by inertia". This, possibly, an answer to the question: What is the mechanism of formation of the right branch of a squat? Hence, we assume that the right branch of the squat develops by the mechanism of transverse shear under the car wheels.

In Fig. 4.51, we also present schematic diagrams of small branches originating from the left branch of the defect and directed toward the rolling surface. In the general case, they are potentially dangerous because, under certain conditions, they may lead to crumbling of fragments of the material from the rail surface. It is possible to assume that the mechanism of their growth is connected with the negative influence of liquid penetrating into the squat. In Fig. 4.52, it is seen how water penetrates into the cracks and reaches their tips.

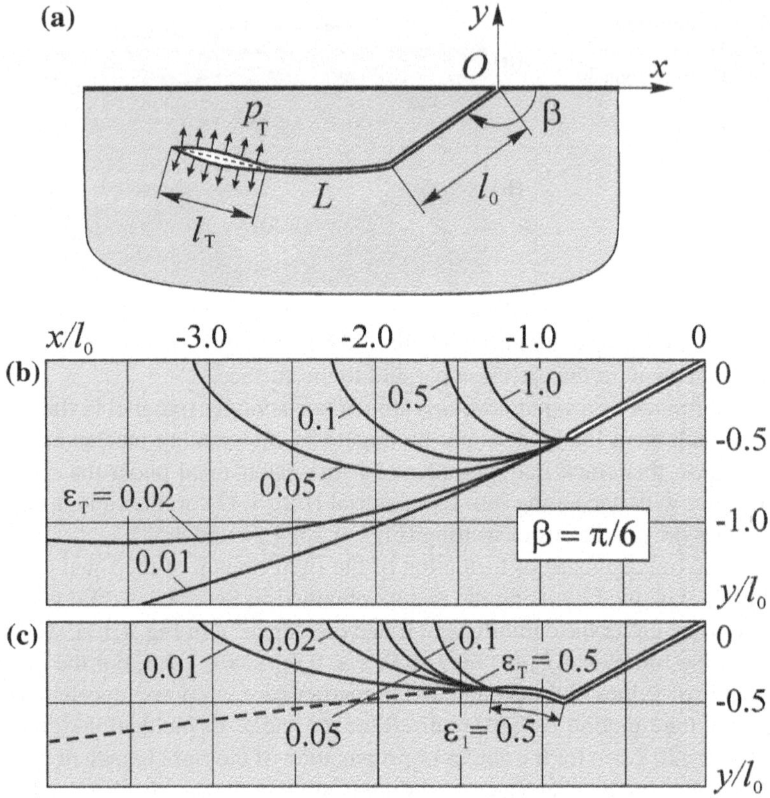

Fig. 4.53 General scheme (**a**); trajectories of propagation of an initially rectilinear crack as functions of the relative length $\varepsilon_T = l_T/l_0$ of a section of action of the liquid (**b**) [28]; trajectories of propagation of an initially curvilinear crack ($l_1 = \varepsilon_1 l_0 = 0.5 l_0$) as functions of the relative length $\varepsilon_T = l_T/l_0$ of the zone (**c**) [26]; the dashed line is the trajectory of propagation of an initially rectilinear crack under the action of contact load (see Fig. 4.48)

We now study the problem of possible influence of the wedging action of liquid on the subsequent propagation of the edge crack. Consider a simplified version [28] in which a section of length l_T (starting from the crack tip) is subjected to action of uniformly distributed pressure with intensity p_T (Fig. 4.53a). In this way, we model the pressure of liquid near the crack tip in the absence of contact load on the boundary.

As already indicated, for small f, the values of $F_{I\theta}^*$ are low. However, the results of numerical calculations indicate that, even in the presence of the lowest pressure upon the crack faces, the picture changes [28]. Indeed, for an initially rectilinear crack and for an initially curvilinear crack formed under the conditions of rolling with wetting (Fig. 4.48), the values of $F_{I\theta}^*$ are much higher than in the absence of the influence of wedging liquid. Moreover, they almost do not decrease as the crack grows even for relatively low pressures. The crack propagates to the boundary more rapidly or more gently depending on the degree of filling of the crack by the liquid (Fig. 4.53b, c).

Thus, it is theoretically shown that the pressure of a liquid at the crack tip plays the decisive role in the formation of pits on the rolling surface.

It is clear that, for a more exact analysis of the development of cracks under the conditions of wetting in the case of contact rolling interaction, it is necessary to consider the combined of action of the load applied on the boundary and the pressure of liquid wedging the crack faces out.

Generalizing our arguments concerning the formation of defects of the squat-type (see Fig. 4.51), we can make the following assertions:

- in the initial stage, the left and right branches of the squat propagate by the shear mechanism (mode II crack growth in Fig. 4.51); it is assumed that the right branch of the defect is initiated and propagates by the mode II mechanism in the process of rolling of the car wheels "by inertia".
- the process of propagation of the left branch is accompanied by the formation of many branches directed both to the rolling surface and into the bulk of the material. The growth of branches to the surface (upward branching cracks) is explained by the wedging action of the pressure of liquid penetrating into the defect. In the course of time, these cracks may appear on the surface and form pits. In combination with the squat, they manifest themselves on the rail surface as a "dark spot";
- the lower branches (downward branching cracks) of the left principal branch propagate under the conditions of dry weather by the opening mechanism. In the course of time, one of these cracks may transform into a through crack (transverse defect) and cause a catastrophe.

4.3.3 Branched Crack

This problem is studied with an aim to investigate the nature of squat-type defects more completely (see Subsection 4.3.2). The damaged driven rolling body is modeled by the elastic isotropic half-plane damaged by an edge two-link crack (Fig. 4.54). For simplicity, we assume that the initial separate links of the branched crack are rectilinear. We also accept that the coefficient of sliding friction f between the rolling bodies varies within a fairly broad range covering the conditions of dry friction and boundary lubrication between the contacting bodies. It is also assumed that, under the conditions of lubrication, solely the level of friction in contact between the rolling bodies decreases, while the medium does not penetrate into the cracks in amounts sufficient to wedge them out. We also believe that the some branches of the crack grow by the opening mechanism. A defect is modeled by a system of two cracks with common mouth on the boundary of the rolling body. Assume that one of these cracks is inclined at an acute ($\beta_1 = 5\ldots30°$) angle to the boundary of the half-plane (in the direction of contact tangential forces) and that the second crack is inclined at an obtuse angle ($\beta_2 = 150°$) (Fig. 4.54). We study the cracks growth paths (trajectories) for both cracks and three typical values of friction coefficient in the wheel-rail system: $f = 0.10, 0.15$ (wet weather), and $f = 0.30$ (dry weather). We assume that the initial length of the second crack (l_2) is constant ($\varepsilon_2 = l_2/a = 1.0$) and vary the length of the first crack ($\varepsilon_1 = l_1/a = 0.2\ldots1.5$). The range of locations of the contact load

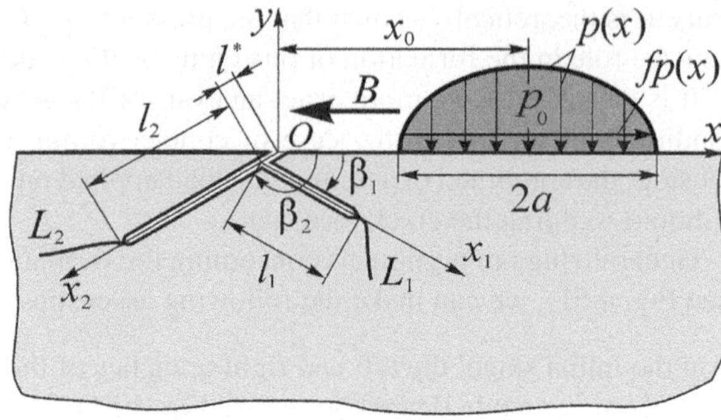

Fig. 4.54 General scheme of the problem; *B* is the direction of motion of the counterbody

guaranteeing the propagation of cracks by the opening mechanism is specified by the condition $\lambda = x_0/a \geq 1.0$ (with the exception of some cases). We study the case where the first crack is inclined at a very acute angle to the surface and its tip is close to the rolling surface, i.e., the case where a neck of length $\delta^* = l^*/a$ is formed between the boundary of the half-plane and the mouth of the first crack (Fig. 4.54). This neck is regarded as a separate crack with the same orientation as the second crack.

The numerical analyses of the paths of propagation of these systems of cracks (Fig. 4.54) are carried out within the framework of the model formulated in Chap. 2. In this case, we assume that the onset of crack propagation occurs when the SIF ($K_{I\theta}$) takes a certain value $K^*_{I\theta}$ for a given load. In each rolling cycle, in the process of motion of the zone of contact loading along the boundary of the half-plane (for variable λ), the parameter $K_{I\theta}$ changes for the tips of each crack taking its maximum values $K^{*(i)}_{I\theta}$ $(i = 1, 2)$ for the corresponding values of $\lambda = \lambda^{*(i)}$ and $\theta = \theta^{*(i)}$. We now assume that each of two cracks grows only for $\lambda = \lambda^{*(i)}$ in the direction determined by the angle $\theta^{*(i)}$ according to the σ_θ-criterion. If the contact of faces is realized in one crack, then we suppose that its presence insignificantly affects the stressed state of the body. The presence of contact of the crack faces is controlled by using relation (3.218) (or (4.13)).

To determine the SIF in each stage of construction of the paths, we solved the system of SIE (3.141) with the additional conditions (3.147) for the first basic problem of the theory of elasticity for a half-plane with curvilinear edge cracks (every time for new cracks lengths). The system of SIE is solved numerically by the method of mechanical quadratures (see Chap. 3 and [91, 96]). Note that we, in fact, solve a system of 3 N linear complex algebraic equations, where $N = 120\ldots130$; the step of motion of the contact load $\Delta\lambda = 0.02\ldots0.05$ and the crack increment $h = 0.02\ldots0.1$. The crack growth rates are computed by the Paris formula (2.12). The calculations were performed for the characteristics of cyclic crack growth resistance of 75KhGST rail steel with a structure of lamellar pearlite (ΔK_{th} = 4.3 MPa·(m)$^{1/2}$, ΔK^* = 19.7 MPa·(m)$^{1/2}$, n = 3.48) (Table 4.4). The crack lengths increase proportionally to the rates of propagation of their tips in a given material for the given stress-strain state (relation (2.32)).

The results of calculations are presented in Figs. 4.55, 4.56, 4.57, 4.58, 4.59, 4.60. Comparing the paths of the first and second cracks with each other and with the case of a single crack (Figs. 4.46, 4.47 and 4.48), we conclude that, in the case where a squat is modeled by two cracks with common mouth on the boundary, the presence of the first crack weakly affects the shape of the path of the second crack for all values of the parameters (f, ε_1, β_1), whereas the path of the second crack insignificantly differs from the case of a single crack. As for the first crack, it produces a branch in the direction of the boundary for small friction coefficients ($f = 0.10$) in the major part of cases (ε_1, β_1). The initially short cracks grow into the depth of the half-plane, whereas the cracks with larger lengths develop to the side of the boundary (Figs. 4.55 and 4.56). For the first crack, the trend of its propagation to the boundary becomes more pronounced as the angle of its inclination increases. In the plots, we show the branches directed into the depth of the half-plane only for some values of ε_1. The cracks that are shorter than the cracks presented in the figures always develop into the bulk of the material. Note that, in Figs. 4.55, 4.56, 4.57, 4.58, 4.59, 4.60, the growth paths of the first and second cracks obtained for a given value of ε_1 are depicted by lines of the same type.

A trend to the propagation of the first crack into the bulk of the material becomes more pronounced as the friction coefficient increases ($f = 0.15$) (Figs. 4.57 and 4.58). For $f = 0.3$, both cracks propagate in this direction for all considered parameters (Fig. 4.59). In this case, if the contact friction between the bodies decreases or increases, the cracks grow slower or faster, respectively. In other words, the corresponding $K^*_{\mathrm{I}\theta}$ along the crack paths increase with the level of friction.

In the case where the first crack starts to develop to the side of the boundary of half-plane, it almost never leads to crumbling of the rolling surface and stops due to the contact of crack faces. In other words, its subsequent propagation by the opening mechanism is impossible. Quite frequently, prior to the stop, the first crack turns to the direction of action of tangential contact forces (see, e.g., Fig. 4.55 for $\varepsilon_1 = 1.2$;

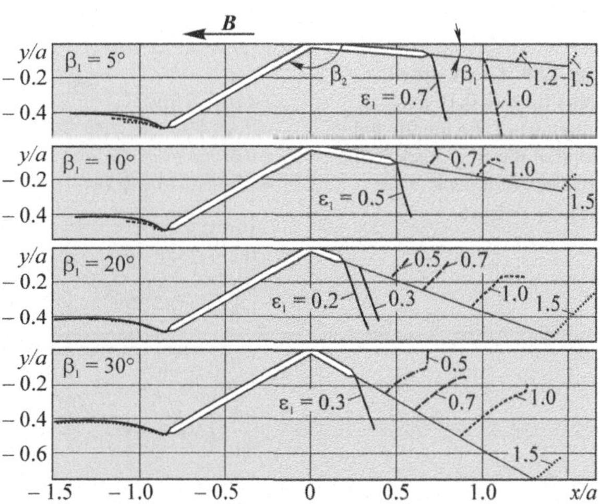

Fig. 4.55 Crack growth trajectories for $f = 0.1$ as functions of the length and slope of the first crack in the case where the cracks have the common mouth ($\delta^* = 0$); $\beta_2 = 5\pi/6$

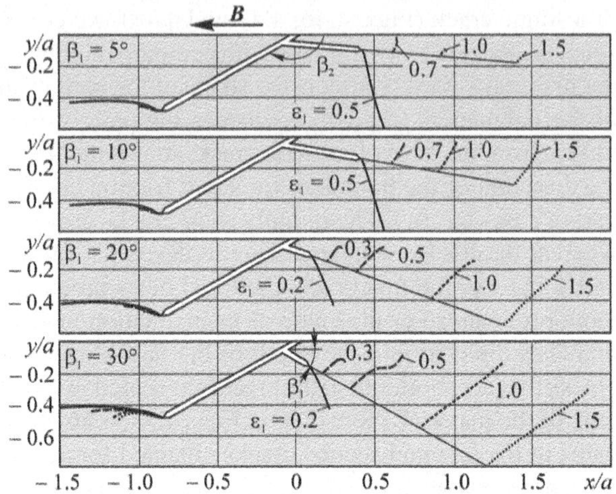

Fig. 4.56 Crack growth paths for $f = 0.1$ as functions of the length and slope of the first crack in the presence of a neck ($\delta^* = 0.1$); $\beta_2 = 5\pi/6$

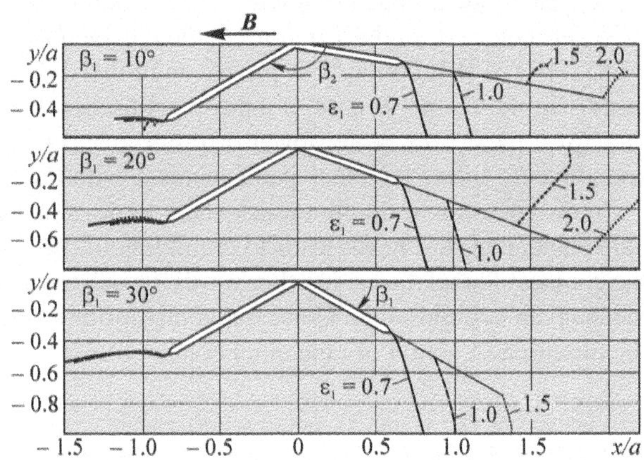

Fig. 4.57 Crack growth paths for $f = 0.15$ as functions of the length and slope of the first crack in the case where the cracks have the common mouth ($\delta^* = 0$); $\beta_2 = 5\pi/6$

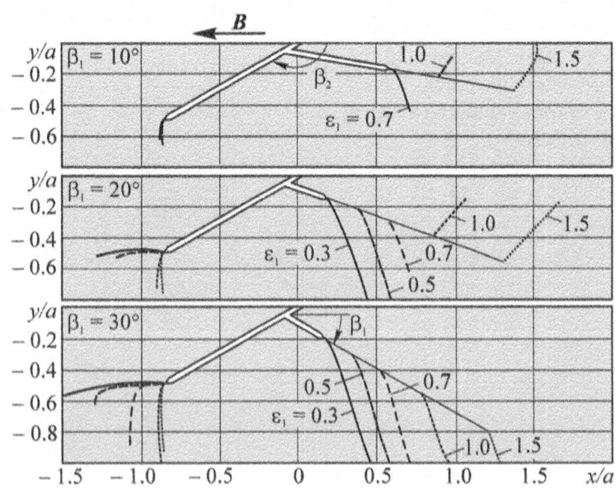

Fig. 4.58 Crack growth paths for $f = 0.15$ as functions of the length and slope of the first crack in the presence of a neck ($\delta^* = 0.1$); $\beta_2 = 5\pi/6$

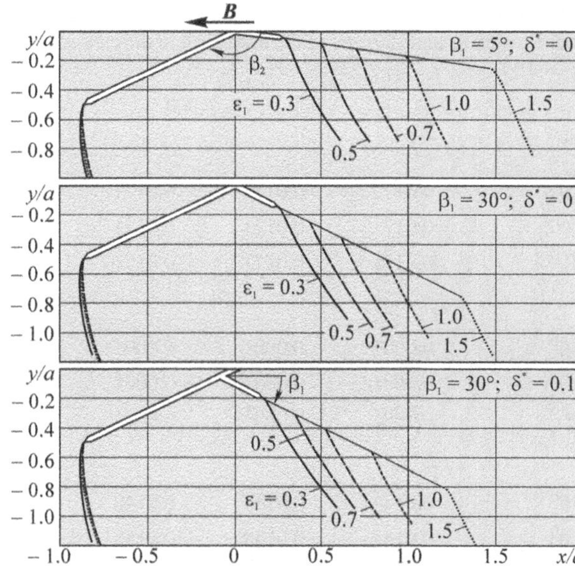

Fig. 4.59 Crack growth paths for $f = 0.3$ as functions of the length and slope of the first crack; $\beta_2 = 5\pi/6$

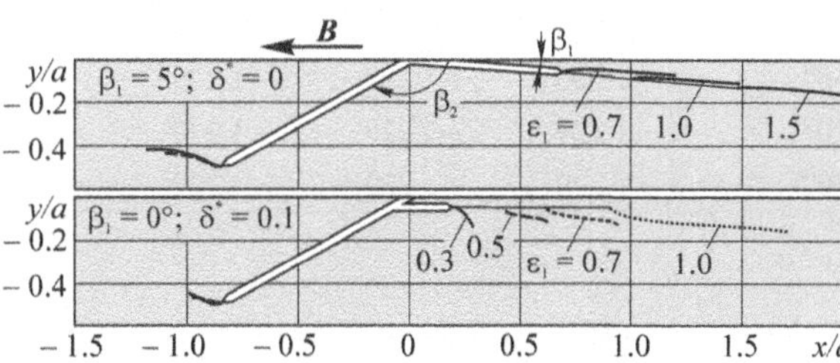

Fig. 4.60 Crack growth paths for $f = 0.1$ in the case where the load is located to the left of the crack mouth ($\lambda < 0$ depending on the length of the first crack; $\beta_2 = 5\pi/6$

1.5). In the presence of a neck, the trend to development of the first crack to the side of the boundary becomes more pronounced (see, Figs. 4.55, 4.56, 4.57 and 4.58).

Under the assumptions made in this section, the crack growth paths are formed by the values of $K_{I\theta}$ maximum in a loading cycle, i.e., when the contact load moves only within the range $\lambda = x_0/a \geq 1.0$. In this case, we observe no branches of the first crack parallel to the edge of the body. The additional investigation of the shapes of trajectories for a certain collection of values of the angles β_2 of initial inclination

Table 4.16 Values of the initial SIF and the angles θ^* of the cracks deviation for $f = 0.1$ depending on the length ε_1 and slope β_1 of the first crack in the case where the cracks have the common mouth ($\delta^* = 0$)

ε_1	0.2	0.5	1.0	0.2	0.5	1.0
β_1	$F_{I\theta}^{*(2)}$			$F_{I\theta}^{*(1)}$		
10^0	0.0172	0.0171	0.0171	0.0092	0.0085	0.0108
20^0	0.0171	0.0171	0.0171	0.0097	0.0081	0.0082
30^0	0.0171	0.0171	0.0171	0.0091	0.0067	0.0056
	$F_{I}^{*(2)}$			$F_{I}^{*(1)}$		
10^0	0.0001	0.0001	0.0001	0.0063	0.0064	0.0001
20^0	0.0001	0.0001	0.0001	0.0076	0.0007	0.0002
30^0	0.0001	0.0001	0.0001	0.0078	0.0001	0.0001
	$F_{II}^{*(2)}$			$F_{II}^{*(1)}$		
10^0	0.0148	0.0148	0.0148	0.0043	0.0035	−0.0093
20^0	0.0148	0.0148	0.0148	0.0037	−0.0067	−0.0070
30^0	0.0148	0.0148	0.0148	0.0028	−0.0057	−0.0048
	$\theta^{*(2)}$			$\theta^{*(1)}$		
10^0	−1.230	−1.229	−1.229	−0.810	−0.742	1.229
20^0	−1.229	−1.229	−1.229	−0.697	1.197	1.222
30^0	−1.229	−1.229	−1.229	−0.573	1.224	1.225

of the second crack and the values of its relative length ε_2 also do not reveal any branches of this kind. Most likely, the horizontal branch of the first crack grows by the shear mechanism. However, some results indicating that the first crack can develop horizontally by the opening mechanism were obtained for the case where the load is located to the left of the crack mouth ($-2.0 \geq \lambda \geq -1.0$), $K_{I\theta} < K_{I\theta}^*$, and the first crack is inclined at a very acute angle (Fig. 4.60). However, we note that the convergence of values of the SIF along the path is not satisfactory in this case.

In Tables 4.16, 4.17, 4.18, 4.19 and 4.20, we present the values of the SIFs and the angles of deviation of the crack tip θ^* for both cracks corresponding to the values of the parameters ε and β in Figs. 4.55, 4.56, 4.57, 4.58, 4.59 and 4.60. These data indicate that if both cracks have the common mouth, then the values of SIF and θ^* for the second crack are practically independent of the length and slope of the first crack. In the presence of a neck, this influence can be detected but it is insignificant: For the second crack, the value of $K_{I\theta}^*$ somewhat increases with the length and slope of the first crack. Note that, in all cases where the crack starts toward the boundary, the values of K_I^* at its tip are very small ($K_I^* = 0\ldots0.0007$), whereas the values of $|K_{II}^*|$ in these cases are much larger than in the case where the crack propagates into

Table 4.17 Values of the initial SIF and the angles θ^* of the crack deviation for $f = 0.1$ as functions of the length ε_1 and slope β_1 of the first crack in the presence of a neck ($\delta^* = 0.1$); $\beta_2 = 5\pi/6$

ε_1	0.2	0.5	1.0	0.2	0.5	1.0
β_1	$F^{*(2)}_{\mathrm{I}\theta}$			$F^{*(1)}_{\mathrm{I}\theta}$		
10^0	0.0172	0.0167	0.0084	0.0083	0.0060	0.0111
20^0	0.0171	0.0158	0.0088	0.0080	0.0087	0.0079
30^0	0.0171	0.0167	0.0094	0.0075	0.0066	0.0054
	$F^{*(2)}_{\mathrm{I}}$			$F^{*(1)}_{\mathrm{I}}$		
10^0	0.0000	0.0002	0.0023	0.0065	0.0042	0.0006
20^0	0.0001	0.0003	0.0022	0.0067	0.0001	0.0002
30^0	0.0001	0.0002	0.0020	0.0068	0.0001	0.0000
	$F^{*(2)}_{\mathrm{II}}$			$F^{*(1)}_{\mathrm{II}}$		
10^0	0.0149	0.0144	0.0061	0.0032	0.0028	−0.0094
20^0	0.0148	0.0135	0.0065	0.0026	−0.0075	−0.0067
30^0	0.0148	0.0144	0.0071	0.0019	−0.0057	−0.0047
	$\theta^{*(2)}$			$\theta^{*(1)}$		
10^0	−1.230	−1.227	−1.108	−0.694	−0.803	1.211
20^0	−1.230	−1.222	−1.121	−0.602	1.225	1.220
30^0	−1.230	−1.227	−1.139	−0.489	1.227	1.229

the bulk of the material. This is especially well visible in the presence of the neck (Table 4.17). This also enables us to believe that, in the cases where $\theta^* \approx 1.2$ for the start of the first crack (toward the boundary), this crack may grow by the shear mechanism.

Summarizing, we can say that the branches directed toward the boundary of the rolling body on the right branch of a squat may be formed by the opening mechanism in the case of wet weather ($f \approx 0.10$). In this case, the left branch grows along the boundary in the direction of motion of the counterbody. Under the conditions of dry weather ($f \approx 0.3$), the branches originating from both branches of the squat develop into the bulk of the material.

If the contact load moves solely within the range $\lambda = x_0/a \geq 1.0$, then no branches of the first crack parallel to the boundary of the body are revealed within the framework of formulated model. These branches appear only if the load is located to the left of the crack mouth ($-2.0 \geq \lambda \geq -1.0$) and the first crack is inclined at a very acute angle.

Table 4.18 Values of the initial SIF and the angles θ^* of the crack deviation for $f = 0.15$ as functions of the length ε_1 and slope β_1 of the first crack in the case where the cracks have the common mouth ($\delta^* = 0$); $\beta_2 = 5\pi/6$

ε_1	0.2	0.5	1.0	0.2	0.5	1.0
β_1	$F_{I\theta}^{*(2)}$			$F_{I\theta}^{*(1)}$		
10^0	0.0191	0.0191	0.0191	0.0158	0.0169	0.0152
20^0	0.0191	0.0191	0.0191	0.0179	0.0159	0.0130
30^0	0.0191	0.0191	0.0191	0.0178	0.0147	0.0118
	$F_{I}^{*(2)}$			$F_{I}^{*(1)}$		
10^0	0.0076	0.0076	0.0076	0.0106	0.0123	0.0115
20^0	0.0076	0.0076	0.0076	0.0136	0.0128	0.0106
30^0	0.0076	0.0076	0.0076	0.0150	0.0127	0.0101
	$F_{II}^{*(2)}$			$F_{II}^{*(1)}$		
10^0	0.0127	0.0127	0.0127	0.0076	0.0073	0.0062
20^0	0.0127	0.0127	0.0127	0.0073	0.0058	0.0046
30^0	0.0127	0.0127	0.0127	0.0059	0.0044	0.0037
	$\theta^{*(2)}$			$\theta^{*(1)}$		
10^0	−1.041	−1.041	−1.041	−0.826	−0.762	−0.732
20^0	−1.041	−1.041	−1.041	−0.726	−0.665	−0.651
30^0	−1.041	−1.041	−1.041	−0.607	−0.559	−0.581

Table 4.19 Values of the initial SIF and the angles θ^* of the crack deviation for $f = 0.15$ as functions of the length ε_1 and slope β_1 of the first crack in the presence of a neck ($\delta^* = 0.1$); $\beta_2 = 5\pi/6$

ε_1	0.2	0.5	1.0	0.2	0.5	1.0
β_1	$F_{I\theta}^{*(2)}$			$F_{I\theta}^{*(1)}$		
10^0	0.0145	0.0112	0.0109	0.0157	0.0148	0.0152
20^0	0.0164	0.0108	0.0110	0.0155	0.0135	0.0131
30^0	0.0174	0.0112	0.0116	0.0149	0.0125	0.0114
	$F_{I}^{*(2)}$			$F_{I}^{*(1)}$		
10^0	0.0080	0.0087	0.0085	0.0119	0.0115	0.0002
20^0	0.0078	0.0084	0.0094	0.0128	0.0111	0.0004
30^0	0.0077	0.0084	0.0092	0.0131	0.0109	0.0097
	$F_{II}^{*(2)}$			$F_{II}^{*(1)}$		
10^0	0.0082	−0.0044	−0.0043	0.0065	0.0059	−0.0131
20^0	0.0100	−0.0042	−0.0035	0.0054	0.0047	−0.0111
30^0	0.0110	0.0046	−0.0043	0.0042	0.0038	0.0035

(continued)

Table 4.19 (continued)

ε_1	0.2	0.5	1.0	0.2	0.5	1.0
β_1	$F_{I\theta}^{*(2)}$			$F_{I\theta}^{*(1)}$		
	$\theta^{*(2)}$			$\theta^{*(1)}$		
10^0	−0.936	0.711	0.706	−0.736	−0.710	1.227
20^0	−0.989	0.698	0.586	−0.635	−0.636	1.219
30^0	−1.012	−0.737	0.680	−0.528	−0.562	−0.580

Table 4.20 Values of the initial SIF and the angles θ^* of the crack deviation for $f = 0.3$ as functions of the length ε_1 and slope β_1; $\beta_2 = 5\pi/6$

ε_1	0.2	0.5	1.0	0.2	0.5	1.0
β_1	$F_{I\theta}^{*(2)}$			$F_{I\theta}^{*(1)}$		
10^0	0.0420	0.0422	0.0422	0.0454	0.0532	0.0534
30^0	0.0415	0.0422	0.0422	0.0566	0.0521	0.0474
	$F_{I}^{*(2)}$			$F_{I}^{*(1)}$		
10^0	0.0308	0.0310	0.0310	0.0299	0.0377	0.0393
30^0	0.0323	0.0309	0.0309	0.0478	0.0454	0.0412
	$F_{II}^{*(2)}$			$F_{II}^{*(1)}$		
10^0	−0.0180	−0.0181	−0.0181	0.0221	0.0240	0.0229
30^0	−0.0163	−0.0181	−0.0181	0.0184	0.0154	0.0141
	$\theta^{*(2)}$			$\theta^{*(1)}$		
10^0	0.759	0.760	0.760	−0.839	−0.788	−0.757
30^0	0.704	0.760	0.760	−0.602	−0.555	−0.556

4.3.4 Conclusions

On the basis of the comparison of the obtained results with the available engineering and experimental data, we determine [28, 29, 37, 87] some causes and regularities of the formation of squat ("dark spot") defects typical of rails and backup-rolls of the rolling mills. Thus, it is shown that:

- squat defects (specific branched systems of nearsurface cracks; see Fig. 4.51) are induced, first of all, by frequent changes in the friction coefficient between the contacting rolling bodies (e.g., the alternation of dry and rainy weather, violations in the technology of lubrication, etc.);
- in the initial stage, the left and right branches of the squat propagate by the shear mechanism (in Fig. 4.51; this if mode II crack growth); we assume that the right branch of the defect is formed and propagates by the mechanism of transverse shear in the process of rolling of car wheels in the mode of motion "by inertia";

- the process of propagation of the left branch is accompanied by the formation of numerous new branches directed both to the rolling surface and into the bulk of the material; the process of growth of branches to the surface (upward branching cracks) is connected with the wedging action of the pressure of liquid penetrating into the defect; in the course of time, these cracks can appear on the surface, and create pits; together with the squat, they are visible on the surfaces of rails in the form of "dark spots";
- lower branches (downward branching cracks) originating from the left main branch propagate under the conditions of dry weather by the opening mechanism; in the course of time, one of these cracks (transverse defect) may turn into a through crack, and finally cause a catastrophe.

4.4 Evolution of Edge Parallel Cracks System. Checks

Systems of parallel edge (surface) cracks are often formed on the races of elements of the rolling couples [76]. In particular, this is typical of the wheel-rail systems (Figs. 4.61 and 4.62) [76, 102]. Therefore, for the evaluation of the contact strength and lifetime of the rolling bodies, it is important to determine the stress-strain state or the stress intensity factors at the tips of the analyzed systems of cracks. In order to model contact interaction in the process of rolling, it is customary to use the model proposed by Keer and Brayant [67]. In this model, a body damaged by cracks is simulated (in the two-dimensional case) by an elastic half-plane with cuts, while the action of counterbody is modeled by the Hertz (elliptic) pressure translationally and unidirectionally moving along the edge of the half-plane. There are numerous investigations of the SIF of a single edge inclined crack available from the literature (see, e.g., the surveys [25, 48]) both for the case where the crack is open and for the case where the crack faces are in contact in the course of motion of the contact load along the boundary of the body (sliding with friction or sticking). However, there are several works [10, 25, 71] in which the kinetics of contact of the crack faces is analyzed and the algorithms used for its evaluation are described. For a system of

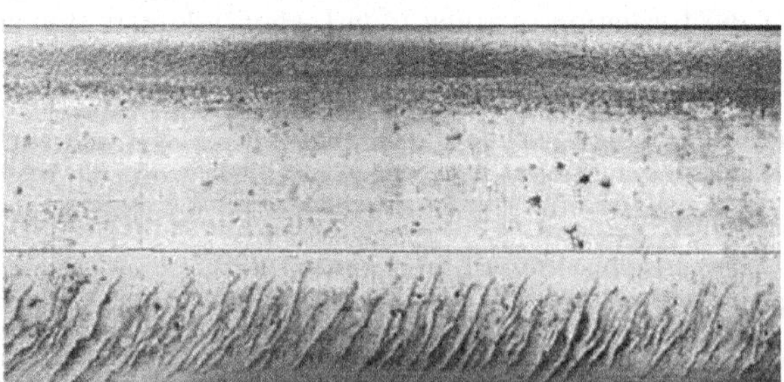

Fig. 4.61 Checks on the surface of rail head are groups of surface cracks on the head corner with a distances 0.5…0.7 mm between them [102]

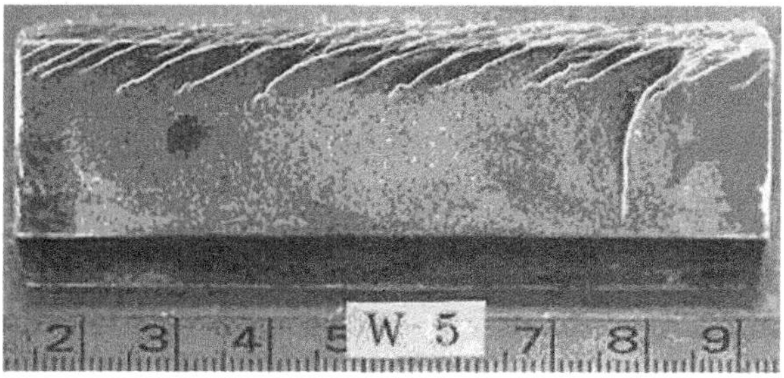

Fig. 4.62 Longitudinal cross section of the defect caused by contact rolling fatigue in standard carbon steel [30, 80]

edge parallel open cracks, the SIF was determined in [30] and some special cases of a system of two parallel cracks were discussed in [57, 58].

In the next subsection, we study the behavior (the parameters of fracture mechanics) for a system of parallel edge cracks under the condition of their propagation by the transversal shear (mode II) mechanism, as well as the growth paths of initially parallel open edge cracks by the normal opening (mode I) mechanism.

4.4.1 Edge Parallel Cracks in Compression Zone

We now write singular integral equations (SIE) for a system of parallel edge rectilinear cracks in the half-plane whose faces are in contact without friction (smooth contact) under the action of the Hertz load moving along the boundary of the half-plane. Solving these equations, we determine the contact sections of the cracks faces and find the stress intensity factors. The numerical results are obtained for the case of two parallel equal inclined cracks.

Hence, we consider a contact problem of the theory of elasticity for a half-plane weakened by a system of N parallel rectilinear edge cracks (cuts) L_j ($j = 1, \ldots, N$) of length l_j inclined to the boundary of the half-plane at an angle β (Fig. 4.63). In the mouth of the first crack, we place the main coordinate system xOy whose abscissa axis coincides with the boundary of the half-plane. All cracks are referred to local

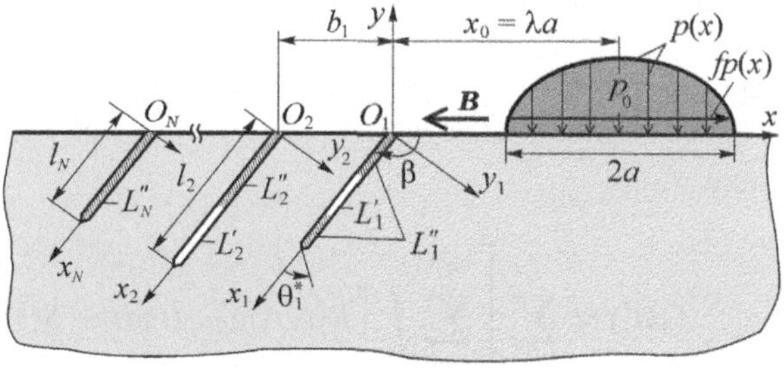

Fig. 4.63 Computational scheme of the model; ***B*** is the direction of motion of the counterbody

coordinate systems $x_jO_jy_j$ whose abscissa axes coincide with the lines of cracks. By b_j we denote the distance between the j-th and $(j+1)$-th cracks.

Assume that the boundary of the half-plane suffers the action of Hertz contact forces with tangential component. Then the conditions of the problem on the boundary of the half-plane can be represented in the form

$$\sigma_y(x) - i\tau_{xy}(x) = -p_0(1+if)\sqrt{a^2-(x-x_0)^2}\Big/a, \quad |x-x_0| \le a, \quad y=0; \tag{4.17}$$

$$\sigma_y(x) - i\tau_{xy}(x) = 0, \qquad |x-x_0| > a, \qquad y=0, \tag{4.18}$$

where x_0 is the coordinate of the center of contact section in the coordinate system xOy, $2a$ is its length, p_0 is the maximum pressure at the center of the contact section, and f is the Coulomb friction coefficient in contact between the bodies in the process of rolling with slip.

Depending on the location of contact load on the boundary of the half-plane, the cracks can be open (the faces are not in contact) or partially (completely) closed. We assume that the cracks faces are in contact without friction (smooth contact). By L'_j, we denote a family of sections $A_{jr}B_{jr}$ $(r=\overline{1, R_j})$ in the crack L_j along which its faces are not in contact (open),

$$L'_j = \bigcup_{r=1}^{R_j} L'_{jr} = \bigcup_{r=1}^{R_j} [A_{jr}B_{jr}], \tag{4.19}$$

Suppose that a smooth contact of the crack faces is realized in the sections $L''_j = L_j \backslash L'_j$. Hence, the boundary conditions on the crack faces take the form

$$\begin{gathered} N(t)^{\pm} + iT(t)^{\pm} = 0, \quad t \in L'_j; \\ v(t)^{+} - v(t)^{-} = 0, \quad T(t)^{\pm} = 0, \quad t \in L''_j; \quad j = 1, \ldots, N. \end{gathered} \tag{4.20}$$

We use the approach proposed in [32, 96]. If the boundary conditions (4.17), (4.18), and (4.20) are satisfied, then the problem is reduced to the solution of a system of singular integral equations for the derivatives of the normal and tangential components of the discontinuities of displacements $g'_n(t)$ and $g'_\tau(t)$ along the cracks in the half-plane

$$\text{Re}\,\Omega(t') = 0, \quad t' \in L'_j; \quad \text{Im}\,\Omega(t') = 0, \quad t' \in L_j; \quad j = 1, \ldots, N, \tag{4.21}$$

where

$$\Omega(t') = \sum_{j=1}^{N}\left[\sum_{r=1}^{R_j}\int_{L'_{jr}}\left[R(t,t')g'_{n\,jr}(t)dt + S(t,t')\overline{g'_{n\,jr}(t)d\bar{t}}\right] + \right.$$

$$+\int_{L_j}\left[R(t,t')g'_{\tau j}(t)dt+S(t,t')\overline{g'_{\tau j}(t)d\overline{t}}\right]\Bigg]-2\pi P(t'), \tag{4.22}$$

and the kernels of the SIE, $R(t,t')$ and $S(t,t')$, and the function $P(t')$ are given by relations (3.123) and (3.208).

In the system of SIE (4.21), we pass to dimensionless coordinates by using the parametric equations for the crack contours L_j and their open sections L'_{jr}:

$$t=w_j(\eta),\quad |\eta|\le 1,\quad t\in L_j;\qquad t=w_{jr}(\eta),\quad |\eta|\le 1,\quad t\in L'_{jr}. \tag{4.23}$$

Then the required functions take the form

$$\varphi_{\tau j}(\eta)=g'_{\tau j}(t)\,w'_j(\eta),\qquad \varphi_{n\,jr}(\eta)=g'_{n\,jr}(t)\,w'_{jr}(\eta),\qquad |\eta|\le 1. \tag{4.24}$$

Assume that the shear component of the required function has a square-root singularity at the ends of the cracks:

$$\varphi_{\tau j}(\eta)=\frac{u_{\tau j}(\eta)}{\sqrt{1-\eta^2}},\qquad |\eta|\le 1. \tag{4.25}$$

The functions $g'_{n\,jr}(t)$ have the same singularities at the ends of the open sections $A_{jr}B_{jr}$:

$$\varphi_{n\,jr}(\eta)=\frac{u_{n\,jr}(\eta)}{\sqrt{1-\eta^2}},\qquad |\eta|\le 1,\qquad r=1,\ldots,R_j, \tag{4.26}$$

where $u_{\tau j}(\eta)$ and $u_{njr}(\eta)$ are real functions continuous on the segment $[-1, 1]$.

We now supplement the system of Eq. (4.21) by the conditions [96]:

$$\int_{-1}^{1}\varphi_{n\,jr}(\eta)d\eta=0\quad \text{for the inner sections}(x_j(B_r)>0); \tag{4.27}$$

$$u_{n\,jr}(-1)=0\quad \text{for the end sections } (x_j(B_r)=0); \tag{4.28}$$

$$u_{\tau j}(-1)=0; \tag{4.29}$$

$$\mathrm{Im}\varphi_{n\,jr}(\eta)=0;\qquad \mathrm{Re}\varphi_{\tau j}(t)=0. \tag{4.30}$$

To determine the boundaries of the open sections L'_{jr} measured from the crack tips, we use the condition of vanishing of the SIF K_{I} at the ends of the intervals $A_{jr}B_{jr}$, namely:

$$K_{\mathrm{I}}(A_{jr}) = 0, \quad r = \begin{cases} 1, \ldots, R_j, & \text{if the first open section does not border the tip;} \\ 2, \ldots, R_j, & \text{if the first open section borders the tip.} \end{cases} \tag{4.31}$$

$$K_I(B_{jr}) = 0, \quad r = \begin{cases} 1, \ldots, R_j, & \text{if the last open section does not border the mouth;} \\ 1, \ldots, R_j - 1, & \text{if the last open section borders the mouth.} \end{cases} \tag{4.32}$$

In the open sections of the cracks faces, the following condition should be additionally satisfied:

$$v^+(t) - v^-(t) = -\frac{1+\kappa}{2G}\mathrm{Re}\left[\left(g_{nj}(t) + g_{\tau j}(t) \right) \frac{d\bar{t}}{dt} \right] > 0, \qquad t \in L'_j. \tag{4.33}$$

where κ and G are the elastic constants. In the sections with slip, we require that

$$N(t)^{\pm} = \mathrm{Re}[\Omega(t)] < 0, \qquad t \in L''_j. \tag{4.34}$$

Here, $\Omega(t)$ is given by relation (4.22).

We solve the system of integral Eq. (4.21) with additional conditions (4.27)–(4.30) by the method of mechanical quadratures (see Sect. 3.5 and [96]).

In order to solve the system of Eqs. (4.21), (4.27)–(4.30) in which the limits of integration are unknown, we apply the method of successive approximations. In order to find the zero-order approximation in the case of a single crack, we solve the problem for a crack open along the entire its length and use condition (4.33) in order to clarify where the crack faces are indeed open. In the case where the entire crack is closed, we solve the problem of smooth contact along the entire its length and use condition (4.34) to determine, where the crack faces are indeed slipping. We make this to reveal open sections if they are encountered in the crack. Note that these actions are performed for all cracks simultaneously and the appearance of several open sections in each crack is possible.

If we approximately find the boundaries of open sections in the crack with the help of conditions (4.33) and (4.34), then we use conditions (4.31) and (4.32) for their correction. Note that, in this case, conditions (4.33), (4.34) are satisfied sufficiently exactly but the convergence in terms of these conditions is quite poor.

The solution of the system of SIE (4.21) and relations (4.27)–(4.30) yield the normalized SIFs at the crack tips and at the ends of intervals $A_{jr}B_{jr}$:

$$F^+_{\mathrm{II}j} = \sqrt{\pi\left|\omega'_j(+1)\right|}\frac{u_{\tau j}(+1)}{\omega'_j(+1)}, \quad F^{\pm}_{\mathrm{I}jr} = \mp\sqrt{\pi\left|\omega'_{jr}(\pm 1)\right|}\frac{u_{n\,jr}(\pm 1)}{\omega'_{jr}(\pm 1)}, \tag{4.35}$$
$$r = 1, \ldots, R_j; \qquad j = 1, \ldots, N.$$

Numerical results: SIF, maps of contact of the crack faces, checks. The calculations were carried out for two identical parallel cracks ($N = 2$; $l_1 = l_2 = l_0$) with typical values of the operating parameters for the wheel-rail system ($f = 0.1$ and $f = 0.3$) and the characteristic features of defects ($\beta = 5\pi/6$; $\varepsilon = l_0/a = 1.0$) for various relative distances between the cracks $\delta = b/a$.

We use a system of dimensionless coordinates according to relation (4.23): $\eta_j = 2x_j/l{-}1$ ($j = 1.\ 2$). Thus, in Figs. 4.64 and 4.65, the values $\eta = -1$ and $\eta = +1$ correspond to the crack mouth and the crack tip, respectively. Note that,

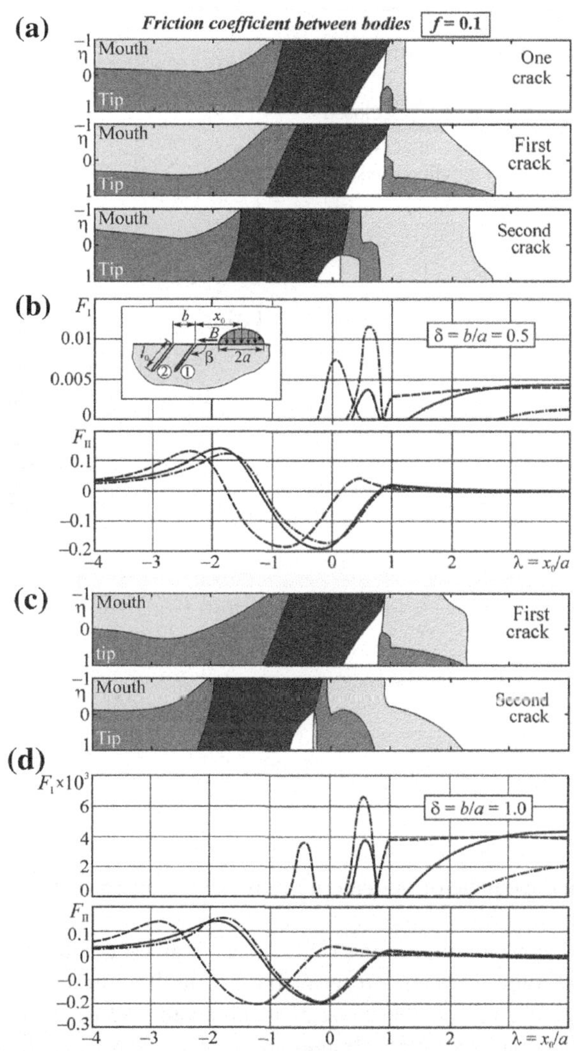

Fig. 4.64 Dependences of the length of contact sections of the cracks faces (**a**, **c**) and the normalized SIF $F_{\mathrm{I,II}} = K_{\mathrm{I,II}}/(p_0\sqrt{\pi\, a})$ (**b**, **d**) on the parameter λ of location of the contact load for one and two cracks for coefficient of friction between rolling bodies $f = 0.1$ ($\varepsilon = 1.0$; $\beta = 5\pi/6$): —— one crack; – • – • – first crack; - – – – second crack; [open crack symbols] open crack with directions of slip of the faces; [closed crack symbols] closed crack (smooth contact of the faces) with directions of slip of the faces

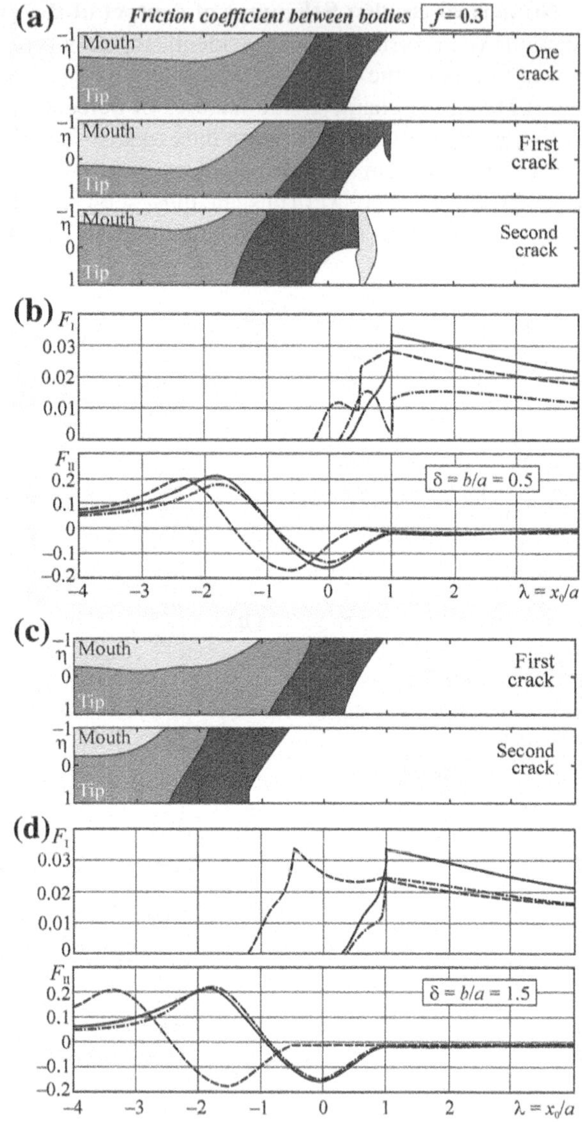

Fig. 4.65 Dependences of the length of contact sections of the cracks faces (**a, c**) and the normalized SIF $F_{I,II} = K_{I,II}/(p_0\sqrt{\pi a})$ (**b, d**) on the parameter λ of location of the contact load for one and two cracks for coefficient of friction $f = 0.3$ (= 1.0; $\beta = 5\pi/6$): —— one crack; – • – • – fir st crack; - - - - second crack; open crack with the directions of slip of the faces; closed crack (smooth contact of the faces) with the directions of slip of the faces

in the course of our calculations, several open sections appeared in each crack but at most one section remained at the end of iterative process.

For the correct analysis of the obtained results, we introduce the parameters of fracture mechanics responsible for the fracture by the mode I mechanism ($\max K_{I\theta}(\lambda, \theta)$) and by the mode II mechanism (ΔK_{II}). In the case of motion of the contact load along the boundary of the half-plane, the normalized mixed-type SIF $F_{I\theta}(\lambda, \theta) = K_{I\theta}(\lambda, \theta)/\left(p_0\sqrt{\pi a}\right)$ determined according to the σ_θ-criterion [84] takes its maximum values for λ^* and θ^*, i.e., $\max F_{I\theta}(\lambda, \theta) = F_{I\theta}(\lambda^*, \theta^*)$ [14]. By F_I^* and F_{II}^*, we denote the values of F_I and F_{II} for λ^*. The range of the normalized SIF $F_{II} = K_{II}/\left(p_0\sqrt{\pi a}\right)$ in the course of motion of the contact load in a contact cycle is defined as follows: $\Delta F_{II} = \max F_{II}(\lambda)–\min F_{II}(\lambda)$.

The accumulated results indicate that the behavior of the curves $F_{II}(\lambda)$ for each crack in the system of two cracks is similar to the case of a single crack. This is observed for various distances between cracks and for different friction coefficients (Figs. 4.64 and 4.65).

In Fig. 4.65a, b, we see that, for the high coefficient of friction ($f = 0.3$), if the load moves from the right to the left and starts to cover the mouth of one crack ($\lambda \leq 1.0$) or the mouths of each of two cracks in turn, the crack faces start to close from the mouth. At the same time, the crack tips are open and $F_I(\lambda) > 0$ in a fairly large range of λ ($0.3 \leq \lambda \leq 1.0$). In the case of low contact friction ($f = 0.1$) realized prior to the closure of crack mouth ($\lambda > 1$) or the mouths of two cracks (in turn) by the load, they insignificantly (without instability) close starting from the tip. However, as soon as the load covers the cracks mouths, the cracks start to close from the mouth, the tips start to open, and the SIF $F_I(\lambda) > 0$ in a quite large range of values of λ, as in the case of large f. This result is quite important because it reveals the possibility of crack growth by the opening mechanism when the contact load is located over the crack. Further, if the center of the contact section is located over the mouth of one crack or of the first crack, then they are completely closed ($F_I(\lambda) = 0$). Later, they start to open from the mouth as soon as the contact load opens the mouths ($\lambda < -1$). However, in this case, the crack tip is closed and $F_I(\lambda) = 0$.

It is worth noting that the parameter $F_{I\theta}^* = \max F_{I\theta}(\lambda, \theta)$ that controls the mode I fracture may have two maxima in a contact cycle for each crack depending on the friction coefficient f: One maximum is low and is observed prior to the closure of the crack mouth by the load, while the second maximum is high and appears when the crack is partially closed from the mouth. The second maximum of $F_{I\theta}$ is attained due to the high values of $|F_{II}|$ observed for $F_I > 0$ (Figs. 4.64 and 4.65). Hence, if the contact load is located directly over the crack (cracks), the fracture may occur not only by the shear mechanism but also by the opening mechanism. In order to reveal the predominant mechanism of crack propagation in the contact zone, it is necessary to perform more extensive investigations with regard for the friction of the crack faces, their initial lengths, the characteristics of crack growth resistance of the materials in transverse shear and normal opening, etc.

The value of the parameter $\Delta F_{II} = \max F_{II}(\lambda)–\min F_{II}(\lambda)$ corresponding to the fracture by shear strongly depends on the distance between the cracks (Table 4.21). Thus, at distances comparable with the length of the contact section ($b \approx a...2a$), we get the most pronounced weakening of the rolling surface (ΔF_{II} is maximum). This conclusion agrees with the experimental data about features of checks-type damages presented by Zerbst et al. [102].

The numerical results presented in Table 4.21 reveal a significant mutual influence of two parallel cracks in the contact zone of rolling bodies: For small distances between the cracks, we observe a noticeable hardening of the rolling surface (ΔF_{II} are small) as compared with the case of a single crack. At the same time, the increase

Table 4.21 Dependence of the normalized SIF on the friction coefficient of f and the relative distance between the cracks δ

f	$\delta = b/a$	λ^*		$F^*_{I\theta}$		F^*_{II}		ΔF_{II}	
		First	Second	First	Second	First	Second	First	Second
0.1	0.25	0.18	−0.06	0.141	0.135	−0.122	−0.117	0.244	0.301
	0.5	0.21	−0.26	0.160	0.127	−0.138	−0.110	0.296	0.317
	1.0	0.23	−0.71	**0.162**	0.129	**−0.140**	−0.112	0.350	0.342
	1.5	0.27	−1.21	0.144	0.146	−0.125	−0.127	**0.350**	0.343
	2.0	0.30	−1.70	0.134	0.143	−0.116	−0.124	0.346	0.346
	Single	0.30		0.133		−0.1154		0.3318	
0.3	0.25	0.22	−0.04	0.112	0.133	−0.096	−0.115	0.253	0.339
	0.5	0.14	−0.26	0.143	0.139	−0.124	−0.120	0.309	0.359
	1.0	0.32	−0.72	0.125	0.149	−0.108	−0.129	0.361	0.381
	1.5	0.34	−1.20	0.120	0.153	−0.104	−0.133	0.368	**0.382**
	2.0	0.32	−1.74	0.126	**0.157**	−0.109	**−0.136**	0.375	0.375
	2.5	0.32	−2.24	0.127	0.150	−0.110	−0.130	0.372	0.370
	Single	0.28		0.138		−0.120		0.367	

in the distance leads to softening (ΔF_{II} increase). Moreover, the presence of one closed crack in a system of two cracks in the process of motion of the contact load may strongly affect the SIF $F_I(\lambda)$ at the tip of the open crack (see, e.g., Fig. 4.64b, d for F_I ($\lambda = 0.6$)). The maximum values of the SIF $F_{I\theta}(\lambda,\theta)$ (depending on the distance between cracks) are mainly attained on the first crack for $f = 0.1$ and on the second crack for $f = 0.3$.

The maximum and minimum of $F_{II}(\lambda)$ are attained in the case where both cracks are in contact over the entire length. Hence, the partial opening displacements of the cracks weakly affect the values of ΔF_{II}. The plots of $F_{II}(\lambda)$ presented in Figs. 4.64 and 4.65 are very similar to the plots obtained under the conditions of smooth contact over the entire crack length. This enables one to determine ΔF_{II} by using solely the solution of the problem of contact of the cracks over the entire length.

Note that the comparison of the results of calculations performed in the present work with the results obtained by Bower [10] for $\beta = 155°$ ($\pi - \beta = 25°$), $\varepsilon = 0.5$ and $f = \pm 0.05$ reveals their satisfactory agreement.

4.4.2 Edge Parallel Cracks Growth by Normal Opening Mechanism. Crumbling

In the present subsection, on the basis of singular integral Eq. (3.141) (see Chap. 3) for a system of curvilinear cracks in the elastic half-plane subjected to the action of

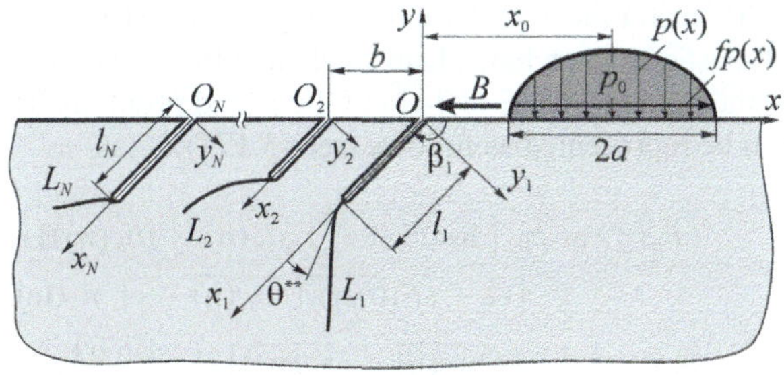

Fig. 4.66 Scheme of the problem; ***B*** is the direction of motion of the counterbody

model contact loads on the boundary of the plane, we compute the stress intensity factors (SIF) for 2–5 parallel edge rectilinear cracks corresponding to various values of the parameters of the problem. We construct growth paths for a system of (two or three) parallel edge initially rectilinear cracks. The contact of the crack faces is not taken into account.

Stress intensity factors [30]. Consider an elastic isotropic half-plane weakened by a system of N parallel edge rectilinear cracks. It is assumed that the cracks have the same length l_0 are inclined at the angle β to the boundary of the half-plane, and the distance between two neighboring cracks is equal to b (Fig. 4.66). The crack nearest to the load is called the first crack. The crack located to the left of this crack it is called the second crack, etc. In a section of length $2a$ on the boundary of the half-plane, we impose a Hertz contact load, i.e., a normal pressure $p(x)$ and tangential forces $q(x)$. They are distributed according to the elliptic law and obey the Amontons–Coulomb law $q(x) = fp(x)$, where f is the friction coefficient between the contacting bodies.

We refer the half-plane to the main coordinate system xOy whose origin is located in the mouth of the first crack. The Ox-axis is directed along the boundary of the half-plane. The contours of the cracks (cuts) L_n ($n = 1, \ldots, N$) are referred to local coordinate systems $x_nO_ny_n$, whose origins are located in the mouths of these cracks. The O_nx_n-axes are directed along the cracks and α is the angle of inclination of the O_nx_n–axes to the Ox-axis ($\beta_1 = \beta_2 = \ldots = \beta_n = \beta = -\alpha$).

On the half-plane boundary, we impose the boundary conditions of problem (4.17, 4.18). It is assumed that the crack faces are load free. The corresponding boundary conditions take the form

$$\sigma^{\pm}(x_k) - i\tau^{\pm}(x_k) = 0, \qquad 0 \le x_k \le l_0\,, \qquad y_k = 0\,, \qquad k = 1, \ldots, N; \tag{4.36}$$

the superscripts "+" and "–" mark the boundary values of stresses on the left and right crack faces, respectively. Relations (4.36) correspond to the case where the cracks enter the zone of tension, open, and develop by the mode I mechanism. These conditions are realized due to the action of sufficiently high tangential (friction) forces in the contact zone directed from the crack in the case where the contact load is located behind the crack (see Fig. 4.66). We now determine (under the formulated conditions), the stress-strain state in the vicinity of the crack tips.

By using the singular integral equations (SIE) of the plane problem of the theory of elasticity for a half-plane weakened by a system of curvilinear cracks, we arrive at the system of SIE (3.141) of the posed problem. The right-hand sides of the SIE can be represented as follows (see (3.171)):

$$\begin{aligned} P_n(\eta) = p_0 \Big\{ & \mathrm{Re}[(1+if)\,(a_n(\eta) - ib_n(\eta))]\, w'(\eta) \\ & - \big[(1-if)\big(\overline{b_n(\eta)}/\overline{a_n(\eta)} - i\big) \times i\,\mathrm{Im}\big(\varepsilon w(\eta) e^{i\alpha} + \gamma_n\big) \\ & - if\big(\overline{a_n(\eta)} + i\overline{b_n(\eta)}\big)\big] e^{-2i\alpha\overline{w'(\eta)}} \Big\}, \qquad n = 1, \ldots, N, \end{aligned} \tag{4.37}$$

where $a_n(\eta) = \sqrt{1 - b_n^2(\eta)}$; $b_n(\eta) = \varepsilon w(\eta)\, e^{i\alpha} / l_0 + \gamma_n - \lambda$; $\varepsilon = l_0/a$; $\lambda = x_0/a$; $\gamma_n = z_n^0 / a$.

For numerical calculations, the parameters of location of the cracks and the friction coefficient between the contacting bodies are chosen mainly with regard for the conditions of rolling (and fretting fatigue). Thus, the investigations are carried out for the angles of orientation of cracks $\beta = \pi/6$, $\pi/2$, $5\pi/6$, their relative lengths $\varepsilon = l_0/a = 0.2$, 0.5, and 1.0, and the relative distances between the cracks $\delta = b/l_0 = 0.1$, 0.5, and 1.0, depending on the distance of the contact load from the mouth of the first crack characterized by the parameter $\lambda = x_0/a$ (Fig. 4.66). We take the friction coefficient $f = 0.3$. Note that the surface cracks are initiated and propagate under the conditions of rolling at the angle $\beta \approx 5\pi/6$ to the boundary. At the same time, under the conditions fretting fatigue, the surface cracks are initiated and propagate within the range of angles $\pi/6 \le \beta \le \pi/2$.

In Fig. 4.67, we present the plots of the normalized stress intensity factors $F_{\mathrm{I,II}}(\lambda, \beta, \varepsilon, \delta, f) = K_{\mathrm{I,II}}(\lambda, \beta, \varepsilon, \delta, f) / \left(p_0\sqrt{\pi a}\right)$ for two cracks as functions of the distance from the contact load to the mouth of the first crack for fixed values of the parameters β, ε, δ, and f. The plots of the SIF $F_{\mathrm{I,II}}(\lambda)$ for a single crack and the same values of the indicated parameters are presented by the dash-dotted curves. Note that the plots of $F_{\mathrm{II}}(\lambda)$ are not shown within the range of values of λ, where $F_{\mathrm{I}}(\lambda) < 0$ (Fig. 4.67b, d). Moreover, in the case of two cracks, the plots of $F_{\mathrm{I}}(\lambda)$ are not presented for the second crack if this parameter is negative for the first crack (Fig. 4.67a, c). The analysis of the presented plots enables us to make the following conditions:

- The absolute values of the SIFs and their maximum values for two cracks are, as a rule, lower than in the case of a single crack (Fig. 4.67a–c, e, f).
- For the orientations of cracks typical of fretting fatigue ($\beta = \pi/2$, $\beta = \pi/6$, Fig. 4.67a–d), the SIFs and, in particular, their maximum moduli for the first of two cracks (closer to the load) are higher than for the second crack. Hence, the first crack is responsible for the subsequent fracture.
- For the orientations of the cracks typical of rolling under the conditions of dry friction (Fig. 4.67e, f), the values of K_{I} for the second crack are much higher than for the first crack. For this crack, K_{II} is somewhat higher than K_{II} for the second crack. Hence, in the course of propagation of these cracks, we may observe either

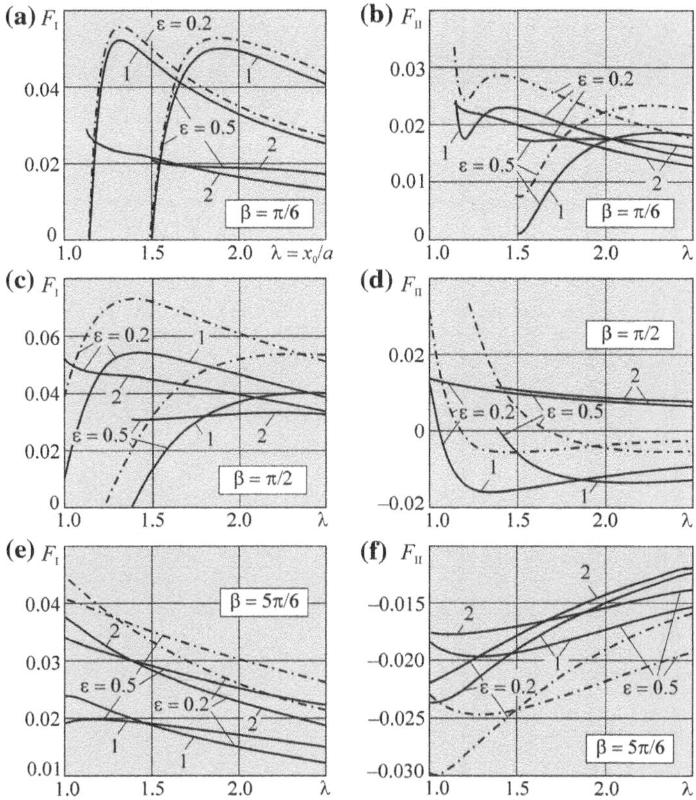

Fig. 4.67 Normalized stress intensity factors F_I and F_{II} for two cracks as functions of the distance of from the contact load to the mouth of the first crack for the angles of orientation of the cracks $\beta = \pi/6$, $\pi/2$, and $5\pi/6$, their relative lengths $\varepsilon = l_0/a = 0.2$ and 0.5, and the relative distance between the cracks $\delta = b/l_0 = 0.5$; 1 —— first crack, 2 —— second crack, - - - - single crack

the predominant development of the second crack (located at a greater distance from the load) or the simultaneous development of both cracks.

- For the angle $\beta = \pi/6$, the contact of crack faces is terminated (i.e., the values of K_I become positive) as soon as the load opens the mouth of the first crack moving in the positive direction of the Ox-axis. For $\beta = \pi/2$, this happens for somewhat higher λ and, for $\beta = \pi/6$, for even higher values of λ. This trend becomes more pronounced as the lengths of the cracks increase;
- The sign of K_{II} corresponding to the direction of tangential stresses in the vicinity of the crack tips strongly depends on the angle of inclination of the cracks to the half-plane boundary.
- The variations of the SIF for the first crack are the same, as for the case of a single crack with the same location.

Table 4.22 Influence of the distance between two cracks on the maximum values of the stress intensity factors at the tip of the first crack for $\varepsilon = l_0/a = 0.2$

β	F_I^*								
	Single crack	$\delta = b/l_0$ (two cracks)							
		0.05	0.1	0.5	1.0	2.0	3.0	4.0	5.0
$\pi/6$	5.59	5.34	5.31	5.24	5.26	5.40	5.46	5.51	5.54
$\pi/2$	7.30	5.25	5.26	5.43	5.64	6.03	6.41	6.70	6.88
$5\pi/6$	4.50	1.78	1.87	2.40	2.90	3.81	4.22	4.36	4.43
β	F_{II}^*								
	Single crack	$\delta = b/l_0$ (two cracks)							
		0.05	0.1	0.5	1.0	2.0	3.0	4.0	5.0
$\pi/6$	3.36	2.26	2.37	2.45	2.64	2.96	3.15	3.24	3.29
$\pi/2$	$\frac{-0.57}{3.19}$	$\frac{-1.79}{0.63}$	$\frac{-1.77}{0.78}$	$\frac{-1.62}{1.26}$	$\frac{-1.45}{1.42}$	$\frac{-1.13}{2.38}$	$\frac{-0.87}{2.78}$	$\frac{-0.73}{2.98}$	$\frac{-0.66}{3.07}$
$5\pi/6$	−2.98	−1.86	−1.98	−2.36	−2.60	−2.74	−2.86	−2.91	−2.94

In Table 4.22, we present the maximum absolute values of the normalized SIFs $F_{\mathrm{I}}^*(\lambda)$ and $F_{\mathrm{II}}^*(\lambda)$ for the cases of a single crack and of the first of two cracks depending on the location of the second crack. The maximum values of the SIF are most frequently used to predict the fracture of contacting bodies. Note that, for the angle $\beta = \pi/2$, the function $F_{\mathrm{II}}(\lambda, \delta)$ changes its sign in the course of motion of the load (see Fig. 4.67d). Hence, in this case, for any curve $\delta = \mathrm{const}$, we present the maximum negative and positive values of $F_{\mathrm{II}}^*(\lambda, \delta)$ in the numerator and denominator, respectively.

The analysis of the data presented in Table 4.22 enables us to conclude that:

- as two cracks becomes closer, the values of F_{I}^* and $|F_{\mathrm{II}}^*|$ at the tip of the first crack decrease to the asymptotic values (for the chosen parameters β, ε, and δ) values; these values are lower than in the case of a single crack;
- as the distance between the cracks increases, the maximum values of the SIFs of the first crack approach the values of the SIFs for a single crack; even for $\delta = 5.0$, the difference between the values of F_{I}^* and F_{II}^* for one crack are lower than 2% for $\beta = \pi/6$ and $5\pi/6$; for $\beta = \pi/2$, the difference between the values of F_{I}^* and F_{II}^* is smaller than 4% and 6%, respectively.

In Table 4.23, we present the maximum absolute values of the normalized SIF $F_{\mathrm{I,II}}^*$ at the tip of the first crack for the cases of one, two, three, or five cracks in the half-plane, fixed sizes and locations of the cracks and variable location of the contact load. The analysis of the results obtained for the chosen values of the parameters β, ε, and δ enables us to conclude that:

- A "double-crack" defect (two identical parallel cracks) is less hazardous than a single crack; in the case of appearance of the second crack, the values of F_{I}^* at the tip of the first crack decrease for $\beta = \pi/6$, $\pi/2$, and $5\pi/6$ by 4–6%, 2–28%, and 36–62%, respectively, whereas the values of F_{II}^* decrease by 13–64%;
- As the number of cracks increases further, the values of the SIF $F_{\mathrm{I,II}}^*$ at the tip of the first crack (closest to the load) approaches an asymptotic value for a given orientation of cracks, their lengths, and the distance between them. The asymptotic values of F_{I}^* and $|F_{\mathrm{II}}^*|$ are lower than the analogous values in the case of a single crack. On the average, for F_{I}^*, this difference is equal to 10%, 30%, and 60% for $\beta = \pi/6$, $\pi/2$, and $5\pi/6$, respectively; for $|F_{\mathrm{II}}^*|$, the difference constitutes 40%.
- As the crack lengths increase and they approach each other, the values of F_{I}^* and $|F_{\mathrm{II}}^*|$ decrease; this trend is violated only for $|F_{\mathrm{II}}^*|$ at $\beta = \pi/2$ as far as the influence of distance between the cracks is concerned.

In the case of five cracks (Fig. 4.68), for the chosen values of the parameters β, ε, and δ, we observe the trends similar to the case of two cracks (Fig. 4.67):

- The character of variations of the SIF for the first crack in the systems of two and five cracks are similar to the case of a single crack; the character of variations of the SIFs of 2–5th cracks in the system of five cracks are similar to the case of the second crack in the system of two cracks.

Table 4.23 Maximum values of the normalized stress intensity factors F_{I}^* and F_{II}^* at the tip of the first crack in the cases of one, two, three, or five cracks

β	One		Two		Three		Five	
	F_{I}^*	F_{II}^*	F_{I}^*	F_{II}^*	F_{I}^*	F_{II}^*	F_{I}^*	F_{II}^*
	$\delta = b/l_0 = 0.1;\ \varepsilon = l_0/a = 0.2$							
π/6	5.59	3.36	5.31	2.37	4.89	2.10	4.77	1.85
π/2	7.30	3.19	5.26	−1.77	4.98	−1.75	4.69	−1.73
5π/6	4.50	−2.98	1.87	−1.98	1.65	−1.80	1.49	−1.66
	$\delta = 0.1;\ \varepsilon = 0.5$							
π/6	5.31	2.32	5.08	1.76	4.97	1.61	4.85	1.49
π/2	5.40	3.36	3.94	−1.43	3.73	−1.41	3.53	−1.40
5π/6	4.09	−2.47	1.55	−1.66	1.37	−1.51	1.24	−1.39
	$\delta = 0.1;\ \varepsilon = 1.0$							
π/6	4.38	1.76	4.21	1.33	4.13	1.22	4.03	1.12
π/2	3.96	1.69	2.89	−1.08	2.74	−1.06	2.59	−1.06
5π/6	3.34	−1.88	1.18	−1.27	1.05	−1.16	0.95	−1.07
	$\delta = 0.5;\ \varepsilon = 0.2$							
π/6	5.59	3.36	5.24	2.45	5.12	2.23	5.02	2.10
π/2	7.30	3.19	5.43	−1.62	5.18	−1.58	4.96	−1.57
5π/6	4.50	−2.98	2.40	−2.36	2.27	−2.22	2.17	−2.13
	$\delta = 0.5;\ \varepsilon = 0.5$							
π/6	5.31	2.32	5.01	1.85	4.91	1.76	4.82	1.69
π/2	5.40	3.36	4.03	−1.35	3.85	−1.32	3.69	−1.31
5π/6	4.09	−2.47	2.00	−1.97	1.89	−1.84	1.81	−1.77
	$\delta = 0.5;\ \varepsilon = 1.0$							
π/6	4.38	1.76	4.15	1.40	4.07	1.33	4.00	1.28
π/2	3.96	1.69	2.95	−1.03	2.82	−1.01	2.70	−1.00
5π/6	3.34	−1.88	1.53	−1.50	1.44	−1.42	1.38	−1.37
	$\delta = 1.0;\ \varepsilon = 0.2$							
π/6	5.59	3.36	5.26	2.64	5.17	2.52	5.11	2.44
π/2	7.30	3.19	5.64	−1.45	5.43	−1.42	5.27	−1.41
5π/6	4.50	−2.99	2.90	−2.60	2.80	−2.51	2.61	−2.44
	$\delta = 1.0;\ \varepsilon = 0.5$							
π/6	5.31	2.32	5.03	1.95	4.96	1.90	4.90	1.86
π/2	5.40	3.36	4.18	−1.22	4.03	−1.20	3.91	−1.20
5π/6	4.09	−2.47	2.50	−2.16	2.42	−2.09	2.37	−2.04
	$\delta = 1.0;\ \varepsilon = 1.0$							
π/6	4.38	1.76	4.16	1.48	4.10	1.43	4.06	1.41
π/2	3.96	1.69	3.06	−0.94	2.95	−0.93	2.86	−0.92
5π/6	3.34	−1.88	1.92	−1.64	1.85	−1.59	1.81	−1.56

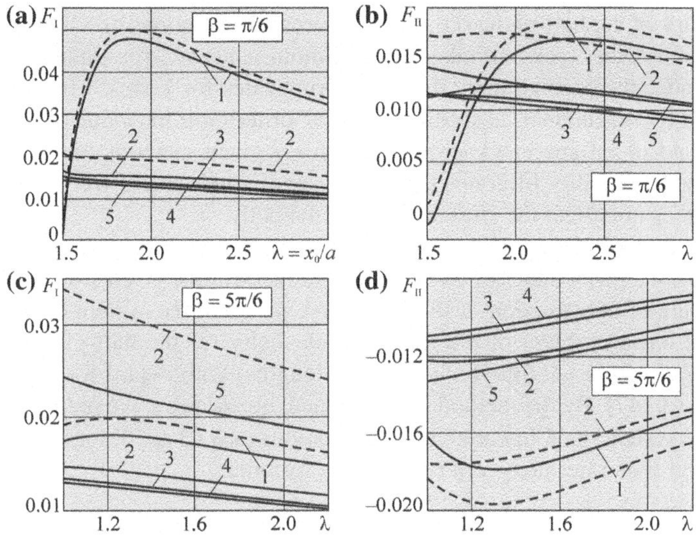

Fig. 4.68 Normalized stress intensity factors for two and five cracks as functions of the distance from the contact load to the mouth of the first crack for the relative crack length $\varepsilon = l_0/a = 0.5$ and the relative distance between cracks $\delta = b/l_0 = 0.5$: ——— five cracks, _ _ _ two cracks; 1, 2, 3, 4, and 5 correspond to the first, second, third, fourth, and fifth cracks, respectively

- The absolute values of the SIFs of the first and second cracks in the system of five cracks are lower than for the SIFs of the first and second crack in the system of two cracks, respectively; in general, the defect formed by five cracks is less dangerous than one crack or a "double" crack.
- If the cracks are inclined to the edge of the half-plane at the angle $\beta = \pi/6$ (Fig. 4.68a, b), then the first crack (closest to the load) is responsible for the fracture; the SIFs and their maximum values for other cracks are lower.
- The SIFs and their maximum absolute values for the fifth crack are larger, in some cases (Fig. 4.68b, c), than for the other cracks in the family of five cracks. In particular, if the cracks are inclined at the angle $\beta = 5\pi/6$ (Fig. 4.68c, d), then the fifth crack can be responsible for fracture.

In conclusion, we note that, in a special case of single edge crack, the obtained results coincide with the results obtained in [45]. In the case of two cracks perpendicular to the boundary, the results are in good agreement with the results from [56]. The calculations were carried out in the case where the order of the system of complex algebraic equations $M = 80$.

Growth paths (trajectories) for a system of parallel edge cracks depending on the level of friction between the rolling bodies. Crumbling. According to the accepted concept [36, 86], we assume that the process of crack growth is controlled by the parameter $K_{I\theta}$ specifying the intensity of normal circumferential stresses at the crack tip. In each loading cycle, in the course of motion of the contact load along the boundary of the half-plane (i.e., in the process of variations of λ), the parameter $K_{\mathrm{I}\theta}$ varies for each crack and takes its maximum value $K_{\mathrm{I}\theta}^*$ for some $\lambda = \lambda^*$ and $\theta = \theta^*$. Under these conditions, the crack propagates for $\lambda = \lambda^*$ in the direction determined by the angle θ^* according to the σ_θ-criterion if the value of $K_{\mathrm{I}\theta}^*$ exceeds the threshold of fatigue crack growth $K_{\mathrm{I}th}$ for a given material. Here, we do not consider the possibility of contact of the faces of some crack; we believe that its presence does not affect the stressed state of the body.

Assume that the crack lengths increase proportionally to the rates of propagation of their tips in a given material (see relations (2.32)) for the given stress-strain state. In each stage of construction of the trajectory, we solve the system of SIE (3.141) of the first basic problem of the theory of elasticity for the half-plane with edge curvilinear cracks (each time for new crack lengths) with regard for the additional conditions (3.147). By the method of mechanical quadratures, we reduce the system of SIE to a system of M linear algebraic equations (see Chap. 3, formulas (3.145)). The crack growth rates are given by the Paris formula

$$v_i = C\left(K_{\mathrm{I}\theta i}^*\right)^n, \quad i = 1, 2, \ldots, N, \tag{4.38}$$

where N in the number of cracks.

The calculations were carried out with regard for the characteristics of cyclic crack growth resistance of 75KhGST rail steel with the structure of lamellar pearlite (Table 4.4).

We constructed the trajectories for the systems of two (Figs. 4.69, 4.70, and 4.71) and three (Fig. 4.72) cracks with identical and different lengths. The initial angle of inclination of the cracks to the boundary of the half-plane (to the direction of tangential forces in contact) $\beta = 5\pi/6$ is chosen according to the experimental data [9, 75, 76, 79]. The calculations were performed for the friction coefficients between the contacting bodies $f = 0.05, 0.10$ and 0.30 corresponding, in particular, to different conditions of operation of the wheel-rail engineering couples (dry or damp weather and lubrication).

Parallel cracks for small values of the coefficient of friction in contact between the rolling bodies. In the case of low values of the friction coefficient ($f \le 0.1$), two initially identical parallel cracks ($l_1 = l_2 = l_0$) grow parallel to the contact boundary in the direction of motion of the counterbody for relatively large distances between the cracks ($\delta = b/a > 1.0$) (Fig. 4.69a). This agrees with the trend of growth of a single crack (Fig. 4.69b) [39, 87]. For smaller distances between the initial cracks ($\delta \le 1.0$), the first crack immediately begins to propagate into the bulk of the material (Fig. 4.69b, c).

In Figs. 4.70 and 4.71, we illustrate the process of growth of two cracks with different initial lengths for low friction coefficients. The obtained results demonstrate that if the first crack is shorter than the second crack, then it mainly propagates toward the second crack. In this case, for any fixed length of the second crack and a given length of the first crack, there exists a certain critical distance δ^* between them. The first crack propagates toward the second crack for $\delta > \delta^*$ and into the bulk of the

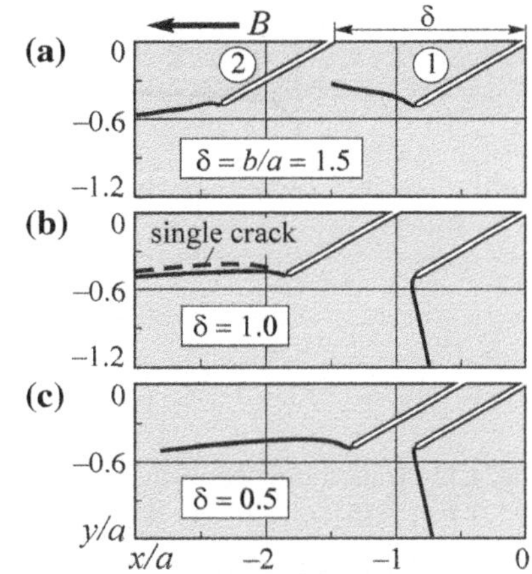

Fig. 4.69 Trajectories of growth of two equal parallel cracks as functions of the relative distance δ between them; $f = 0{,}1$; $\varepsilon = l_0/a = 1{,}0$; B is the direction of motion of counterbody

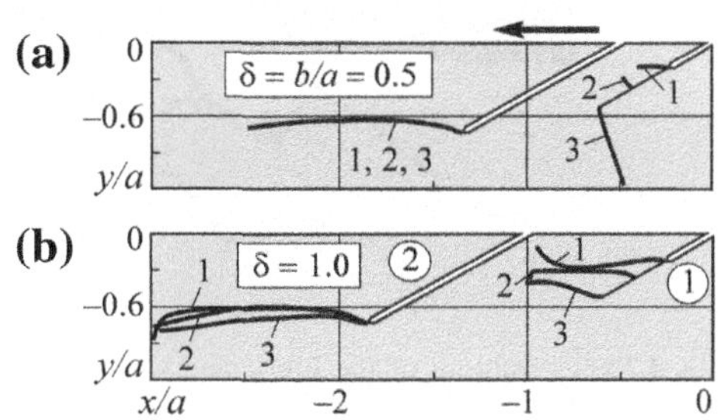

Fig. 4.70 Growth paths of two unequal parallel cracks as functions of the length of the first crack; $f = 0.1$; $\varepsilon_2 = l_2/a = 1.0$; $1 — \varepsilon_1 = l_1/a = 0.3$; $2 — \varepsilon_1 = 0.5$; $3 — \varepsilon_1 = 0.7$

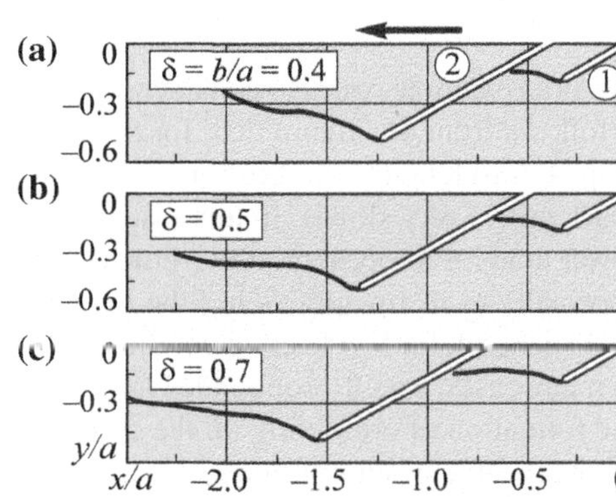

Fig. 4.71 Trajectories of growth of two unequal cracks as functions of the distance between them; $f = 0.05$; $\varepsilon_1 = l_1/a = 0.4$; $\varepsilon_2 = l_2/a = 1.0$

material for $\delta < \delta^*$. The value of δ^* increases with the length of the first crack. Note that the family of analyzed values of the parameters f, ε_1,. ε_2, and δ includes the values for which the first crack closely approaches the second crack and creates the hazard of crumbling (curve 1 in Fig. 4.70b). As the friction coefficient f decreases, the trend toward the development of crumbling increases, especially in the case of short ($\varepsilon_1 < 0.5$) first crack (Fig. 4.71).

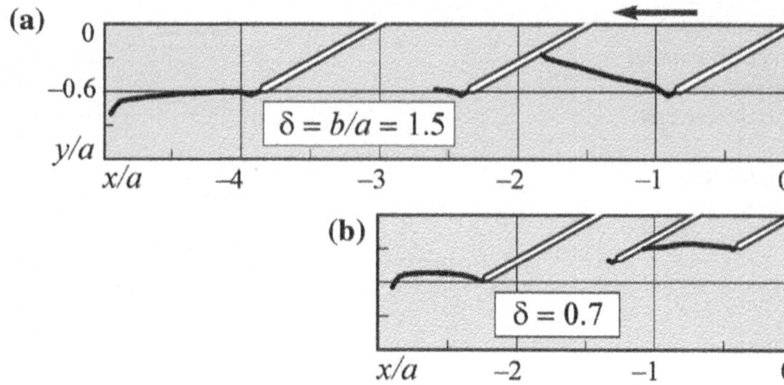

Fig. 4.72 Trajectories of growth of three equal (**a**) and unequal (**b**) cracks; $f = 0.1$; **a** $\varepsilon_1 = \varepsilon_2 = \varepsilon_3 = 1.0$; **b** $\varepsilon_1 = 0.5$; $\varepsilon_2 = 0.7$; $\varepsilon_3 = 1.0$

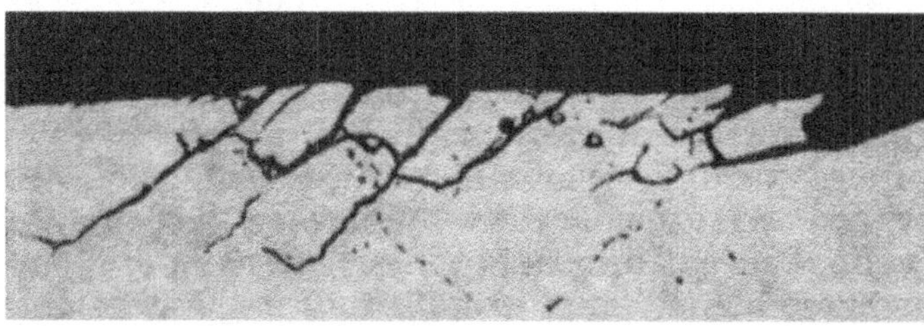

Fig. 4.73 Crumbling of the lubricated surface of a roll made of rail steel in testing for contact fatigue [75]

As an example, we also consider a system of three parallel edge cracks (Fig. 4.72). It follows from this figure that, for three cracks with equal (Fig. 4.72a) and different (Fig. 4.72b) lengths, the extreme cracks develop quite intensely, whereas the middle crack grows very slowly. In this case, the values of the SIFs along the paths are much lower than for the system of two cracks. For the first crack, these values are unstable, especially as its tip approaches the neighboring crack.

In general, for low friction coefficients in contact between the bodies, we observe, in many cases, a well-pronounced trend in the development of parallel edge cracks to the formation of crumbling on the contact surface. In particular, this trend becomes stronger for a system of unequal cracks if the first crack encountered on the path of motion of the counterbody is shorter than the next crack. The obtained numerical results are in good agreement with the experimental data (Fig. 4.73).

Two cracks for high values of the contact friction coefficient. For high values of the contact friction coefficient in ($f = 0.3$), the cracks of the same length propagate into the bulk of the material (Fig. 4.74) as in the case of a single crack [38, 87]. After the appearance of the second crack, the first crack changes the direction of its propagation. Then the first crack stops (except the value $\delta = b/a = 0.5$) and the second crack begins to propagate as the main crack.

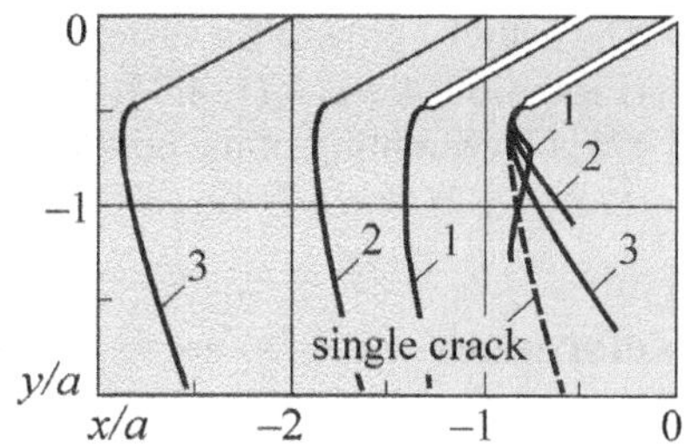

Fig. 4.74 Crack growth paths for two parallel equal cracks as functions of the distance between them; $f = 0.3$; $\varepsilon_1 = \varepsilon_2 = 1.0$: 1 — $\delta = 0.5$; 2 — $\delta = 1.0$; 3 — $\delta = 2.0$

The numerical calculations show that, for the friction coefficient $f = 0.3$, the behaviors of cracks for the other angles of their initial inclination to the boundary of the body ($\beta = \pi/2$ and $\pi/6$) are similar.

4.4.3 Conclusions

We study the stress intensity factors and the kinetics of propagation of one and two parallel edge cracks in the elastic half-plane. The cracks appear on the boundary of the half-plane depending on the motion of the model contact load (Hertz pressure with tangential component) along the boundary. We studied the cracks inclined at the angle $\beta = 5\pi/6$ to the direction of tangential contact forces and, hence, at a small angle $\bar{\beta} = 30° = \pi - \beta$ to the direction of motion of the contact load (Fig. 4.66). The proposed scheme models the contact interaction of rolling bodies with slip in the wheel-rail system, while the value of the angle β is characteristic of a system of surface cracks in the rails. The numerical results are obtained for two values of the friction coefficient $f = 0.1$ and $f = 0.3$ in contact between the rolling bodies, relative crack length $\varepsilon = l_0/a = 1.0$, and various distances between the cracks. It is established that:

- For a low coefficient of sliding friction between the rolling bodies ($f = 0.1$), the cracks begin to close (under the conditions of smooth contact between the faces) from the tips in the course of motion of the contact load (counterbody) prior to covering of the crack mouth by the counterbody, which is not observed for higher values of the friction coefficient ($f = 0.3$). However, if the counterbody covers the crack mouths, then they start to close from the mouth independently of the value of the friction coefficient.
- The maximum value of the range ΔK_{II} in a contact cycle (for variable $\lambda = x_0/a$) corresponding to fracture by the mode II mechanism strongly depends on the distance between the cracks. For the distances between two parallel cracks comparable with the length of the contact section ($d \approx a\ldots 2a$), ΔK_{II} attains its highest value. In other words, the maximum softening of the rolling surface is attained under this condition (see Table 4.20); this result agrees with the experimental data obtained in [102];

- The paths of growth for systems of (2–5) parallel cracks of different lengths agree (Figs. 4.69, 4.70, 4.71, 4.72, 4.73 and 4.74) with the available experimental data [75] on crumbling of the contact surfaces of rolls made of rail steels.

References

1. Akama M, Mori T (2002) Boundary element analysis of surface initiated rolling contact fatigue cracks in wheel/rail contact systems. Wear 253:35–41
2. Andreikiv AE, Darchuk AI (1992) Ustalostnoye razrusheniye i dolgovechnost' konstrukciy (Fatigue Fracture and Durability of Structures). Naukova Dumka, Kiev
3. Beghini M, Bertiny L, Fontanari V (2005) Parametric study of oblique edge cracks under cyclic contact loading. Fatigue Fract Eng Mater Struct 28(1/2):31–40
4. Beghini M, Bertiny L, Fontanari V (1999) Stress intensity factors for an inclined edge crack in semiplane. Eng Fract Mech 62:607–613
5. Benuzzi D, Bormetti E, Donzella G (2003) Stress intensity factor range and propagation mode of surface cracks under rolling-sliding contact. Theor Appl Fract Mech 40:55–74
6. Beynon JH, Brown MW, Kapoor A (1999) Initiation, growth and branching cracks in railway track. In: Beynon JH, Brown MW, Lindley TC et al. (eds) Engineering against fatigue, pp. 461–472, A.A. Balkema Publ., Rotterdam
7. Bogdanski S (2002) The behaviour of kinked cracks in contact. In: Proceedings of 14th Bienniel conference on Fracture "Fracture mechanics beyond" (ECF 14), Vol. I/III, pp. 289–296, EMAS Publication
8. Bogdanski S, Olzak M, Stupnicki J (1996) Numerical stress analysis of rail rolling contact fatigue cracks. Wear 191:4–24
9. Bold PE, Brown MW, Allen RJ (1991) Shear mode crack growth and rolling contact fatigue. Wear 144:307–317
10. Bower AF (1988) The influence of crack face friction and trapped fluid on surface initiated rolling contact fatigue cracks. J Tribol Trans ASME 110(4):704–711
11. Cannon DF, Edel KO, Grassie SL, Sawley K (2003) Rail defects: an overview. Fatigue Fract Eng Mater Struct 26(10):865–886
12. Clayton P, Su X (1996) Surface initiated fatigue of pearlitic and bainitic steels under water lubricated rolling/sliding contact. Wear 200:63–73
13. Datsishin OP, Marchenko GP (1991) Edge-crack growth. Sov Mater Sci 27(5):465–471
14. Datsishin OP, Marchenko GP, Panasyuk VV (1993) Theory of crack growth in rolling contact. Mater Sci 29(4):373–383
15. Datsyshin AP, Panasyuk VV, Savruk MP (1976) General method for the solution of two-dimensional problems of the theory of cracks. In: Abstracts of the 4 Vsesoyuznogo Kongressa po teoreticheskoy i prikladnoy mekhaniki (4th All-Union Congress on Theoretical and Applied Mechanics). Naukova Dumka, Kiev
16. Datsyshyn OP (1996) Fracture and wear processes simulating under cyclic contact of solid bodies. In: Petit J (ed) ECF-11, vol II. Mechanism and mechanics of damage and failure. EMAS LTD, Warley, pp 1411–1416
17. Datsyshyn OP (2005) Service life and fracture of solid bodies under the conditions of cyclic contact interaction. Mater Sci 41(6):709–733
18. Datsyshyn OP, Glazov AYu (2004) Durability prognosis of rolling bodies by pitting development. In: Panasyuk VV (ed) Mekhanika ruynuvannya materialiv ta mitsnist' konstruktsiy (Fracture Mechanics of Materials and Strength of Structures), pp 243–248, Lviv
19. Datsyshyn OP, Glazov AYu (2006) Edge cracks paths in rolling bodies under action of linear pressure on the crack faces. In: Kit HS, Kushnir RM (eds) Matematychni problemy mekhaniky neodnoridnykh struktur (Mathematical Problems of Mechanics of Inhomogeneous Structures): Proceedings of the 7th international science conference Vol 2, pp 43–45, Lviv

20. Datsyshyn OP, Glazov AYu (2009) On some specific features of typical contact-fatigue defects formation in rolling bodies, In: Abstracts of the international Scientific conference "Suchasni problemy mekhaniky" ("Modern Problems of Mechanics"). p 34, Lviv
21. Datsyshyn OP, Glazov AYu (2013) Rolling bodies durability evaluation by formation of typical contact fatigue damages—pitting and spalling. Visnyk Ternopil'skogo Natsional'nogo Tekhnichnogo Universytetu, No. 3, pp 75–87
22. Datsyshyn OP, Glazov AYu (2010) Some features of subsurface cracks development in contact zone of wheel steels. Mashynoznavstvo 5:3–8
23. Datsyshyn OP, Glazov AYu, Khrunyk RA (2005) Prediction mill roller life by pitting development. In: Tezy 7 Mizhnarodnogo sympoziumu ukrayins'kyh inzheneriv-mekhanikiv u Lvovi (Abstracts of the 7th Int. Symposium of Ukrainian Mechanical Engineers in Lviv), pp 116–117, Lviv
24. Datsyshyn OP, Glazov AYu, Lenkovs'kyi TM (2016) Kinetics of spalling formation in the nearsurface area of railway rail. In: Problemy ta perspektyvy rozvytku zaliznychnogo transportu (Problems and Prospects of Development of Railroad Transport): Abstracts of the 76th international scientific-practical conference, pp 161–163, Dnipropetrovs'k
25. Datsyshyn OP, Hlazov AYu, Levus AB (2014) Specific features of contact of the faces of an edge crack under moving Hertzian loads. Mater Sci 49(5):589–601
26. Datsyshyn OP, Kopylets' MM (2003) Prediction of the service life of rolling bodies according to the development of a subsurface crack. Mater Sci 39(6):765–779
27. Datsyshyn O, Kuzin M, Glazov A, Kravchuk O (2014) Features of subsurface cracks propagation and spalling formation in rolling bodies, In: Panasyuk VV (ed) Mekhanika ruynuvannya i mitsnist' konstruktsiy (Fracture Mechanics of Materials and Strength of Structures): Proceedings of the international conference pp 179–186, Lviv
28. Datsyshyn OP, Levus AB (2003) Propagation of an edge crack under the pressure of liquid in the vicinity of the crack tip. Mater Sci 39(5):754–757
29. Datsyshyn OP, Levus AB (2009) Some formation features of squat defects in rail steels under rolling contact. In: Panasyuk VV (ed) Mekhanika ruynuvannya i mitsnist' konstruktsiy (Fracture Mechanics of Materials and Strength of Structures): Proceedings of the 4th international scientific conference, pp. 903–910, Lviv
30. Datsyshyn OP, Levus AB (2000) Stress intensity factors for a system of edge parallel cracks in a half-plane due to Hertzian pressure on its boundary. Mashynoznavstvo 11:9–15
31. Datsyshyn OP, Levus AB, Glazov AYu, Marchenko HP (2009) On some development features of pitting, spalling, cracking and dark-spot damages in rail steels under rolling contact. In: Proceedings of the 8th international conference on contact mechanics and wear of rail/wheel systems. Vol 1, pp 325–326, AB Editore, Firenze
32. Datsyshyn OP, Marchenko HP (2003) Estimation of mode II surface crack growth under rolling contact. Mashynoznavstvo 7:17–23
33. Datsyshyn OP, Marchenko HP, Glazov AYu (2019) On the special angle of surface cracks propagation in the railway rail heads. Eng Fract Mech 206:452–462
34. Datsyshyn OP, Marchenko HP, Hlazov AYu, Levus AB (2015) Influence of compressive stresses on the propagation of surface shear cracks in railroad rails. Mater Sci 51(2):235–243
35. Datsyshyn OP, Marchenko HP, Hlazov AYu, Levus AB (2004) One approach to the evaluation of durability of solid bodies. Mater Sci 40(4):484–490
36. Datsyshyn OP, Marchenko HP, Levus AB (2001) Investigation of rolling surface cracking under dry friction. In: Franek F, Bartz WJ, Pauschitz A (eds) Papers, Posters and Abstracts from 2nd World Tribology Congress (WTC–2001). On CD (file M-12-P64-603-Datsyshyn.pdf), Vienna (2001)
37. Datsyshyn OP, Marchenko HP, Levus AB (2001) Edge cracks growth paths during rolling under dry friction. Mashynoznavstvo 4–5, 38–41
38. Datsyshyn OP, Marchenko HP, Zynyuk OD, Hrytsyshyn PM (1995) Residual durability of the surface layer of ShKh15 bearing steel. Mater Sci 31(2):192–199
39. Datsyshyn OP, Panasyuk VV (2001) Pitting of the rolling bodies contact surface. Wear 251:1347–1355

40. Datsyshyn OP, Panasyuk VV, Glazov AYu (2011) Modelling of fatigue contact damages formation in rolling bodies and assesment of their durability. Wear 271(1–2):186–194
41. Datsyshyn OP, Panasyuk VV, Glazov AYu (2016) The model of the residual lifetime estimation of trybojoint elements by formation criteria of the typical contact fatigue damages. Int J Fatigue 83(2):300–312
42. Datsyshyn OP, Panasyuk VV, Glazov AYu (2009) The model of fatigue contact damages formation in rolling bodies and estimation of their durability. In: Proceedings of the 8th international conference on contact mechanics and wear of rail/wheel systems. Vol 1, pp 35–43, AB Editore, Firenze
43. Datsyshyn OP, Panasyuk VV, Pryshlyak RE (2014) Effect of rounding the edges in the base of a rider on the stress intensity factors in a body with edge cracks. Mater Sci 50(1):1–13
44. Datsyshyn OP, Pryshlyak RYe (2001) Investigation of the rolling surface pitting under boundary lubrication conditions. In: Franek F, Bartz WJ, Pauschitz A (eds) Papers, Posters and Abstracts from 2nd World Tribology Congress (WTC–2001). On CD (file M-12-P65-605-Datsyshyn.pdf), Vienna
45. Datsyshyn OP, Pryshlyak RE, Prykhods'ka SV, et al (1998) Influence of the shape of model contact load on the stress intensity factors for an edge crack. Problemy Trybologii 3:3–16
46. Datsyshyn OP, Tkachov VI, Hlazov AYu, Khrunyk RA (2006) Prediction of the contact durability of back-up rolls of forge-rolling mills in the process of development of pitting. Mater Sci 42(6):823–836
47. Donzella G, Faccoli M, Ghidini A, Mazzu A, Roberti R (2005) The competitive role of wear and RCF in rail steel. Eng Fract Mech 72:287–308
48. Ekberg A, Kabo E (2005) Fatigue of railway wheels and rail under rolling contact and thermal loading—an overview. Wear 258:1288–1300
49. Fletcher DI, Beynon JH (1999) A simple method of stress intensity factors calculation for inclined surface-breaking crack with crack face friction under contact loading. Proc Inst Mech Engrs, Part J, J Eng Tribol, 213:481–486
50. Fletcher DI, Franklin FJ, Kapoor A (2009) Rail surface fatigue and wear. In: Lewis R, Olofsson U (eds) Wheel-rail interface handbook. Woodhead Publishing, pp 280–310
51. Frolish MF, Fletcher DI, Beynon JH (2002) A quantitative model for predicting the morphology of surface initiated rolling contact fatigue cracks in back-up roll steels. Fatigue Fract Eng Mater Struct 25:1073–1086
52. Glazov AYu (2007) Evaluation of the residual contact durability of rail steels, In: Proceedings of the open scientific-technical conference of young scientists and specialists of the Karpenko Physicomechanical Institute of the Ukrainian National Academy of Sciences, pp 54–58, Lviv
53. Glazov AYu (2009) Kinetics of development of subsurface cracks in wheel steels, In: Proceedings of the Conference "Problemy koroziyno-mekhanichnogo ruynuvannya, inzheneriya poverhni, diagnostychni systemy" ("Problems of Corrosion-Mechanical Fracture, Surface Engineering, and Diagnostic Systems"), pp 76–79
54. Glazov AYu (2003) Residual life of 9KhF roll steel under the conditions of rolling contact fatigue. In: Proceedings of the Open Scientific-Technical Conf. of Young Scientists and Specialists of the Karpenko Physicomechanical Institute of the Ukrainian National Academy of Sciences, pp 142–145, Lviv
55. Glodež S, Flasker J, Ren Z (1997) A new method for the numerical determination of pitting resistance of gear teeth flanks. Fatigue Fract Eng Mater Struct 20(1):71–83
56. Godet M, Berthier Y, Dubourg MC, Vincent L (1992) Contact mechanics: needs for broader applications. J Phys D Appl Phys 25:A273–A278
57. Goshima T (2003) Thermomechanical effects on crack propagation in rolling contact fatigue failure. J Thermal Stresses 26:615–639
58. Goshima T, Kamishima Y (1996) Mutual interference of two surface cracks in a semi-infinite body due to rolling contact with frictional heating by a rigid roller. JSME Int J Ser A 39(1):26–33
59. Goshima T, Soda T (1997) Stress intensity factors of a subsurface crack in a semi-infinite body due to rolling/sliding contact and heat generation. JSME Int J, Ser A, 40(3):263–270

60. Heyer R (2002) The new UIC catalogue of rail defects. Eisenbahningenieur 52(9):94–109
61. Ishida M, Abe N (1996) Experimental study on rolling contact fatigue from the aspect of residual stress. Wear 191:65–71
62. Johnson KL (1985) Contact mechanics. Cambridge University Press, Cambridge
63. Kaneta M, Matsuda K, Murakami K et al (1995) Reproduction of rail dark-spot defects. Trans JSME C 61:3402–3409
64. Kaneta M, Matsuda K, Murakami Y, Nishikawa H (1998) A possible mechanism for rail dark spot defects. Trans ASME J Tribol 120:304–309
65. Kaneta M, Murakami Y (1987) Effects of oil hydraulic pressure on surface crack growth in rolling/sliding contact. Tribology Int 20(4):210–217
66. Kaneta M, Murakami Y (1991) Propagation of semi-elliptical surface crack in lubricated rolling/sliding elliptical contact. J Trib ASME 113:270–275
67. Keer LM, Bryant MD (1983) A pitting model for rolling contact fatigue. Trans ASME: J Lubric Technol 105(2):198–205
68. Keer LM, Bryant MD, Haritos GK (1982) Subsurface and surface cracking due to Hertzian contact. Trans ASME: J Lubric Technol 104(3):347–351
69. Komvopoulos K, Cho S-S (1997) Finite element analysis of subsurface crack propagation in a half-space due to a moving asperity contact. Wear 209:57–68
70. Lenkovs'kyi TM (2014) Determination of the characteristics of cyclic crack resistance of steels under transverse shear (a survey). Mater Sci 50(3):340–349
71. Levus AB (2007) Kinetics of contact of the crack faces in rolling, In: Proceedings of the conference "Problemy koroziyno-mekhanichnogo ruynuvannya, inzheneriya poverhni, diagnostychni systemy" ("Problems of Corrosion-Mechanical Fracture, Surface Engineering, and Diagnostic Systems"), pp 79–82
72. Levus AB, Datsyshyn OP (2013) System of parallel edge cracks in the contact zone of rolling bodies, Visnyk Ternopil'skogo Natsional'nogo Tekhnichnogo Universytetu 4:42–52
73. Levus AB, Glazov AYu, Datsyshyn OP (2008) Contact of faces of an edge crack under the action of Hertzian loading. In: Suchasni problemy mekhaniky ta matematyky (Modern Problems of Mechanics and Mathematics): Proceedings of the 2nd International science. conference, Vol 2, pp 56–59, Lviv
74. Lundberg G, Palmgren A (1947) Dynamic capacity of rolling bearing. Acta Polytechnica, Ser. Mech Eng RSAEE 1(3):50
75. Mašin A (1985) Přispěvek k porušeni kolejnic kontaktni únavou. Strojirenstvi 35(8):447–451 (1985)
76. Miller KJ (2001) Structural integrity—whose responsibility?. The 36th John Player Memorial Lecture presented at an Ordinary Meeting of the Institution of Mechanical Engineers, London: Inst. Mech. Engn, 24
77. Morozov NP, Nikolaev VA, Polukhin VP, Legun AM (1977) Proizvodstvo I ekspluatatsiya krupnykh opornyhkh valkov (Production and Operation of Large Backup Rolls). Metallurgiya, Moscow
78. Murakami Y (1987) Stress intensity factors handbook. Oxford, UK: Pergamon Press, 1456 p
79. Murakami Y., Sakae C., Hamada S.: Mechanism of rolling contact fatigue and measurement of ΔK_{IIth} for steels. In: Beynon JH, Brown MW, Lindley et al (eds) Engineering against fatigue. A.A. Balkema Publ., Rotterdam, pp 473–485
80. Mutton PJ, Tan M, Bartle P, Kapoor A (2009) The effect of severe head wear on rolling contact fatigue in heavy haul operations. In: Proceedings of the 8th international conference on contact mechanics and wear of rail/wheel systems, Vol 2, pp 397–407, AB Editore, Firenze
81. Orringer O, Morris JM, Steele RK (1984) Applied research on rail fatigue and fracture in the United States. Theor Appl Fracture Mech 1:23–49
82. Ostash OP, Andreiko IM, Kulyk VV, Uzlov IH, Babachenko OI (2007) Fatigue durability of steels of railroad wheels. Mater Sci 43(3):403–414
83. Otsuka A, Sugawara H, Shomura M (1996) A test method for mode II fatigue crack growth relation to a model for rolling contact fatigue. Fatigue Fract Eng Mater Struct 19(10):1265–1275

84. Panasyuk VV, Berezhnits'kyi LT (1964) Determination of the limit forces in tension of a plate with an arc-shaped crack. In: Karpenko GV (ed) Voprosy mekhaniki real'nogo tverdogo tela (Problems of Mechanics of Real Solid Body). Naukova Dumka, Kiev, pp 3–19
85. Panasyuk VV, Datsyshyn OP, Glazov AYu (2007) Prediction of the contact durability of rails by pitting development. Mashynoznavstvo 3:3–10
86. Panasyuk VV, Datsyshyn OP, Levus AB (2002) Evolution of a system of edge cracks in the region of rolling bodies cyclic contact. In: Neimitz A et al (ed), ECF-14, Fracture mechanics, Beyond 2000, V. I/III, UK: EMAS Publishing, Sheffield, pp 609–616
87. Panasyuk VV, Datsyshyn OP, Marchenko HP (2001) Crack growth in rolling bodies under the conditions of dry friction and wetting. Mater Sci 37(1):1–11
88. Panasyuk VV, Datsyshyn OP, Marchenko HP (1995) To crack propagation theory under rolling contact. Eng Fract Mech 52(1):179–191
89. Panasyuk VV, Ostash OP, Andreiko IM, et al (2012) Norms for steels aimed at preventing in-service defects on the rolling surfaces of whole-rolled high-strength railroad wheels. In: Problyemy resursu i bezpeky ekspluatatsiyi konstruktsiy, sporud ta mashyn (Problems of the Service Life and Operating Safety of Structures, Installations, and Machines), pp 594–598, Kyiv
90. Panasyuk VV, Ostash OP, Datsyshyn OP, Andreiko IM, et al (2009) Determination of the fatigue durability of railroad wheels made of high-strength steel. In: Problyemy resursu i bezpeky ekspluatatsiyi konstruktsiy, sporud ta mashyn (Problems of the Service Life and Operating Safety of Structures, Installations, and Machines), pp 659–663, Kyiv
91. Panasyuk VV, Savruk MP, Datsyshyn AP (1976) Raspredeleniye napryazheniy okolo treshchin v plastinakh i obolochkakh (Distribution of Stresses near Cracks in Plates and Shells). Naukova Dumka, Kiev
92. Pinegin SV (1969) Kontaktnaya prochnost' i soprotivleniye kacheniyu (Contact Strength and Rolling Resistance). Mashinostroenie, Moscow
93. Ringsberg JW, Bergkvist A (2003) On propagation of short rolling contact fatigue cracks. Fatigue Fract. Engng Mater. Struct. 26(10):969–983
94. Romaniv ON, Shur EA, Simin'kovich VN, Tkach AN, Kiseleva TN (1983) Crack resistance of pearlitic eutectoid steels. II: Failure of steels in cyclic loading. Sov Mater Sci 19(2):111–118
95. Sato M, Anderson PM, Rigney DA (1993) Rolling-sliding behavior of rail steels. Wear 162–164:159–172
96. Savruk MP (1981) Dvumernyye zadachi uprugosti dlya tel s treshchinsmi (Two-Dimensional Problems of Elasticity for Bodies with Cracks). Naukova Dumka, Kiev
97. Smith RA (2003) The wheel-rail interface–some recent accidents. Fatigue Fract Engng Mater Struct 26(10):901–907
98. Vakulenko IO, Anofriev VH, Hryshchenko MA, Perkov OM (2009) Defekty zaliznychnykh kolis (Defects of Railroad Wheels). Dnipropetrovs'k
99. Way S (1935) Pitting due to rolling contact. J Appl Mech Trans ASME 2:A49–A58
100. Yarema SYa, Mikitishin SI (1976) Analytical description of the fatigue-failure diagrams of materials. Sov Mater Sci 11(6):660–666
101. Zang WL, Gudmundson P (1991) Frictional contact problems of kinked cracks modeled by a boundary integral method. Int J Numer Meth Eng 31:427–446
102. Zerbst U, Madler K, Hintze H (2005) Fracture mechanics in railway applications—an overview. Eng Fract Mech 72(2):163–194

Chapter 5
Fretting Fatigue Fracture

Abstract This chapter deals with the investigation of contact interaction of two bodies under the conditions of fretting fatigue. We study the fracture processes in the material (and construct the paths of crack propagation) in the zone of cyclic contact of two bodies under the conditions of fretting fatigue, in particular, depending on the friction coefficient and stick/slip conditions between the bodies, the form of the base of counterbody (the type of modeling contact loading), etc. We also present some examples of evaluation of the residual service life for turbine blades of the GTE (gas-turbine engine) made of TS-5 (TC-5) titanium alloy.

5.1 Foreword Theses

The phenomenon of fretting is the process of accumulation of defects on the contact surfaces two bodies pressed to each other in the course of their cyclic mutual displacements with small amplitude. This phenomenon has been extensively studied for about 100 yr. First, it was regarded as the phenomenon of surface damage, which could be now called fretting wear. The first information about fretting wear in the literature appeared in 1911 in the work by Eden et al. [24]. They revealed fretting particles (powder and fragments) of iron oxide between the contacting metal surfaces. The contact studied in the cited was realized under the conditions of sliding of one of the pressed surfaces along the other surface (Fig. 5.1) with microscopic oscillatory displacements. At present, the phenomenon of fretting contact (contact interaction) is defined just in this way, although the authors did not mention this feature. Purposeful investigations of fretting were originated in 1927 by Tomlinson [61] who introduced the term "fretting corrosion" for the first time, thus emphasizing the presence of oxidation processes accompanying the accumulation of defects. Since that time, numerous researchers and engineers revealed various types of defects (including micro- and macrocracks) in the course of contact of bodies under the conditions of fretting and started to give much attention to the processes of fatigue fracture, connected with this contact. It was established that, under the conditions of fretting fatigue, the reliability of contacts in machine units is violated, the quality of the

O. Datsyshyn and V. Panasyuk, *Structural Integrity Assessment of Engineering Components Under Cyclic Contact*, Structural Integrity 9,
https://doi.org/10.1007/978-3-030-23069-2_5

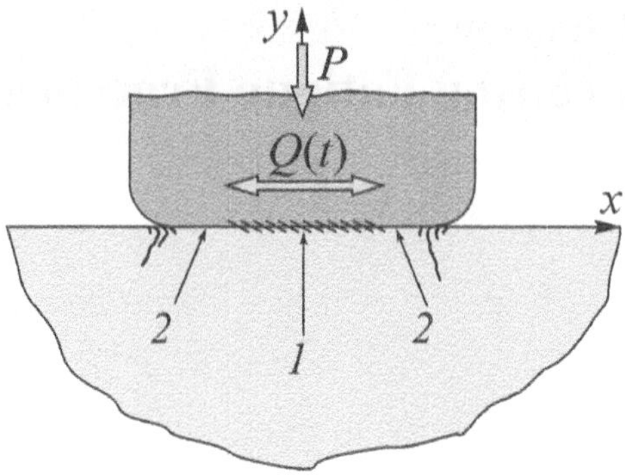

Fig. 5.1 Scheme of realization of fretting fatigue; (*1*) sticking, (*2*) slipping

surfaces of workpieces is deteriorated, and their fatigue strength and lifetime noticeably decrease.

At present, it is customary to distinguish [7, 66] three types of processes and defects developed in the course of fretting:

- the process of fretting fatigue realized under the combined action of fretting and fatigue; it is accompanied by the appearance and development of micro- and macrocracks, pits, cavities as well as by the insignificant formation of powder-like wear products;
- fretting corrosion, which is a kind of fracture and wear caused by fretting when the process is accom-panied by the chemical reactions between the materials and the medium, mostly of oxidizing nature;
- the process of fretting wear accompanied by the appearance of powder-like wear products; as a result of this process, the sizes of the workpieces change, which leads to the appearance of gaps and stress concentrators.

The process of fretting fatigue is most often realized in machine units and structures under the action of cyclic loads or in-service vibrations in the contact zones of junctions of the bodies and casings of various vehicles (cars, aircrafts, rockets, ships), in numerous industrial structures (turbines of nuclear power plants, gas-turbine engines, oil platforms, bridges, and pipelines), and also in various spline, bolt, and key connections. The processes of fretting fatigue in the nearsurface zone of elements of the units are often accompanied by the formation of cracks (Fig. 5.2). Thus, fretting in the railroad wheel axes was found by Maxwell et al. [40] in 1967. The experimental data and engineering practice demonstrate [3, 7, 30, 40, 43, 56, 59, 66] that the cracks are mainly localized on both sides of the contact zone. They are initiated and propagate under the action of cyclic contact and (possibly) bulk forces (tension-compression and bending) and most often grow into the bulk of the material (through the element of the unit). In this way, they substantially shorter the lifetime of some units and machines.

Collins [7] noted that “the loss of the serviceability of machine elements caused by fretting fatigue is one of the most dangerous kinds of fracture both due to the high frequency of its appearance and heaviness of its consequences.” Waterhouse

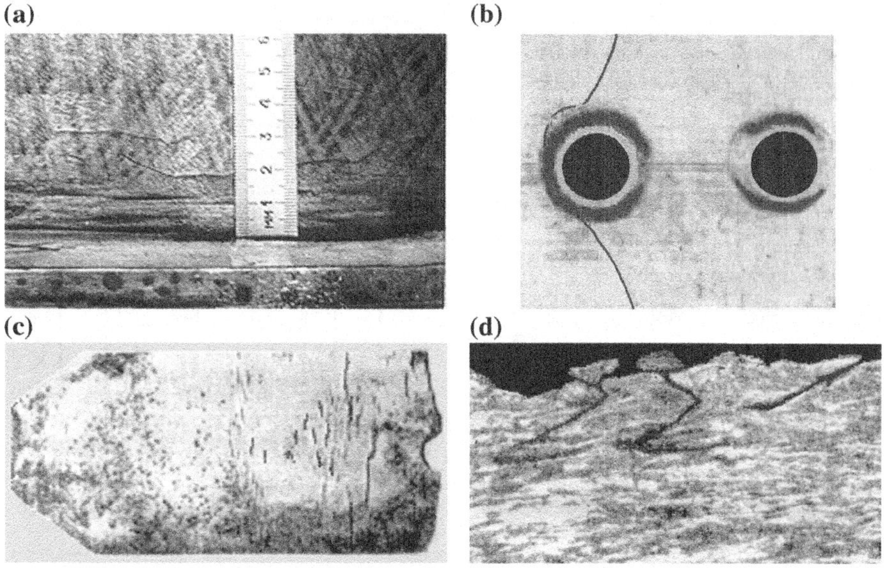

Fig. 5.2 Fatigue cracks initiated from the site of damage by fretting: **a** in the axle set of the shaft [3], **b** under the rivet head [66], **c** on the surface of the plate of car spring [56], **d** on the surface of a rope wire [66]

[66] emphasized: "It is especially important to understand the role of fretting in the initiation of fatigue fracture and establishing one of the causes why the fatigue resistance of structural units is lower than the fatigue resistance of separate parts."

There are numerous publications devoted to the investigation of fretting fatigue (fretting corrosion) of structural materials. These are, in particular, the monographs by [2, 3, 29, 31, 33, 49, 56, 66], selected chapters of the monographs by Collins [7] and Johnson [34], and the surveys [6, 43, 59]. In recent years, in connection with the development of fracture mechanics, the researchers started to extensively apply its approaches to the investigation of fretting fatigue. The surveys and generalizations of these investigations can be found in [6, 30, 33, 37, 43, 59, 63]. Note that the major part of these works is experimental and, at present, their methodology and, in particular, the schemes of testing (Fig. 5.3) are thoroughly developed [31, 33, 59, 66].

It is clear that the final aim of all these investigations is the development of numerical methods aimed at the prediction of lifetime and fracture of the elements of joints under the conditions of fretting fatigue. The work by Rooke and Jones [52] was one of the first publications where a computational model was proposed for the estimation of the lifetime of structural materials according to their crack resistance. In this method and in the major part of procedures developed later, the crack growth paths are predicted in a simplified way, on the basis of the experimental data. The stress intensity factors at the crack tip along the crack path are determined approximately by using the solution of a problem of the theory of elasticity for a

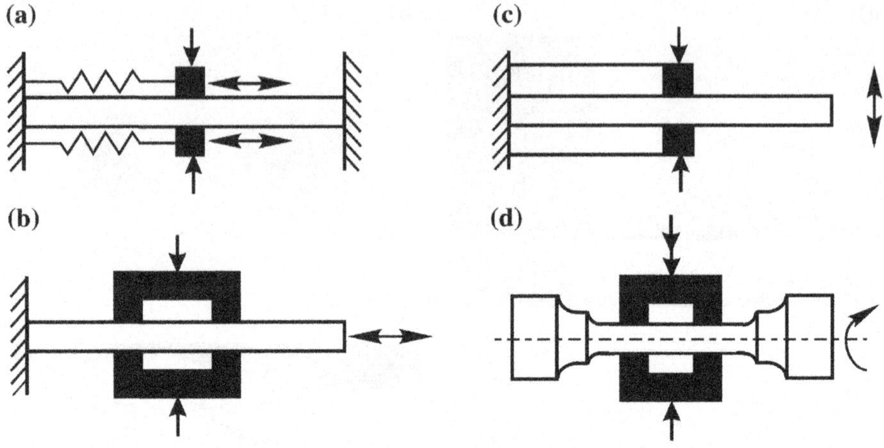

Fig. 5.3 Typical schemes of specimens loading under fretting fatigue test

half-plane with edge rectilinear crack either perpendicular [24, 52, 64] or inclined [28, 33, 39] to the boundary of the half-plane under loads modeling the force schemes of fretting fatigue. By using the SIF along the crack paths and the methods of fatigue fracture mechanics (see, e.g. [44], vol. 4), the researchers estimate the crack growth rate and the lifetime of the nearsurface layer of the analyzed product. In a different approach proposed in [27], the crack path is predicted according to the stress-strain state of the continuous half-plane (without cracks).

In the present work, the computational model used for the investigation of fracture processes and estimation of the lifetime of elements of the fretting couples (according to the crack growth resistance of material) is based on the concepts of the general model described in Chap. 2. It was successively developed in [9, 14, 15, 19, 47]. In what follows, we propose a description of this model.

Assume that one of the bodies is damaged by cracks. As the simplest version of model domain for this body, we choose an elastic half-plane weakened by a system of curvilinear cracks (Fig. 5.4). We replace the contact influence of the counterbody either by the action of a punch (Fig. 5.4) or by the action of normal static forces $p(x, \lambda)$ and alternating (in a contact cycle. as functions of time t) tangential forces $q(t, x, \lambda)$ distributed in a certain way over a part of the boundary (Fig. 5.5). These forces depend on the shape and sizes of the counterbody, the specific features of contact interaction in a cycle, the mechanical characteristics of the materials and contact surfaces, etc. As the most extensively used force schemes of model loads $p(x, \lambda)$, we can mentioned concentrated forces, contact pressure, or elliptic (Hertzian) distribution of forces [23, 28, 33, 39, 42, 52, 64]. Under conditions of full slip between the contacting bodies, the model tangential forces are expressed via the normal forces by the Amontons–Coulomb law, whereas in the presence of the sticking region, it is customary to use more complicated dependences between the tangential and normal forces [33, 34, 42]. If the body is subjected to the nominal (bulk) cyclic loading

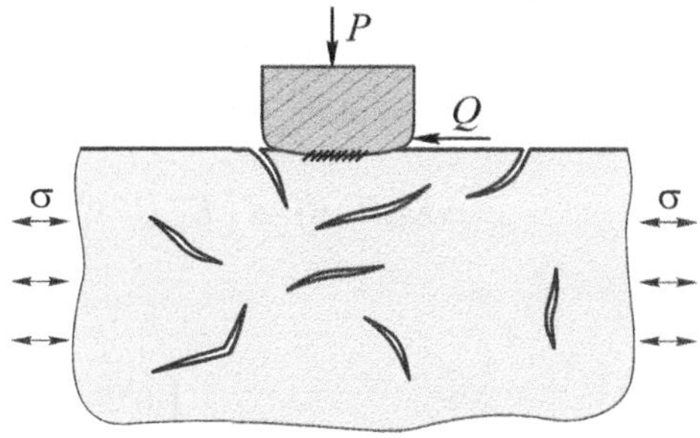

Fig. 5.4 General scheme of the model

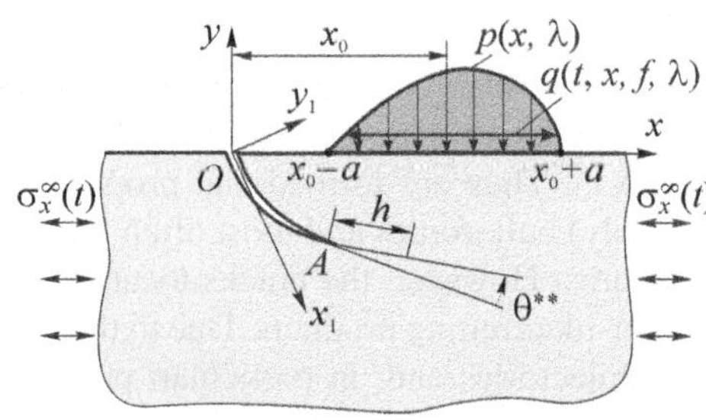

Fig. 5.5 Computational scheme of the model

(Fig. 5.3b, c), then the scheme of model contact loading is supplemented by the loading of the half-plane at infinity σ_x^∞ (Figs. 5.4 and 5.5).

In view of the alternating character of displacements of the counterbody, we approximate the model tangential forces as follows:

$$q(x, f, \lambda, t) = q^{\pm}(x, f, \lambda) = \pm q(x, f, \lambda). \tag{5.1}$$

The parameter $\lambda = x_0/a$ is regarded as fixed. Thus, in order to construct the crack growth paths (trajectories) under the conditions of fretting fatigue (unlike the conditions of rolling), we use only the main step h of their increments (see Fig. 5.5). The tangential forces $q^{+}(x, f, \lambda)$ and $q^{-}(x, f, \lambda)$ (identically distributed but with opposite directions) form a contact cycle and correspond to its positive and negative phases. Under the conditions of fretting fatigue, we may observe both the full slipping of the bodies and their sticking with partial slipping. We assume that, under the conditions of full slipping, the tangential forces are given by the formula

$$q^{\pm}(x, f, \lambda) = \pm f p(x, \lambda), \tag{5.2}$$

where f is the friction coefficient between the bodies.

We now study the growth paths of the edge cracks the under the conditions of realization of the opening mechanism. Hence, we can assume that the trajectory is formed by the maximal values of the SIF $K_{I\theta}$ in a contact cycle (see condition (2.24)). As shown above, they are given by the formula

$$K_{I\theta}^{\max}(l) = \max\{K_{I\theta}^{+}(l),\ K_{I\theta}^{-}(l)\}. \tag{5.3}$$

Hence, we represent the condition of crack growth (2.26) takes the form

$$\Delta K_{I\theta}(l,\ \theta) = \left| K_{I\theta}^{+}(l,\ \theta^{**}) - K_{I\theta}^{-}(l,\ \theta^{**}) \right| \geq \Delta K_{Ith}, \tag{5.4}$$

where

$$\theta^{**} = \begin{cases} \theta^{*+} & \text{for } K_{I\theta}^{+} > K_{I\theta}^{-}; \\ \theta^{*-} & \text{for } K_{I\theta}^{-} > K_{I\theta}^{+}. \end{cases} \tag{5.5}$$

The experimental data and engineering practice show that, under the conditions of fretting fatigue, the cracks are mainly localized on both sides of the contact zone (Fig. 5.1). They are formed and propagate under the action of cyclic contact and (possibly) bulk forces and most often grow into the bulk of the material, which lead to failures. However, the cracks located under the counterbody may be filled with powder-like fretting products. Due to their wedging action, these cracks may change their trajectories and, in particular, propagate toward the surface of the body and cause its crumbling. In what follows, we study both cases of crack growth in the fretting zone. The case where the conditions of partial sticking are realized in the contact zone is also analyzed.

5.2 Edge Crack Mouth Outside the Contact Section. Full Slipping Between Contacting Bodies

Stress intensity factors for the edge cracks and the paths of their growth. It is known from the engineering practice and experimental data that the cracklike defects observed under the conditions of fretting fatigue mainly have the surface character and are localized near the edges of the counterbody (contact zone). Therefore, prior to estimating the lifetime of the material of these most sensitive zones, we determine the stress-strain state (SIF) and the crack growth paths in the elastic half-plane near the ends of the contact section (Fig. 5.6).

We first investigate and analyze the SIF for an edge rectilinear crack of length l_0 inclined at an arbitrary angle β to the boundary of the half-plane such that the crack mouth appears on the boundary beyond the contact section (Fig. 5.6). We model the contact load by a constant pressure ($p(x) = p = \text{const}$) with tangential component ($q(x) = fp$) in a part of the half-plane boundary of length $2a$ at a distance x_0 from the crack mouth. Without loss of generality of formulation of the problem, we assume that the contact load is located to the right of the crack mouth. Here, the positive direction of the tangential forces coincides with the direction of the Ox-axis. We also suppose that the crack faces do not contact and are free of loads. Then the boundary conditions of our problem can be written as follows:

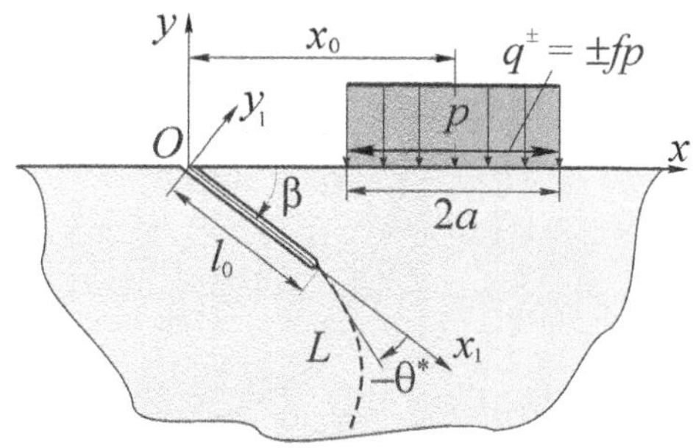

Fig. 5.6 Simplified computational scheme of the problem

$$\begin{aligned}\sigma_y(x) \pm i\tau_{xy}(x) = -p \pm iq = -p(1 \mp if); \quad |x - x_0| \le a, \quad y = 0; \\ N^{\pm}(t) + iT^{\pm}(t) = 0, \quad t \in L.\end{aligned} \tag{5.6}$$

In this case, the problem of evaluation of the SIF is reduced to the solution of the singular integral Eq. (3.151) with the right-hand side (3.168), namely,

$$\begin{aligned}P\,(\eta) &= P_0(\eta) \\ &= \frac{p}{\pi}\left\{\frac{w'(\eta)}{2}\left[(i-f)\ln\frac{\lambda+1-b(\eta)\varepsilon}{\lambda-1-b(\eta)\varepsilon} - (i+f)\ln\frac{\lambda+1-\overline{b}(\eta)\varepsilon}{\lambda-1-\overline{b}(\eta)\varepsilon}\right]+\right. \\ &\left. + \frac{\overline{w'}(\eta)}{\exp(2i\alpha)}\left[\frac{(f+i)\,\varepsilon\left(\overline{b}(\eta)-b(\eta)\right)}{\left(\lambda-\varepsilon\,\overline{b}(\eta)\right)^2-1} + f\ln\frac{\lambda+1-\overline{b}(\eta)\varepsilon}{\lambda-1-\overline{b}(\eta)\varepsilon}\right]\right\},\end{aligned} \tag{5.7}$$

where $b(\eta) = [z_{01} + w(\eta)\exp(i\alpha)]/l$; $\varepsilon = l/a$; $\lambda = x_0/a$; $\alpha = -\beta$, l is a parameter characterizing the crack length (in the general case, curvilinear), and $t = w(\eta)$ is the parametric equation of the crack contour L.

If the half-plane is additionally subjected to cyclic tension at infinity with uniformly distributed forces $\sigma_x^{\infty} = \sigma$, then one more term appears on the right-hand side of the SIE (3.151) (see (3.154)):

$$P_0^{\infty}\,(\eta) = \sigma\left[\overline{w'}(\eta)\exp(-2i\alpha) - w'(\eta)\right]/2. \tag{5.8}$$

We solve the singular integral equation (3.151) by the method of mechanical quadratures (see Sect. 3.5).

The preliminary numerical analysis of the dependences of the SIF for an initially rectilinear crack on the parameters $\varepsilon = l_0/a$, β, λ, and f shows that, in the negative phase of a contact cycle, where $q = q^- = -fp$ (Fig. 5.6), the crack is mainly located in the compression zone ($K_{\mathrm{I}} < 0$) and, in its positive phase, where $q = q^+ = +fp$, it is located in the tension zone ($K_{\mathrm{I}} > 0$). Hence, in what follows, we investigate the SIF and trajectories for the positive phase of the contact cycle.

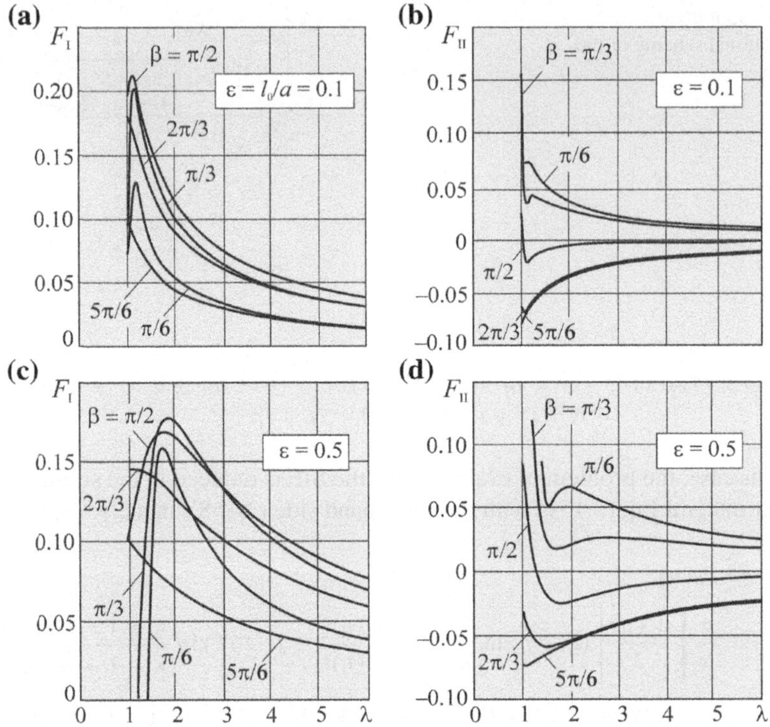

Fig. 5.7 Normalized SIFs $F_{\mathrm{I}} = K_{\mathrm{I}}/(p\sqrt{\pi a})$ **a**, **c** and $F_{\mathrm{II}} = K_{\mathrm{II}}/(p\sqrt{\pi a})$ **b**, **d** for an edge rectilinear crack inclined at an angle β to the boundary of the half-plane depending on the relative distance $\lambda = x_0/a$ of the contact load from the crack mouth; $f = 0.5$

In Figs. 5.7 and 5.8, we present the plots of the normalized SIF $F^{+}_{\mathrm{I,II}}(\lambda,\ \beta,\ \varepsilon) = K^{+}_{\mathrm{I,II}}(\lambda,\ \beta,\ \varepsilon)/\left(p\sqrt{\pi a}\right)$ on the angle β of crack orientation and the relative distance ($\lambda = x_0/a$) of the contact load from its mouth for the values of friction coefficient f = 0.5 and f = 0.8 and (mainly) for two values of the relative crack length $\varepsilon = l_0/a$ = 0.1 and $\varepsilon = 0.5$. The range of the parameter λ is chosen with regard for the direction of tangential forces q to guarantee that the crack faces do not contact. In the course of these calculations, the absence of contact between the crack faces was checked by satisfying the condition $K_{\mathrm{I}}(\lambda) > 0$ and also by the sign of the discontinuity of normal displacements along the crack (see formula (4.11)).

The data presented in Figs. 5.7 and 5.8 and the results of other additional calculations enable us to make the following conclusions:

- The values of the SIF $K_{\mathrm{I}}(\lambda)$ and max $K_{\mathrm{I}}(\lambda)$ strongly depend on the friction coefficient f, i.e., on the value of contact tangential forces. The quantity max K_{I} strongly increases with f. Moreover, the range of λ in which the crack is open (its faces are not in contact) enlarges. Here, the values of $K_{\mathrm{II}}(\lambda)$ and max $|K_{\mathrm{II}}(\lambda)|$ weakly vary as functions of f. However, the direction of tangential contact forces determines,

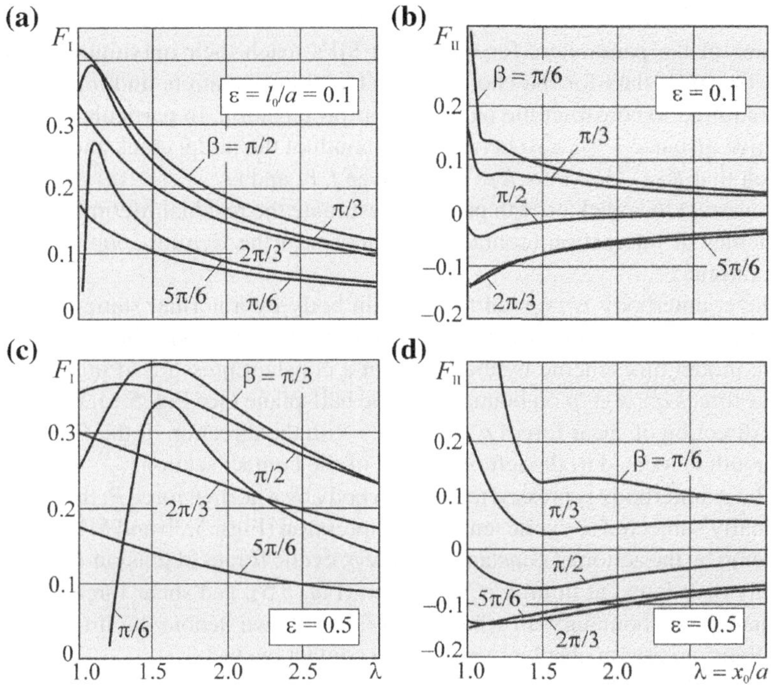

Fig. 5.8 Normalized SIFs $F_{\mathrm{I}}(\lambda, \beta, \varepsilon)$ and $F_{\mathrm{II}}(\lambda, \beta, \varepsilon)$ for the edge rectilinear crack inclined to the boundary; $f = 0.8$

to a large extent, the direction of tangential stresses along the crack faces near the tip.

- In the case of short cracks ($\varepsilon = 0.1$), for all considered values of f, the values max $K_{\mathrm{I}}(\lambda)$ are attained for $\beta = \pi/2$ if the edge of the load is located near the crack mouth ($\lambda \approx 1.0$). For longer cracks ($\varepsilon = 0.5, 1.0, 2.0$), the values max $K_{\mathrm{I}}(\lambda)$ shift to the side of $\beta = \pi/6$ and are attained when the distance between the contact load and the crack mouth is much larger (see, in particular, Fig. 5.8a, c).
- The character of changes in $K_{\mathrm{II}}(\lambda)$, including its sign, strongly depends on the crack orientation (see Fig. 5.8b, d).
- The maxima of $|K_{\mathrm{II}}(\lambda)|$ in a broad range of the lengths of (open) cracks and the values of the friction coefficient $f = 0.3\ldots0.5$ are attained for the orientations $\beta = \pi/3$. The increase in the friction coefficient in contact between the bodies implies that max $|K_{\mathrm{II}}(\lambda)|$ is attained for cracks inclined to the boundary at a smaller angle ($\beta = \pi/6$; Fig. 5.8b, d). These results are in good agreement with the experimental data on the orientation of shear cracks in the initial stage of their propagation [28, 59, 64].
- The results of additional calculations show that the character of changes in the mixed-type SIF $K_{\mathrm{I}\theta}(\lambda)$ is, in general, very similar to $K_{\mathrm{I}}(\lambda)$.

Summarizing these results, we note that the data on the SIFs and, especially, on the values of the parameters for which the SIFs reach their maximum values are used as the input data for the choice of the length, orientation, and location of the cracks required to construct the paths of their propagation. In particular, we choose the relative distance $\lambda^* = x_0^*/a$ between the mouth of the initial crack and the contact load such that $K_{I\theta}(\lambda^*) = \max K_{I\theta}(\lambda)$ for given f, β, and ε.

To construct the crack growth paths and estimate the residual lifetime, we choose two schemes of interaction (contact) of elements of the fretting couples typical of fretting fatigue:

I. The counterbody is pressed to the main body by a normal static force P and realizes cyclic reciprocating displacements relative to main body (Figs. 5.1, 5.3a and 5.4). We model this scheme by the action of a constant pressure of intensity p and the shear forces $q^{\pm} = \pm fp$ on boundary of the half-plane (see Fig. 5.6). It is assumed that the direction of shear forces q^+ coincides with the direction of the Ox-axis if the crack mouth is located to the left of the end of the contact section;

II. The counterbody is pressed to the main body by a normal force P; the main body is nominally subjected to cyclic tension-compression (Figs. 5.3b and 5.4). We model this scheme by the action of constant pressure p, cyclic forces of tension-compression uniformly distributed at infinity $\sigma_x^{\infty}(t) = \pm\sigma$ (Fig. 5.5), and shear forces $q^{\pm} = \pm fp$ on the half-plane boundary. In what follows, by $\sigma^{\pm}$ we denote the limits of forces $\sigma_x^{\infty}(t)$ (their maximum absolute values) in a contact cycle.

Note that the shear forces $q^{\pm}$ formed in the contact zone in the case of contact scheme I differ from the shear forces $q^{\pm}$ formed in the contact scheme II. In the first case, they appear under the simultaneous action of the pressure p and alternating displacements $s(t)$, while in the second case, under the action of the forces $\pm\sigma$ and pressure p. Clearly, for a more precise modeling of the contact schemes I and II, the dependences $q(s(t), x)$ and $q(\sigma_x^{\infty}(t), x)$ should be determined either experimentally or as a result of the solution of quite complicated nonstationary contact problems.

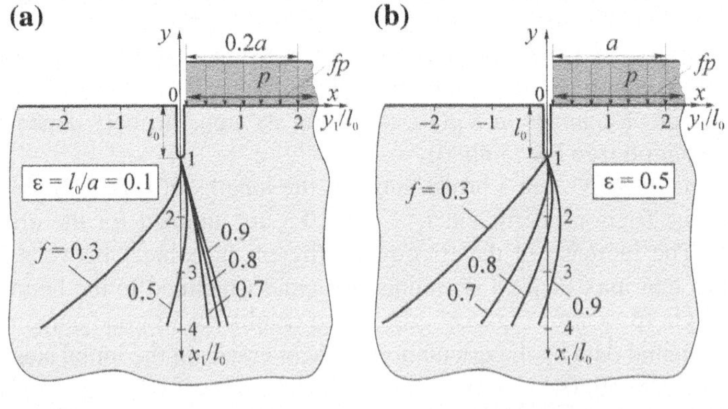

Fig. 5.9 Dependences of the propagation trajectories of an initially rectilinear crack perpendicular to the half-plane boundary on the friction coefficient f in contact between the bodies

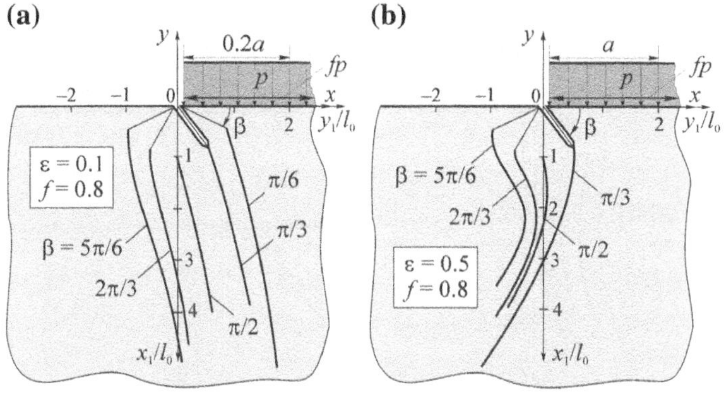

Fig. 5.10 Dependences of the propagation trajectories of an initially rectilinear crack on the angle β of its inclination to the half-plane boundary

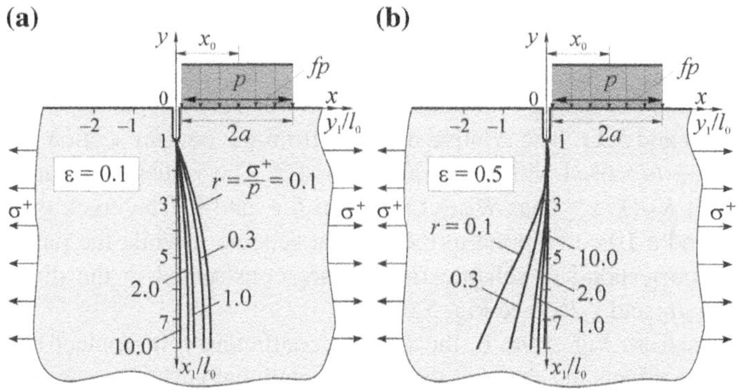

Fig. 5.11 Dependences of the shapes of the paths of initially rectilinear cracks perpendicular to the half-plane boundary on the ratio r of the nominal forces σ^+ to the contact pressure $p; f = 0.8$

On the basis of the proposed computational model and the algorithm of step-by-step construction of the crack growth paths, we studied the paths (trajectories) of propagation of the most dangerous edge and initially rectilinear cracks located in the contact zone (the schematic diagrams of these problems are shown in Figs. 5.9, 5.10 and 5.11). For the construction of these trajectories, we solve the SIE (3.151) for a curvilinear edge crack in the elastic half-plane loaded either by forces of the form (5.6) in the contact scheme I or by the forces $\sigma_x^\infty = \sigma$ and forces of the form (5.6) in the contact scheme II. Note that, in the case of scheme I, we construct the paths solely for the positive phase of a cycle ($q = q^+ = +fp$); at the same time, for scheme II, this is done for the zero-to-tension loading mode ($\sigma^- = 0$). In this case, $\max \sigma_x^\infty(t) = \sigma^+ > 0$. It should be emphasized that, in the case of contact of the crack faces either under the action of tangential forces q^- in the negative phase of the contact cycle or under the action of compressive bulk forces ($\sigma_x^\infty = \sigma^- < 0$), the problem can be solved by using singular integral equations deduced in Sect. 3.10 of

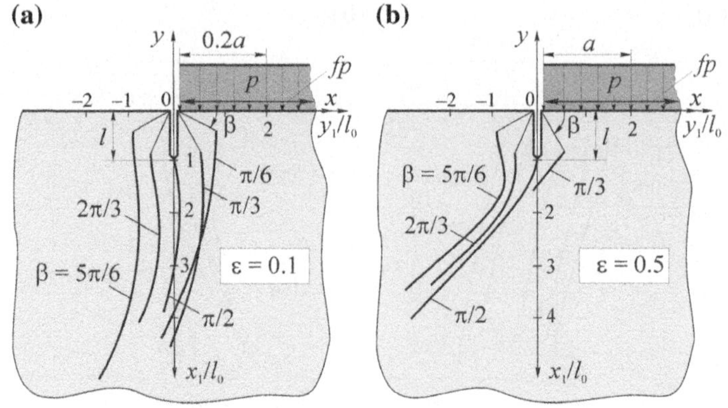

Fig. 5.12 Dependences of the propagation paths of an initially rectilinear crack on the angle β of its orientation for the initial relative crack length $\varepsilon = 0.1$ (**a**) or $\varepsilon = 0.5$ (**b**); $f = 0.5$; the relative distance of the contact load from the crack mouth $\lambda^* = 1.1$ (**a**) and $\lambda^* = 1.2$ (**b**)

the present work. The results of evaluation of the crack growth paths are displayed in Figs. 5.9, 5.10 and 5.11. The relative distance from the contact section to the crack mouth $\lambda^* = x_0^* / a$ varied within the range 1.1–1.2. The values of λ^* are chosen to guarantee that $K_{\mathrm{I}\theta}(\lambda^*) \approx \max K_{\mathrm{I}\theta}(\lambda)$ for fixed f, ε, and β. The crack growth paths in Figs. 5.9 and 5.10 correspond to the contact scheme I, while the paths depicted in Fig. 5.11 correspond to scheme II; they are constructed in the dimensionless coordinates x_1/l_0 and y_1/l_0 (see Fig. 5.6).

As follows from Fig. 5.9a, b, the friction coefficient f in contact between the bodies strongly affects the shape of the crack growth path. The increase in f "drags" the analyzed crack under the contact zone. For lower f, the crack deviates from the contact zone.

By analyzing Fig. 5.10a and b and the trajectories of propagation of the cracks initially inclined at different angles β to the half-plane boundary for $f = 0.5$ (Fig. 5.12a, b), we can conclude that the initial orientation of the crack weakly affects the character of its subsequent development. The cracks with different initial orientations approach the asymptote typical of every value of f.

Under the contact conditions II, when the process of fretting between the elements of fretting couple is caused by the nominal tension-compression of the analyzed body, the crack paths (Figs. 5.11 and 5.13) originating from the edge of the contact section are closer to rectilinear than for the contact scheme I. As the intensity of nominal forces (the parameter $r = \sigma^+/p$) increases, the crack growth paths exhibit a trend to take orientation perpendicular to the direction of these forces.

In Fig. 5.14a, we present the images of microcracks in a thin surface layer near the end face of the fretting bridge [42]. At the same time, in Fig. 5.14b, we see the main crack caused by the complex action of fretting and nominal tension-compression [26]. The comparison of the computed trajectories shown in Fig. 5.10a (curves β =

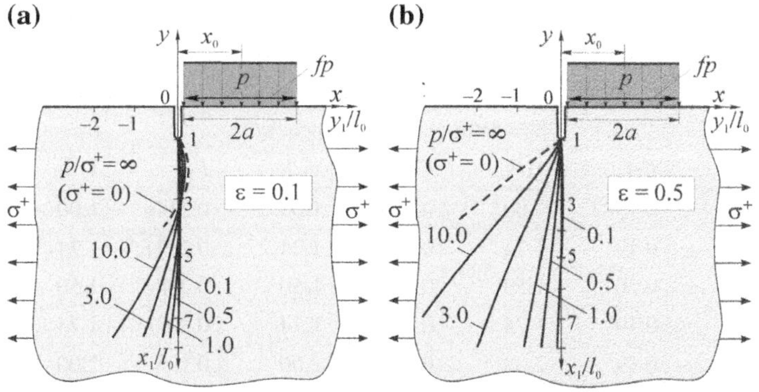

Fig. 5.13 Dependences of the shape of paths of initially perpendicular to the half-plane boundary crack on the ratio p/σ of the normal contact pressure p to the intensity of tensile forces σ at infinity for the $\varepsilon = 0.1$ (**a**) and $\varepsilon = 0.5$ (**b**); $f = 0.5$; the relative distance between the contact load and the crack mouth is $\lambda^* = 1.1$

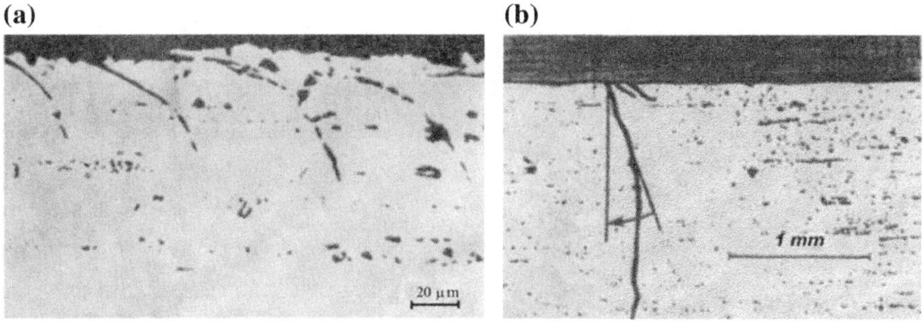

Fig. 5.14 Initiation (**a**) and propagation (**b**) of cracks in aluminum alloy under the conditions of fretting fatigue [37]

$\pi/3$ and $\beta = \pi/6$) and Fig. 5.11a with the experimental data presented in Fig. 5.14 demonstrates that they are in good agreement.

It should be emphasized that, in the absence of the nominal load, the SIF $F_{I\theta}$ decrease along the trajectories (Table 5.1a). With the appearance and growth of the nominal load (increase in $r = |\sigma^+|/p$), the crack paths become perpendicular to the direction of tension at infinity (Figs. 5.11 and 5.13) and the SIF $F_{I\theta}$ increase along the trajectories (Table 5.1b).

Note that the numerical data were mainly obtained for the order of the system of algebraic equations $M = 40$ and the step of construction of the trajectory $h/l = 0.2$.

Estimation of the residual lifetime. We now find the residual lifetime of the contact surface of VT3-1 (BT3-1) titanium alloy damaged by an edge initially rectilinear crack for its locations relative to the counterbody (contact loading) typical of fretting fatigue. We consider scheme II of contact interaction and the stage of crack propagation by the opening mechanism. We perform our calculations on the basis

Table 5.1 Changes in the SIF $F_{I\theta} = K_{I\theta}/\left(p\sqrt{\pi a}\right)$ along the crack paths with relative initial length $\varepsilon = 0.1$

a	$f = 0.5$					$f = 0.9$	
$\beta = \pi/6$		$\beta = 5\pi/6$		$\beta = \pi/2$		$\beta = \pi/2$	
x_1/l_0	$F_{I\theta}$	x_1/l_0	$F_{I\theta}$	x_1/l_0	$F_{I\theta}$	x_1/l_0	$F_{I\theta}$
1.0	0.1583	1.00	0.1341	1.00	0.2146	1.00	0.4686
1.1	0.19	1.24	0.208	1.24	0.2131	1.24	0.488
1.2	0.20	1.50	0.21	1.50	0.2082	1.50	0.502
1.3	0.19	1.74	0.19	1.74	0.2024	1.74	0.50
1.4	0.19	2.00	0.18	2.00	0.195	2.00	0.51
1.5	0.19	2.24	0.16	2.24	0.189	2.24	0.51
1.6	0.19	2.50	0.15	2.50	0.183	2.50	0.51
1.7	0.18	2.74	0.14	2.74	0.178	2.74	0.50
1.8	0.17	3.00	0.14	3.00	0.172	3.00	0.50
1.9	0.16	3.24	0.13	3.24	0.167	3.24	0.49
2.0	0.15	3.50	0.13	3.50	0.162	3.50	0.49
2.1	0.13	3.74	0.12	3.74	0.158	3.74	0.48
		4.00	0.13	4.00	0.153	4.00	0.47

b	$f = 0.5, \beta = \pi/2$		
x_1/l_0	$F_{I\theta}$		
	$r = \sigma^+/p = 10$	$r = 1.0$	$r = 0.1$
1.0	3.75	0.5681	0.2498
1.5	4.55	0.6427	0.2517
2.0	5.21	0.698	0.2462
2.5	5.79	0.744	0.2395
3.0	6.31	0.784	0.233
3.5	6.79	0.821	0.227
4.0	7.23	0.854	0.221
4.5	7.65	0.885	0.217
5.0	8.04	0.915	0.212
5.5	8.42	0.943	0.208
6.0	8.78	0.971	0.205
7.0	9.45	1.022	0.200
8.0	10.08	1.072	0.197

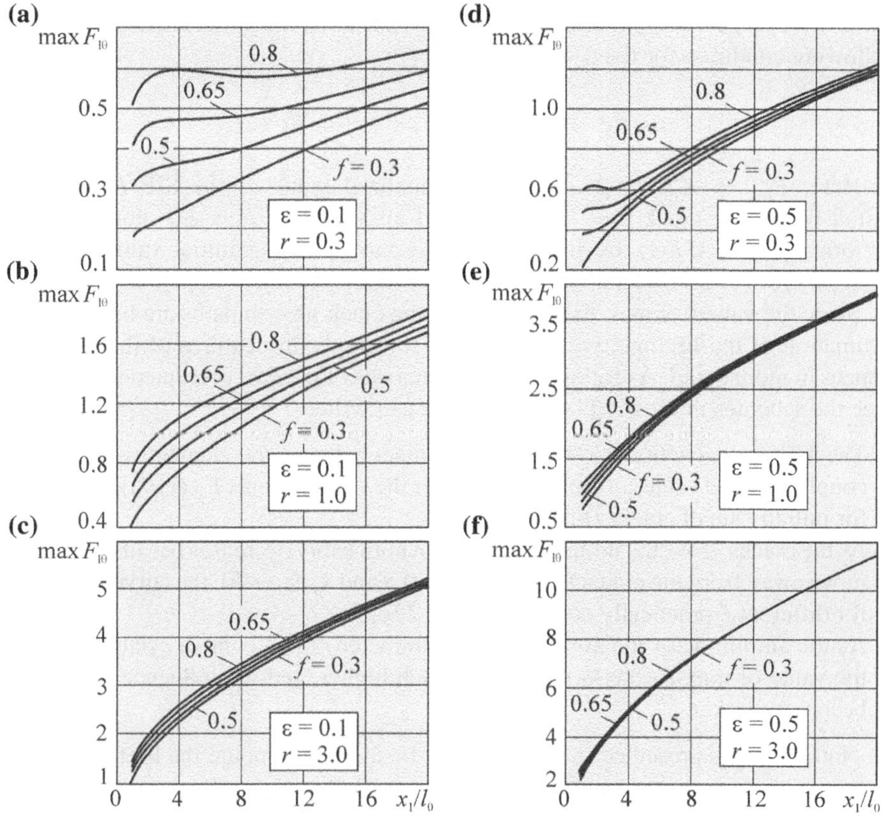

Fig. 5.15 Maximal values of the normalized SIF $F_{I\theta}(x_1/l_0)$ along the trajectories of the edge crack propagation; $\beta = \pi/3$; $\lambda^* = 1.1$; $r = \sigma^+/p$

of the second relation in (2.17) with the use of the values $\Delta K_\sigma = \Delta K_{I\theta} = \max K_{I\theta}$ along the crack growth paths (Fig. 5.15). The crack growth rate is given by the Paris relation [67]:

$$v = C(\Delta K_{I\theta})^n. \tag{5.9}$$

The characteristics of cyclic crack resistance of VT3-1 alloy are as follows [4]: $C = 2.14 \times 10^{-13}$ (MPa)$^{-n}\times$ m$^{1-n/2}$, $n = 4.17$, $K_{I,\,1\text{–}2} = 11.6$ MPa $\sqrt{\text{m}}$ and $K_{I,\,2\text{–}3} = 62.0$ MPa $\sqrt{\text{m}}$. Here, $K_{I,\,1\text{–}2}$ and $K_{I,\,2\text{–}3}$ are the boundaries of Paris section in the FFD [44]. We use the quantity $K_{I,\,2\text{–}3}$ to find the critical values (relation (2.27)) of the crack length $l_{c\sigma} = l_c\ (x_{1c}/l_0)$ and the lifetime $N_{g\sigma}$.

Assume that the initial crack length $l_{0\sigma} = l_0 = 1$ mm and the contact pressure p = 120 MPa. As already indicated, for the onset of propagation of an edge crack, it is necessary that condition (5.4) be satisfied. We guarantee the validity of this condition satisfaction by applying the corresponding load, i.e., specify the maximal nominal force in a cycle $\sigma^+ = \sigma_*^+$ (parameter $r = \sigma^+/p$) such that the initial crack satisfies the following condition for fixed ε, β, λ^*, r, and f :

$$p\sqrt{\pi a} \cdot \max F_{\mathrm{I}\theta}^0(\varepsilon, \beta, f, \lambda^*, r) = K_{\mathrm{I},\,1-2}. \tag{5.10}$$

Here, $F_{\mathrm{I}\theta}^0 = K_{\mathrm{I}\theta}^0/(p\sqrt{\pi a})$ is the normalized value of the SIF $K_{\mathrm{I}\theta}$ for the initial rectilinear crack $\varepsilon = l_0/a$. Note that all values of max $F_{\mathrm{I}\theta}$ along the trajectories (Fig. 5.15a–f) are higher than the corresponding initial values of max $F_{\mathrm{I}\theta}^0(\varepsilon, \beta, f, \lambda^*, r)$. Hence, after their start, the analyzed cracks do not stop.

Since the values of max $F_{\mathrm{I}\theta}(x_1/l_0)$ along the crack growth paths are basic for the estimation of the lifetime, we now discuss some specific features of these dependences in more detail. As follows from the analysis of the data presented in Fig. 5.15 (see the schemes in Figs. 5.11 and 5.13), we can write:

- The higher the friction coefficient f in contact between the elements of a friction couple for fixed values of ε and r, the greater the values of max $F_{\mathrm{I}\theta}(x_1/l_0)$, especially for initially small cracks (Fig. 5.15a–c).
- As the cracks grow, the influence of friction noticeably decreases because their tips move away from the contact zone: for $\varepsilon = 0.5$ and $x_1/l_0 \to 20$, the curves obtained for different f practically coincide (Fig. 5.15e, f).
- As the amplitude of the nominal load (parameter r) in a contact cycle increases, the value of max $F_{\mathrm{I}\theta}$ (x_1/l_0) becomes much higher, and the influence of friction becomes weaker.

Note that data presented in Fig. 5.15 can be used to compute the lifetime of any isotropic elastic material (within the framework of the proposed approach).

In Table 5.2, we present the results of evaluation of the lifetime of the nearsurface layer in the contact zone of VT3-1 titanium alloy according to its crack resistance. In this case, the lifetime $N_{g\sigma}$ is estimated by the number of loading cycles for which the length $l_0 = 1$ mm of the initial edge crack increases to its critical value $l_{c\sigma}$ upon attainment of which the crack begins to grow spontaneously. We also present the approximate values of these critical lengths $l_{c\sigma}/l_0 \approx x_{1c}/l_0$. Moreover, in Table 5.2a, b, we present the intermediate values of the number of loading cycles $N_{p\sigma}$ for which the initial crack grows by about 20 its lengths ($l_p/l_0 \approx x_{1p}/l_0 = 20$).

The quantities $N_{g\sigma}$ and $N_{p\sigma}$ in Table 5.2a were established under the condition that every initial crack for fixed values of the parameters f, ε, β, λ^* and r starts and propagates under the nominal load, determined from equality (5.10). Hence, every initial crack has its own "individual" load $\sigma^+ = \sigma_*^+$. The data presented in Table 5.2a illustrate the dependence of lifetime on the contact friction coefficient and on the ratio of the intensity of nominal cyclic forces σ^+ to the contact pressure p. As already indicated, the angles of initial inclination of the crack to the half-plane boundary chosen for calculations ($\beta = \pi/2$ and $\beta = \pi/3$) correspond to the typical orientations of macrocrack nuclei under the conditions of fretting fatigue. It follows from Table 5.2a that, as the nominal load (parameter r) increases, the lifetime decreases. The intermediate $N_{p\sigma}$ and critical $N_{g\sigma}$ lifetimes for the initial orientations of the crack at the angle $\beta = \pi/3$ are, as a rule, lower than for the angle $\beta = \pi/2$,

Table 5.2 Residual lifetime of the nearsurface layer of VT3-1 (BT3-1) titanium alloy in the contact zone: (**a**) for different values of the nominal forces σ^+ determined from condition (5.10); (**b**) for identical values of the nominal forces σ^+

a	$\varepsilon = 0.1;\ \lambda^* = 1.1$						$\varepsilon = 0.5;\ \lambda^* = 1.1$					
f	$\beta = \pi/2$			$\beta = \pi/3$			$\beta = \pi/2$			$\beta = \pi/3$		
	$N_{p\sigma} \times 10^{-6}$	x_{1c}/l_0	$N_{g\sigma} \times 10^{-6}$	$N_{p\sigma} \times 10^{-6}$	x_{1c}/l_0	$N_{g\sigma} \times 10^{-6}$	$N_{p\sigma} \times 10^{-6}$	x_{1c}/l_0	$N_{g\sigma} \times 10^{-6}$	$N_{p\sigma} \times 10^{-6}$	x_{1c}/l_0	$N_{g\sigma} \times 10^{-6}$
$r = \sigma^+/p = 0.3$												
0.3	0.88	81	0.99	0.65	63	0.69	0.35	40	0.37	0.22	21	0.22
0.5	1.4	88	1.9	1.3	81	1.7	0.67	81	0.71	0.63	62	0.66
0.65	1.6	88	2.7	1.8	85	–	0.96	85	1.1	0.94	80	1.0
0.8	1.8	91	3.5	2.0	85	–	1.1	86	1.3	1.2	81	1.4
$r = 1.0$												
0.3	0.29	49	0.30	0.25	33	0.25	0.21	32	0.21	0.19	23	0.19
0.5	0.35	75	0.39	0.33	55	0.35	0.26	42	0.26	0.24	32	0.24
0.65	0.40	83	0.46	0.38	75	0.43	0.30	51	0.31	0.28	40	0.29
0.8	0.44	86	0.54	0.44	79	0.51	0.34	61	0.36	0.33	49	0.34
$r = 3.0$												
0.3	0.20	35	0.20	0.17	25	0.18	0.18	30	0.18	0.16	22	0.16
0.5	0.21	42	0.22	0.19	31	0.20	0.19	33	0.19	0.17	25	0.17
0.65	0.22	47	0.23	0.21	36	0.21	0.20	35	0.20	0.18	27	0.18
0.8	0.24	53	0.25	0.22	42	0.23	0.21	38	0.21	0.19	30	0.20

b	$\varepsilon = 0.1;\ \beta = \pi/2;\ r = 2.0$	
f	x_{1p}/l_0	$N_{p\sigma} \times 10^{-4}$
0.3	20	24.6
0.5	– ‖ –	14.9
0.7	– ‖ –	9.54

which is confirmed by the experimental data on the hazard of cracks inclined to the boundary of the body at the angle $\beta = \pi/3$. A certain increase in the lifetime with the friction coefficient f can be explained by the fact that, in this case, the cracks start not for the same levels of loading σ^+ but for different ("individual") levels of σ^+.

Comparing the data presented in Table 5.2a for smaller ($\varepsilon = 0.1$) and larger ($\varepsilon = 0.5$) initial crack lengths, we conclude that the lifetime is longer for smaller initial cracks. However, this effect becomes weaker as the nominal load increases: If, for $|\sigma^+|/p = 0.3$ and $\varepsilon = 0.1$, the lifetime $N_{g\sigma}$ exceeds its value for $\varepsilon = 0.5$, on the average, by a factor of 2.8 for the same values of f and β, then, for $|\sigma^+|/p = 3.0$, the lifetime $N_{g\sigma}$ obtained for $\varepsilon = 0.1$ exceeds the value of $N_{g\sigma}$ for $\varepsilon = 0.5$, on the average, only by 15%.

The influence of the coefficient f on the lifetime is well visible from the data in Table 5.2b. In this table, we present the intermediate lifetime $N_{p\sigma}$ under the load σ^+ ($r = 2.0$) identical for all three analyzed cases ($f = 0.3$, 0.5, and 0.7). This load is starting, i.e., it guarantees the validity of condition (5.10) for the case $f = 0.7$ and definitely exceeds the starting values of σ^+ for $f = 0.5$ and $f = 0.3$. Thus, we can conclude that the increment of the half amplitude of the friction coefficient in a contact cycle by 0.2 leads to a significant decrease in the lifetime (by about a factor of 1.6).

5.3 Crack Mouth Under Counterbody. Full Slipping Between Fretting Pair Elements

The experimental data and engineering practice show [3, 30, 66] that, under the conditions of fretting fatigue, the cracks and other defects are mainly localized on both ends of the contact zone. They are formed and propagate under the action of cyclic contact and bulk (tension-compression and bending) forces. Among the surface damages encountered in the contact zone of elements of the fretting couples, one can often encounter crumbling of various sizes and configurations. As typical elements of crumbling, one can mention fairly large dish-shaped, wedge-like, and small needle-shaped pits. The experimental investigations of the processes of fracture and wear carried out by R.B. Waterhouse, J.S. Halliday, A.R. Wayson, and other researchers enable us to assume that the fretting products closed in the contact zone may serve as a possible cause of crumbling. Under the conditions of fretting, the cracks formed in the contact zone under the counterbody can be filled with powder-like fretting products and oil. Due to the wedging action of these products, after a certain penetration into the depth of the material, the cracks may propagate to the surface of the body and lead to its crumbling. Actually, the present section is devoted to the simulation and investigation of the process of pitting formation in the contact zone of the bodies under the conditions of fretting fatigue.

We consider a problem of fretting fatigue of contacting bodies in the case of contact with slip but without sticking. We construct the trajectories of edge cracks whose mouth is located in the contact zone under the counterbody. Moreover, we determine the dependences of the shapes of these paths on the friction coefficient between the bodies, pressure of the filler upon the crack faces, and the location and length of the initial macrocrack. The theoretical results are compared with the experimental data [15, 66]. As a result, we make an attempt to estimate the residual lifetime (according to the formation of pitting) in the nearsurface layer of elements of the fretting couple made of TS-5 titanium turbine alloy.

Stress intensity factors. Assume that pitting is formed as a result of propagation of the surface shear macrocrack preliminarily initiated under the counterbody for the conditions of its cyclic filling and wedging by the fretting products. Within the framework of the outlined approach to modeling the process of pitting formation

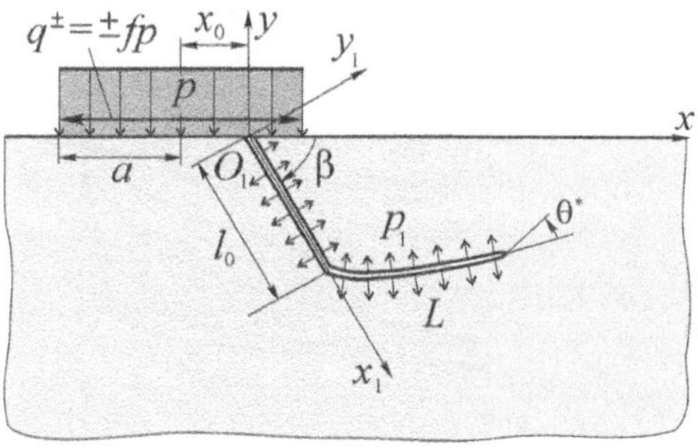

Fig. 5.16 Computational scheme of the problem for the case of uniformly distributed pressure on the half-plane boundary

described above, we use the computational scheme outlined in Fig. 5.16 and the corresponding boundary conditions (5.6):

$$\begin{gathered}\sigma_y \pm i\tau_{xy} = -p(1 \mp if), \quad |x - x_0| \le a; \qquad \sigma_y - i\tau_{xy} = 0, \\ |x - x_0| > a \quad \text{for} \quad y = 0; \\ N^{\pm}(t) + iT^{\pm}(t) = -p_1 = -sp, \qquad t \in L,\end{gathered} \tag{5.11}$$

where $N^{\pm}$ and $T^{\pm}$ are the limit normal and tangential stresses on approaching the crack contour L from the left ("+") and from the right ("–"), respectively, in Muskhelishvili's notation; $0 < s \le 1.0$.

By analogy with [22], we establish the right-hand side of SIE (3.151):

$$\begin{aligned}P(\eta) = \frac{p}{\pi}\Bigg\{ & w'(\eta)\mathrm{Re}\left[(i \mp f)\ln\frac{d(\eta)+1}{d(\eta)-1}\right] + \\ & + \frac{\overline{w'(\eta)}}{e^{2i\alpha}}\left[\frac{2\varepsilon(1 \mp if)\mathrm{Im}b(\eta)}{\overline{d^2(\eta)}-1} \pm f\ln\frac{\overline{d(\eta)}+1}{\overline{d(\eta)}-1}\right]\Bigg\} - spw'(\eta);\end{aligned} \tag{5.12}$$

where $d(\eta) = \lambda - \varepsilon b(\eta)$;$b(\eta) = (z_{01} + w(\eta)\, e^{i\alpha})/l_0$; $\varepsilon = l_0/a$; $\lambda = x_0/a$; l_0 is the length of the initial macrocrack regarded, for the sake of simplicity, as rectilinear and $t = w(\xi)$ is the parametric equation of the crack contour L.

We perform numerical investigations for broad ranges of values of the operating parameters and the parameters specifying the location of an edge initially rectilinear crack. (We assume that this crack is preliminarily initiated by the transverse shear mechanism in the contact zone). The friction coefficient varies within the range $f = 0.3\ldots0.9$. The parameter characterizing the pressure of the filler on the crack faces was $s = p_1/p = 0.5\ldots1.0$. The quantities specifying the length and location of the initial crack (Fig. 5.16) take the following values: the relative initial crack length $\varepsilon = l_0/a = 0.1\ldots1.0$, while the angle of crack inclination to the edge of the body $\beta = \pi/9\ldots8\pi/9$. The parameter of location of the crack mouth in the contact section varies within the range $|\lambda| = |x_0/a| \le 1.0$. The origin of the local coordinate system O_1 coincides with the origin of the main coordinate system O at the crack mouth ($z_{01} = 0$) and the O_1x_1-axis is oriented along the initial crack ($\beta = -\alpha$) (Fig. 5.16).

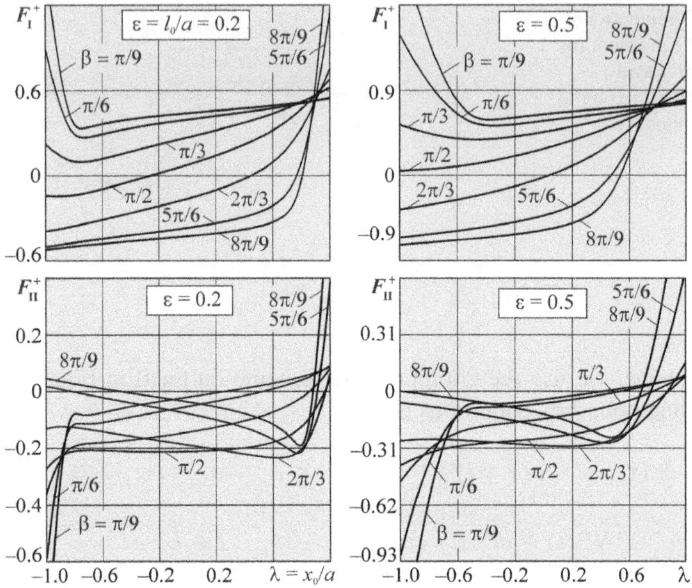

Fig. 5.17 Dependences of the normalized SIFs $F_I^+ = K_I^+/\left(p\sqrt{\pi a}\right)$ and $F_{II}^+ = K_{II}^+/\left(p\sqrt{\pi a}\right)$ in the positive phase of the contact cycle for the edge initially rectilinear crack on the location of its mouth in the contact section; $f = 0.5$; $s = p_1/p = 1.0$

First, for the described ranges of the parameters of the problem (f, s, ε, β, and λ), we compute the SIFs (K_I, K_{II}, and $K_{I\theta}$) and the angle θ^* of propagation (deviation) of an initially rectilinear crack with an aim to choose the basic sets of these parameters for the construction of the crack growth paths. In particular, it is important to determine the sets of parameters f, s, ε, β, and λ favorable for the propagation of cracks to the surface of the body (half-plane) and its crumbling. The characteristic regularities of behavior of the normalized stress intensity factors $F_{I,II} = K_{I,II}/\left(p\sqrt{\pi a}\right)$ and $F_{I\theta} = K_{I\theta}/\left(p\sqrt{\pi a}\right)$ and the angles θ_* are shown in Fig. 5.17 and Table 5.3. The normalized SIF $F_{I,II}^-$ in the negative phase of the contact cycle ($q = q^- = -fp$) can be easily found by using the SIF $F_{I,II}^+$ in the positive phase of a cycle ($q = q^+ = +fp$) on the basis of the dependence $F_{I,II}^-(f, \lambda, \beta) = F_{I,II}^+(f, -\lambda, \pi - \beta)$, which follows from the symmetry of geometric-force scheme of the problem (Fig. 5.16). The analysis of the obtained results enables us to conclude that:

- as the relative initial crack length $\varepsilon = l_0/a$ (Fig. 5.17) and the relative pressure $s = p_1/p$ of the filler upon the crack faces increase (Table 5.3), the values of K_I, $|K_{II}|$, and $K_{I\theta}$ become higher;
- for a given crack orientation (fixed β), the SIFs sharply increase and reach their maximal values as the crack mouth becomes closer to the end of contact section near which the crack is directed toward the load-free surface of the body; the more

Table 5.3 Normalized SIFs and the angle θ^* of initial deviation of the rectilinear crack located near the edge of the contact section for positive ($q = q^+$) and negative ($q = q^-$) phases of the contact cycle; $\lambda = -0.9$; $\beta = \pi/6$

f	F_I^+	F_{II}^+	$F_{I\theta}^+$	θ^{*+}, deg.	F_I^-	F_{II}^-	$F_{I\theta}^-$	θ^{*-}, deg.
			$\varepsilon = l_0/a = 0.2$					
$s = 0.5$								
0.3	−0.007	0.050	–	–	0.010	0.217	0.256	−70
0.5	−0.012	−0.006	–	–	0.015	0.272	0.323	−69
0.7	−0.018	−0.062	–	–	0.021	0.328	0.390	−69
0.9	−0.023	−0.117	–	–	0.026	0.384	0.458	−69
$s = 0.7$								
0.3	0.200	−0.044	0.213	23	0.217	0.123	0.292	−43
0.5	0.195	−0.099	0.252	41	0.222	0.179	0.352	−50
0.7	0.189	−0.155	0.302	50	0.228	0.235	0.415	−54
0.9	0.183	−0.211	0.358	55	0.233	0.290	0.479	−56
$s = 1.0$								
0.3	0.510	−0.184	0.594	33	0.527	−0.017	0.528	4
0.5	0.505	−0.240	0.636	39	0.533	0.039	0.537	−8
0.7	0.499	−0.295	0.683	44	0.538	0.094	0.562	−19
0.9	0.494	−0.351	0.734	47	0.544	0.150	0.599	−27
			$\varepsilon = 0.5$					
$s = 0.5$								
0.3	0.425	−0.193	0.528	38	0.488	−0.034	0.492	8
0.5	0.404	−0.247	0.560	44	0.509	0.019	0.511	−4
0.7	0.382	−0.300	0.597	49	0.531	0.072	0.545	−15
0.9	0.361	−0.353	0.637	53	0.552	0.125	0.591	−23
$s = 0.7$								
0.3	0.752	−0.341	0.934	38	0.815	−0.182	0.872	23
0.5	0.731	−0.394	0.964	42	0.837	−0.129	0.865	17
0.7	0.710	−0.448	0.998	45	0.858	−0.075	0.868	10
0.9	0.688	−0.501	1.035	48	0.879	−0.022	0.880	3
$s = 1.0$								
0.3	1.243	−0.563	1.542	38	1.306	−0.403	1.470	30
0.5	1.222	−0.616	1.572	40	1.328	−0.350	1.452	26
0.7	1.200	−0.669	1.604	42	1.349	−0.297	1.440	23
0.9	1.179	−0.722	1.638	44	1.370	−0.244	1.432	19

Comment: The dashes correspond to $F_I < 0$

acute the angle of crack inclination to the boundary, the higher the maximum values of the SIFs (Fig. 5.17);

- the friction coefficient f weakly affects the SIF K_{I}; however, its value and the direction of tangential contact forces strongly affect the SIF K_{II} and, hence, the angle of initial deviation θ^* and the quantity $K_{\mathrm{I}\theta}$; if the contact tangential forces are directed to the crack mouth and along the direction of its orientation, then, in the process of their growth, the angle θ^* increases, and the susceptibility to crack propagation toward the boundary of the body becomes stronger; on the other hand, if these forces are directed from the mouth and act in the direction opposite to the direction of crack orientation, then the angle θ^* decreases with their growth (Table 5.3); this promotes the process of deepening of the crack into the bulk of the material.

Note that all calculations were performed for the order of the system of complex algebraic equations $M = 60$ and the step $\Delta\lambda = 0.01$. As compared with the results obtained for $M = 90$, the maximal relative error of the values of SIFs does not exceed 1.4% even for small angles of crack inclination to the half-plane edge ($\beta = \pi/9$ and $\beta = 8\pi/9$) and the values of the parameter λ close to ± 1.0, when the convergence of the numerical results becomes much worse.

Crack growth paths (trajectories). Formation of pitting and evaluation of the residual lifetime. In order to study the formation of pitting based the outlined analysis of the values of SIF and angles θ^* for the initial edge rectilinear macrocrack, we place its mouth near the right end of the contact section and assume, as a rule, that $\lambda = x_0/a = -0.9$. The basic values of the other parameters are $\beta = \pi/6$; $\varepsilon = 0.2$; $s = 1.0$; and $f = 0.5$. We constructed the trajectories both for the positive ($q = q^+ = +fp$) and negative ($q = q^- = -fp$) directions of tangential forces in contact between the bodies, i.e., for the positive and negative phases of the contact cycle of elements of the fretting couple. We also take into account the fact that the role of leading crack can be played not only by a crack formed by the positive phase but also by a crack formed in the negative phase (see Table 5.3).

Hence, if we place the mouth of an initial crack near the right end of the contact section (Fig. 5.16), then we can study the influence of crack orientation on the shape of its trajectory (Fig. 5.18). It was discovered that if the crack is oriented toward the load-free surface of the body ($\beta < \pi/2$), then it propagates to the surface and may cause crumbling. The positive phase of the cycle ($q = q^+$) is especially favorable for this. The smaller the angle of crack inclination to the half-plane edge and the larger its length, the faster the process crack growth toward the edge. On the contrary, as the angle of inclination β increases, the paths formed in the positive phase of the cycle become more gently sloping, while the paths formed in the negative phase propagate into the bulk of the material. If the crack is initially directed into the contact zone ($\beta > \pi/2$), then it does not start in the positive phase of cycle ($K_{\mathrm{I}} < 0$) and grows into the bulk of the material in the negative phase.

In view of these considerations, in what follows, we study the trajectories of a crack initially oriented favorably for the formation of pitting; in particular, we take $\beta = \pi/6$. If the initial relative length of this crack is small, then its behaviors in the

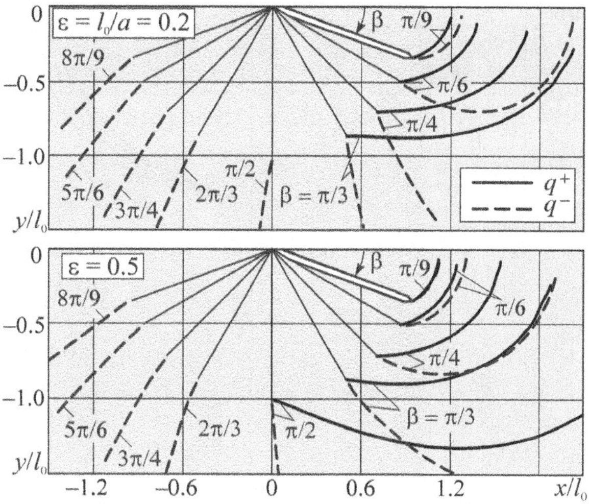

Fig. 5.18 Dependences of the crack growth paths on the initial angle β of its inclination to the body boundary; $\lambda = -0.9; f = 0.5$; $s = 1.0$

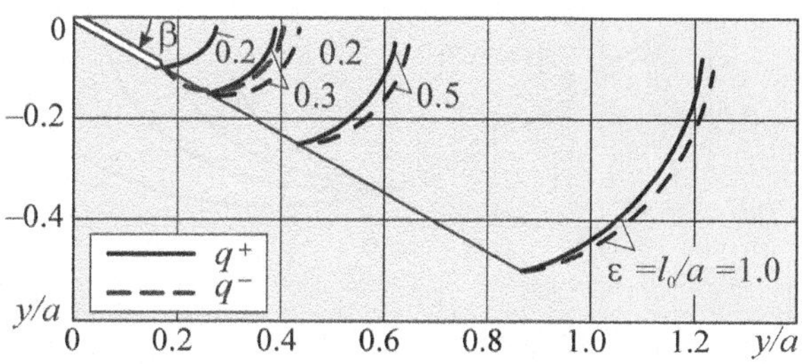

Fig. 5.19 Dependence of the crack growth paths on its initial relative length ε; $\lambda = -0.9; f = 0.5$; $\beta = \pi/6; s = 1.0$

positive and negative phases of a cycle are noticeably different (Fig. 5.19). In the positive phase, the trajectories rapidly approach the edge of the body. In the negative phase, their inclination angles are much lower. Finally, if the initial crack is longer ($\varepsilon = l_0/a \geq 0.5$), then the trajectories of positive and negative phases approach each other: The crack propagates as in the case of its growth solely under the action of pressure on its faces in the absence of contact load [8].

If the friction coefficient f increases, then the crack propagation toward the surface and its crumbling in the positive phase of a cycle becomes somewhat more pronounced ($q = q^{+}$; Fig. 5.20a). On the contrary, in the negative phase ($q = q^{-}$; Fig. 5.20b), this type of behavior becomes much weaker or even changes its sign and the crack starts to propagate into the material.

It is clear that the trajectories of crack growth are very sensitive to the pressure of filler on the faces. As the pressure (the value of $s = p_1/p$) decreases, the trajectories either become less steep (Fig. 5.20c), especially in the negative phase of a cycle (Fig. 5.20d), or the crack does not grow at all.

The analysis of the shapes of crack paths (Figs. 5.18, 5.19 and 5.20) and the functions $F_{\mathrm{I,II}}(\lambda)$ (Fig. 5.17) shows that the cracks initiated under the counterbody in the contact zone may reach the surface of the body outside this zone. Hence, the pits

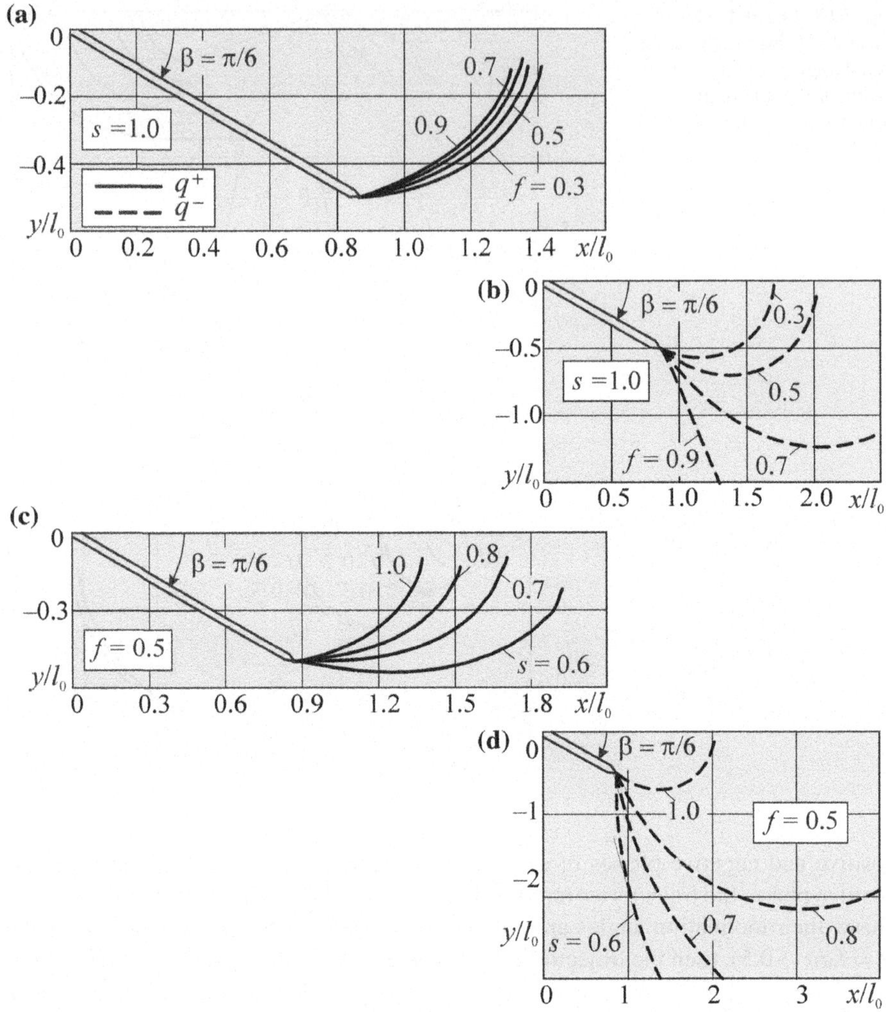

Fig. 5.20 Dependences of the crack growth paths on the half amplitude of the friction coefficient f (**a**, **b**) and the intensity of normal pressure on the crack faces $s = p_1/p$ (**c**, **d**) in the negative and positive phases of the contact cycle; $\lambda = -0.9$; $\varepsilon = 0.2$

are formed mainly on the boundary of the contact zone. It is of interest to answer a question whether a pit can be formed in the contact zone under the counterbody. For this purpose, we performed numerical analyses for initial cracks whose mouths are located closer to the center of the contact zone (Fig. 5.21). It is easy to see that, for the chosen parameters, the cracks grow to the surface up to its crumbling. They can attain the surface both under the counterbody and beyond it, forming dish-shaped pits (cavities). It should be emphasized that, as the distance from the mouth of an initial crack to the edge of the contact zone increases, the values of the SIF along

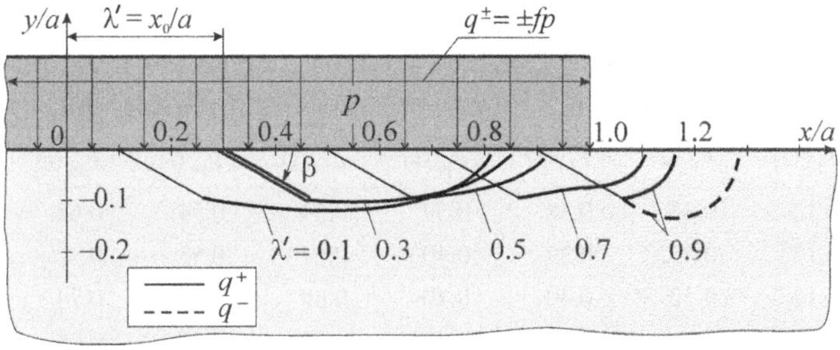

Fig. 5.21 Dependence of the crack growth paths on the location of crack mouth in the contact section; $f = 0.5$; $s = 1.0$; $\varepsilon = 0.2$; $\beta = \pi/6$; $\lambda' = -\lambda$

the crack path become lower and grow slower. The analysis of the SIF for the initial crack (Fig. 5.17 and Table 5.3) enables us to make the following prediction: If the pressure upon crack faces and the friction coefficient decrease and the slope of the initial crack to the surface increases, then it seems likely that a crack formed far from the boundary of the contact zone do not start ($K_{\rm I} < 0$). As the relative initial crack length decreases, the probability of its start also becomes lower.

In general, in the course of crack propagation in the positive phase of the contact cycle, the values of the SIF $F^+_{\rm I\theta}(l)$ along the crack paths become higher (see Table 5.4) and then sharply increase as the crack tip approaches the boundary of the body, which may serve as an indication of the transfer to spontaneous fracture and surface crumbling. The values of SIF $F^-_{\rm I\theta}(l)$ also grow along the paths but much slower, especially if the pressure on the crack faces is low (Table 5.4a and b). Thus, if the crack mouth is located near the right end of the contact section, then the positive phase of the contact cycle is determining in the formation of pitting.

Note that a satisfactory accuracy of our calculations is guaranteed if we use the step of construction the paths $h = 0.02l_0$ and the order of the system of complex algebraic equations $M = 70$.

To illustrate the relationship between the trends of crack propagation and lifetime, we performed the numerical analysis of the residual lifetime of the nearsurface layer of TS-5 titanium turbine alloy. To do this, we use here the second relation in (2.17), where the fatigue crack growth rate is described by the Paris formula [44]:

$$v = C(\Delta K_{\rm I\theta})^n = 10^{-7}\left(\Delta K_{\rm I\theta}/\Delta K^*_{\rm I}\right)^n; \quad \Delta K_{\rm I,\,1\text{-}2} \le \Delta K_{\rm I\theta} \le \Delta K_{\rm I,\,2\text{-}3};$$

here, $\Delta K_{\rm I,1–2}$ and $\Delta K_{\rm I,2–3}$ are the ranges of the stress intensity factor $K_{\rm I}$ corresponding to the lower and upper boundaries of the rectilinear Paris section in the FFD of the material; C, n, and $K^*_{\rm I}$ are the parameters of the Paris equation characterizing the cyclic crack growth resistance of the material. The values of these characteristics are can be found in Table 5.5.

The numerical results are presented in the lower row of Table 5.4. In our calculations, we take $\lambda = -0.9$; $\varepsilon = 0.2$; $\beta = \pi/6$; the half length of the contact section

Table 5.4 Normalized SIF $F_{\mathrm{I}\theta}^{\pm}(l)$ along the crack growth paths in the positive ($q = q^{+}$) and negative ($q = q^{-}$) phases of the contact cycle; $\lambda = -0.9$, $\varepsilon = 0.2$, and $\beta = \pi/6$

k	**(a)** $s = 0.6; f = 0.5$		**(b)** $s = 0.8; f = 0.5$		**(c)** $s = 1.0; f = 0.5$		**(d)** $s = 1.0; f = 0.7$	
	$F_{\mathrm{I}\theta}^{+}(l)$	$F_{\mathrm{I}\theta}^{-}(l)$	$F_{\mathrm{I}\theta}^{+}(l)$	$F_{\mathrm{I}\theta}^{-}(l)$	$F_{\mathrm{I}\theta}^{+}(l)$	$F_{\mathrm{I}\theta}^{-}(l)$	$F_{\mathrm{I}\theta}^{+}(l)$	$F_{\mathrm{I}\theta}^{-}(l)$
1	0.12	0.33	0.38	0.39	0.64	0.54	0.68	0.56
2	0.12	0.33	0.39	0.40	0.67	0.55	0.72	0.58
3	0.12	0.32	0.40	0.40	0.69	0.56	0.74	0.59
4	0.13	0.32	0.42	0.42	0.73	0.57	0.78	0.60
5	0.14	0.33	0.44	0.42	0.76	0.59	0.81	0.61
15	0.26	0.37	0.69	0.48	1.23	0.72	1.31	0.71
30	0.49	0.41	1.68	0.57	4.01	1.03	5.98	0.83
34	0.59	0.43	2.59	0.58	13.36	1.15	12.60[a]	0.86
35	0.60	0.43	2.99	0.59	14.16[a]	1.19	–	0.87
41	0.82	0.44	11.58[a]	0.62	–	1.46	–	0.92
45	1.01	0.46	–	0.63	–	1.71	–	0.96
58	4.11[a]	0.48	–	0.70	–	3.35	–	1.11
60	–	0.49	–	0.70	–	3.75	–	1.13
73	–	0.51	–	0.77	–	13.40[a]	–	1.34
75	–	0.54	–	0.77	–	–	–	1.38
90	–	0.55	–	0.86	–	–	–	1.78
105	–	0.57	–	0.95	–	–	–	2.49
120	–	0.60	–	1.05	–	–	–	3.89
$\theta^{*}(l_0)$	43°	– 61°	40°	– 36°	39°	– 8°	44°	– 19°
$N_{g\sigma}$	crack does not start		1.71×10^5		0.81×10^5		0.46×10^5	

Comment: k is the number of step in the construction of the path

[a]The values of $F_{\mathrm{I}\theta}^{\pm}(l)$ corresponding to the onset of spontaneous fracture

Table 5.5 Characteristics of cyclic crack growth resistance by normal opening for TS-5 (TC-5) turbine alloy [35]

$\Delta K_{\mathrm{I}th}$	$\Delta K_{\mathrm{I},1\text{–}2}$	$\Delta K_{\mathrm{I}}^{*}$	$\Delta K_{\mathrm{I},2\text{–}3}$	$\Delta K_{\mathrm{I},fc}$	C	n
$\mathrm{MPa}\sqrt{\mathrm{m}}$					$(\mathrm{MPa})^{-n}\mathrm{m}^{(1-n/2)}$	
2.5	3.1	10.3	32.0	39.0	3.7×10^{-11}	3.4

$a = 16$ mm ($l_0 = 3.2$ mm), and the contact pressure $p = 130$ MPa. In the first case (Table 5.4a), we have $F^-_{\mathrm{I}\theta}(l) > F^+_{\mathrm{I}\theta}(l)$ and, according to conditions (5.3) and (5.4), the crack propagates into the bulk of the material (Fig. 5.20d) along the path formed by the negative phase of the contact cycle. Thus, the crack grows for 15 steps (this corresponds to the increment of its length of about 1 mm) and then stops because $\Delta K_{\mathrm{I}\theta}$ becomes lower than $\Delta K_{\mathrm{I},1\text{-}2}$. In the second case (Table 5.4b), the crack does not start because, in the initial stage, we have $F^+_{\mathrm{I}\theta}(l_0) \approx F^-_{\mathrm{I}\theta}(l_0)$ and, hence, $\Delta K_{\mathrm{I}\theta}(l) \approx 0$. In the third and fourth cases (Table 5.4c and d), the crack starts and propagates along the trajectory formed by the positive phase of a cycle (Fig. 5.20a) up to the time where $\Delta K^+_{\mathrm{I}\theta}(l) = \Delta K_{\mathrm{I},2\text{-}3}$, i.e., up to the onset of spontaneous fracture prior to the surface crumbling. In this case, the increment of crack length is approximately equal to 1.3 mm. Hence, in this case, the quantitative influence of pressure upon the crack faces (Table 5.4a–c) and contact friction between the bodies (Table 5.4c and d) on their residual lifetime is well pronounced. We also see that the validity of the conditions of fatigue crack growth (5.3) and (5.4) and the characteristics of cyclic crack growth resistance of materials specified by the proposed model significantly improves the possibilities of prediction of the specific features of the fracture process (or its absence) in the bodies subjected to fretting fatigue. In this case, we quantitatively taken into account various important operating characteristics of the fretting couple, such as the size of the contact section and contact pressure.

Finally, it should be emphasized that the numerical values of $N_{g\sigma}$ may become much higher if, for the description of the fatigue crack growth rate, instead of the Paris formula, we use a more exact relation with taking into account the nearthreshold section of the $v - \Delta K_{\mathrm{I}}$ diagram [17]. It is clear that the complete lifetime corresponding to the formation of pitting and estimated with regard for the stages of initiation and growth of an initial edge crack (by the shear mechanism) may exceed the residual lifetime $N_{g\sigma}$ by 1–2 orders of magnitude.

Experimental investigations. For the comparison and illustration of the process of pitting formation caused by fretting fatigue, we analyze the experimental data [15, 50] obtained for specimens of TS-5 (TC-5) titanium turbine alloy and the data taken from the well-known monograph by Waterhouse [66].

The chemical composition of TS-5 alloy is as follows: Ti, 5.0% Al, 3.0% Sn, 2.0% Zr, 2.0% V. Its conditional yield limit $\sigma_y = 722$ MPa, ultimate strength $\sigma_u = 766$ MPa, and relative elongation $\delta = 17.9\%$.

On the surface of prismatic specimens of TS-5 titanium turbine alloy $110 \times 11 \times 5$ mm in size symmetrically clamped between two prismatic counterbodies made of the same material, as a result of its cyclic bending near the edge of the contact zone under the counterbody, we observe the formation of relatively large pits (cavities) (Fig. 5.22a, b) and a series of smaller pits (Fig. 5.22c). The level of slipping in the contact zone estimated by the method of cantilever bending constituted 5...11 μm [60]. The tensile stresses acting along the fretting line are as follows: $\sigma_{\max} = 200$ MPa (Fig. 5.22a), $\sigma_{\max} = 130$ MPa (Fig. 5.22b), and $\sigma_{\max} = 100$ MPa (Fig. 5.22c). For the maximal sizes of the largest pit (Fig. 5.22a), we get a length of 3.5 mm and a depth of 35μm. Along the boundary of the contact zone, smaller pits (Fig. 5.22c) form grooves with a width of 1.0...1.5 mm and a depth of up to 40 μm. These

(a) (b) (c)

Fig. 5.22 Pitting on the surfaces of the specimens of TS-5 (TC-5) titanium turbine alloy: **a** after $N = 1.2\times10^5$ loading cycles; $\sigma_{max} = 200$ MPa; **b** after $N = 4.6\times10^5$ cycles, $\sigma_{max} = 130$ MPa; **c** after $N = 6\times10^6$ cycles, $\sigma_{max} = 100$ MPa

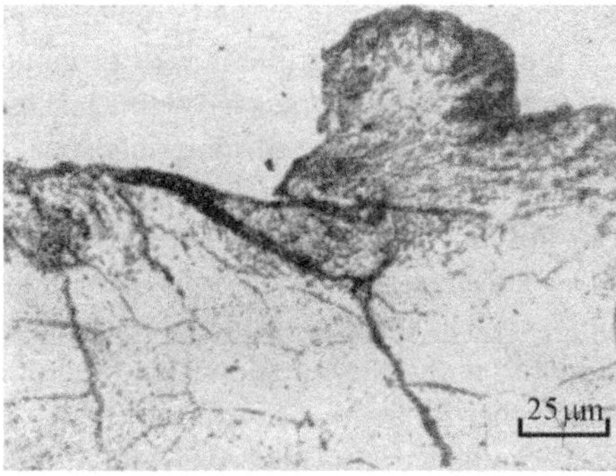

Fig. 5.23 Surface damage in soft steel [66]; $N = 1.2\times10^5$

grooves were observed in numerous specimens subjected to fretting fatigue. In the photographs, we can also clearly see the process of filling of the pits with fretting products. The X-ray structural analysis shows that these products mainly consist of titanium oxides and the traces of titanium particles, which strongly promote the process of wedging of the cracks.

Comparing the cross section of the defect depicted in Fig. 5.23 with the shape of the computed trajectories of growth of the edge cracks (see, e.g., Fig. 5.20), we can assume that these defects are formed as a result of branching of the initial edge crack filled with fretting products, which leads to the formation of two branches (either consecutively or simulta-neously) one of which is controlled by the negative phase of the fretting cycle and propagates into the bulk of the material, while the other branch is controlled by the positive phase of the cycle and grows in the direction of the body boundary.

By using the proposed model, we can explain the formation of pits of wedge-like shape (Fig. 5.24), as indicated by Field and Waters [66]. In our opinion, defects of this type are initiated when the edge crack growing from beneath the counterbody toward the load-free surface of the body wedged out by fretting products crosses another crack propagating from the load-free surface into the bulk of the material. It is known [10, 47, 66] that, in the initial stage, the cracks from the free surface

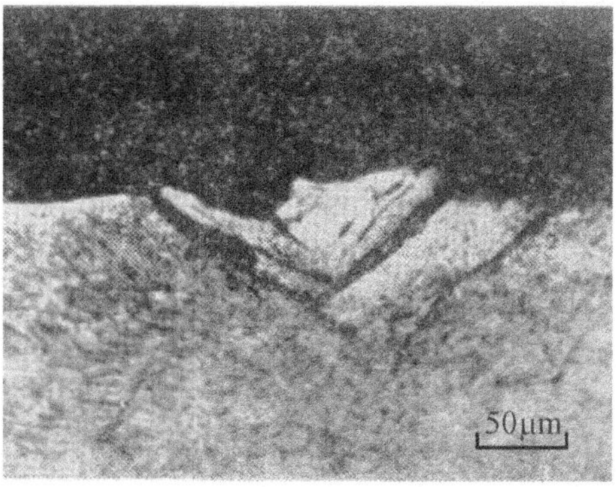

Fig. 5.24 Formation of a free particle on the steel surface [66]

propagate toward the contact zone. As a result, the crumbling pits take a wedge-like shape.

Despite the fact that the loading modes used in the numerical analysis and experimental investigations of the specimens of titanium alloy are somewhat different, the obtained results are in good qualitative agreement and do not have quantitative contradictions in estimating the lifetime. They also agree with the observations and conclusions made by Waterhouse in [66] concerning the formation of typical dish-shaped pits of crumbling in the bodies contacting under the conditions of fretting fatigue.

5.4 Partial Slipping Between Contacting Bodies

In the literature, there are numerous works (see Chap. 1 and also the surveys [33, 47]) devoted to the investigation of the stress intensity factors near the tips of edge or subsurface rectilinear or broken cracks in the elastic half-plane (half-space) under the action of a contact load on the boudary of the half-plane simulating the action of the counterbody under the conditions of full slip between the bodies. In [9, 12, 18, 22, 47], the authors also constructed the trajectories of propagation of edge cracks under the conditions of full slip between the contacting bodies. However, there are numerous practical cases in which sticking is realized between the bodies in a part of the contact section. In [33, 42], one can find examples of evaluation of the SIF for edge rectilinear cracks in the half-plane under the action of contact load modeling the conditions of sticking and slipping.

In the present section, we model and study the process of crack propagation in the presence of fretting fatigue under the conditions of sticking and slipping (partial slipping) in contact between the bodies. We determine the stress intensity factors and construct the trajectories of the propagation of edge initially rectilinear cracks depending on the following operating parameters: friction in contact between the bodies, ratio of the lengths of the sections of sticking and slipping, the location of

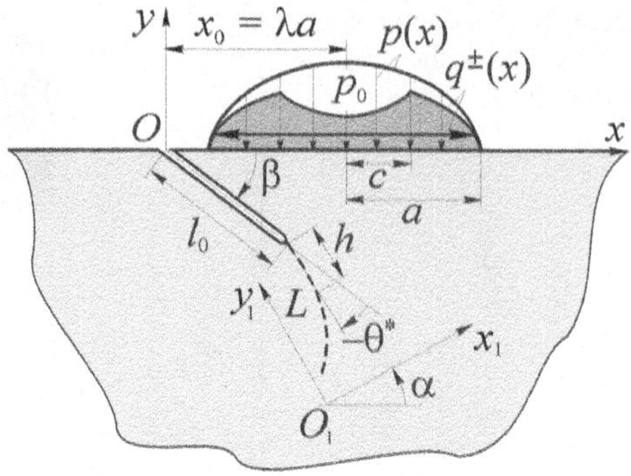

Fig. 5.25 Computational scheme of the problem under action of complicated elliptic load on the half-plane boundary

the initial crack relative to the contact load, etc. These investigations were carried out within the framework of the proposed model for the evaluation of lifetime and analysis of the fracture processes in solid bodies in the course of their cyclic contact (Chap. 2) and, in particular, under the conditions of fretting fatigue [9, 18, 47].

Model loading and stress intensity factors. We assume that one of contacting bodies is damaged by an edge macrocrack whose mouth is located beyond the contact section. We model this body by an elastic half-plane weakened by a curvilinear (rectilinear) crack along a contour L (Fig. 5.25). The contact influence of the other body (counterbody) is modeled by the action of normal $p(x)$ and alternating tangential $q^{\pm}(x)$ forces distributed according to the elliptic (Hertzian) law. Note that the tangential forces have a somewhat more complicated distribution than the normal forces, which enables us to model [33, 34] the presence not only of the sections of slipping but also of the sections of sticking in contact between the bodies. Hence, we represent the normal and tangential components of the model contact load on the boundary of the half-plane ($y = 0$) in the form

$$p(x) = \frac{p_0}{a}\sqrt{a^2 - (x - x_0)^2}, \quad |x - x_0| \le a;$$

$$q^{\pm}(x) = \begin{cases} \pm f\dfrac{p_0}{a}\sqrt{a^2 - (x - x_0)^2}, & -a \le x - x_0 \le -c, \quad c \le x - x_0 \le a; \\ \pm f\dfrac{p_0}{a}\left[\sqrt{a^2 - (x - x_0)^2} - \sqrt{c^2 - (x - x_0)^2}\right], & |x - x_0| \le c, \end{cases} \tag{5.13}$$

where $2a$ is the length of the contact section, p_0 is the maximal pressure in this section, and f is the coefficient of friction between the bodies. The identically distributed tangential forces $q^{+}(x)$ and $q^{-}(x)$ with opposite directions form a contact cycle and correspond to its positive and negative phases. The relative length of the section of sticking c/a is given by the formula $c/a = \sqrt{1 - |Q/fP|}$, where Q and P are, respectively, the tangential and normal components of the vector of external load ($Q \le fP$). Note that the distribution of forces (5.13) in the contact section between two elastic cylinders in the presence of partial sticking was established for the first time in [5].

Assume that, in the contact (compression) zone, the macrocracks first (in the initial stage) propagate by the shear (mode II) mechanism almost rectilinearly at an angle typical of a given fretting couple and then, when the cracks enter the region tension, their growth occurs by the normal opening (mode I) mechanism, and the trajectory becomes curvilinear. We now consider the second stage. The crack paths are constructed step-by-step by using the σ_θ-criterion of local fracture, the characteristics of cyclic crack growth resistance of the material, and the solutions of singular integral equations of the corresponding stationary problems of the theory of elasticity.

As already indicated, in each stage of construction of the crack paths, it is necessary to find the SIF at the crack tip from the corresponding SIE. In our case, this is the SIE (3.151) for the plane problem of the theory of elasticity for the half-plane weakened by an edge curvilinear crack along the contour L under the action of model contact load (5.13) on the half-plane boundary (Fig. 5.25). We assume that this crack is free of loads and its faces are not in contact.

For the boundary conditions formulated above (5.13), the right-hand side of the SIE (3.151) take the form (3.180), i.e.,

$$
\begin{aligned}
P(\eta) = p_0 \Big\{ &\mathrm{Re}[(f - i)(b(\eta) + i a(\eta))]\, w'(\eta) \\
&- i(1 - i f)\left(\frac{\overline{b(\eta)}}{\overline{a(\eta)}} - i\right) \mathrm{Im}\big(w(\eta)\varepsilon e^{i\alpha} + \gamma\big)\times \\
&\times \overline{w'(\eta)} e^{-2i\alpha} - f\left(\overline{b(\eta)} - i\overline{a(\eta)}\right) \overline{w'(\eta)} e^{-2i\alpha} \Big\} \\
&- p_0 f \Big\{ \frac{c}{a} \mathrm{Re}(b_1(\eta) + i a_1(\eta))\, w'(\eta) + \\
&+ i\left(1 + i\frac{\overline{b_1(\eta)}}{\overline{a_1(\eta)}}\right) \mathrm{Im}\big(w(\eta)\varepsilon e^{i\alpha} + \gamma\big)\, \overline{w'(\eta)} e^{-2i\alpha} \\
&- \frac{c}{a}\left(\overline{b_1(\eta)} - i\overline{a_1(\eta)}\right) \overline{w'(\eta)} e^{-2i\alpha} \Big\};
\end{aligned}
\tag{5.14}
$$

here, $t = w(\eta)$ is the parametric equation of crack contour L; $\varepsilon = l_0/a$; $a(\eta) = \sqrt{1 - b^2(\eta)}$; $b(\eta) = e^{i\alpha} w(\eta)/a + \gamma - \lambda$; $\lambda = x_0/a$; $\gamma = z_{01}/a$; $a_1(\eta) = \sqrt{1 - b_1^2(\eta)}$; $\lambda_1 = x_0/c$; $b_1(\eta) = e^{i\alpha} w(\eta)/c + \gamma_1 - \lambda_1$; $\gamma_1 = z_{01}/c$; l_0 is the length of the initial macrocrack (for simplicity, its is regarded as rectilinear), z_{01} is the complex coordinate of the origin of the local coordinate system $x_1 O_1 y_1$ in the main coordinate system xOy, and α is the angle between the O_1x_1-axis and the Ox-axis (Fig. 5.25).

To determine favorable conditions for crack propagation by the mode I mechanism, we study the SIFs (K_{I}, K_{II}, $K_{\mathrm{I}\theta}$) and the angle θ^* for initially rectilinear edge cracks. We analyzed cracks whose mouth is mainly located outside the contact section. The calculations were performed for broad ranges of the operating parameters ($f = 0.3\ldots0.9$; $Q/fP = 0.27, 0.67, 1.00$) and the parameters of location of the

Table 5.6 Threshold values of sticking section length

Q/fP	1,00	0,67	0,27
c/a	0,00	0,57	0,85

crack ($\lambda = x_0/a = 0.8\ldots3.0$; $\varepsilon = l_0/a = 0.1, 0.5$; $\beta = \pi/6, \pi/4, \pi/3, \pi/2, 2\pi/3, 5\pi/6$). For the sake of convenience, we assumed that the origins of the main and local coordinate systems are located in the crack mouth ($z_{01} = 0$) and the O_1x_1-axis is directed along the crack ($\alpha = -\beta$) (Fig. 5.25). Without loss of generality of formulation of the analyzed problem, the crack mouth is placed near the left end of the contact section ($\lambda \geq 0.8$). In Table 5.6, we present the relative lengths c/a of the section of sticking corresponding to the chosen ratios of the vector of tangential component of the contact load Q to its limit value fP corresponding to the full slipping of the contacting bodies.

The analysis of the dependences of SIF (presented in Figs. 5.26, 5.27 and 5.28) on the parameters λ, β, ε, f, and c/a for an initially rectilinear edge crack demonstrates that, in the negative phase of the contact cycle ($q(x) = q^{-}(x)$, Fig. 5.25), the crack is mainly located in the compression zone ($K_{\mathrm{I}}^{-} < 0$). At the same time, in the positive phase ($q(x) = q^{+}(x)$), it is located in the zone of tension ($K_{\mathrm{I}}^{+} > 0$). Hence, we investigate the SIF and the crack growth paths for the positive phase of the contact cycle.

In Figs. 5.26, 5.27 and 5.28, we show the dependences of the normalized SIF $F^{+} = K^{+}/\left(p_0\sqrt{\pi a}\right)$ on the main parameters of problem for the positive phase of cyclic fretting fatigue. Here and in what follows, we omit the superscript "+". As follows from the presented plots, the stress intensity factors strongly depend on the initial crack orientation (Figs. 5.26 and 5.27, Tables 5.7 and 5.8) when there is no sticking in contact. The maximum values of $K_{\mathrm{I}}(\lambda, \beta)$ are, as a rule, reached for cracks inclined to the edge of the half-plane at the angles $\beta = \pi/3\ldots\pi/2$, the maximum $K_{\mathrm{I}\theta}(\lambda, \beta)$ are attained at the angles $\beta = \pi/6\ldots\pi/2$, while max $|K_{\mathrm{II}}(\lambda, \beta)|$ mainly reached at the angles $\beta = \pi/6$ (sometimes, at $\beta = 5\pi/6$). If the relative length of the initial crack is small ($\varepsilon = 0.1$), then the maxima of the SIF $K_{\mathrm{I}\theta}$ (max $K_{\mathrm{I}\theta}(\lambda) = K_{\mathrm{I}\theta}(\lambda^{*}) = K_{\mathrm{I}\theta}^{*}$) are realized for $\lambda = \lambda^{*} \approx 1.0$. As the crack length increases, both under the conditions of full slipping and in the presence of a section of sticking, the maxima of the SIF $K_{\mathrm{I}\theta}$ are realized for higher values of λ^{*}, i.e., the region favorable for crack propagation by the opening mechanism moves away from the end of the contact section. The appearance of the section of sticking leads to a substantial decrease in the SIF (Fig. 5.28, Tables 5.7 and 5.8). Thus, for the collection of the parameters $f = 0.8$; $\beta = \pi/3$; $\varepsilon = 0.1$ ($\lambda^{*} \approx 1.1$), we get $F_{\mathrm{I}\theta}^{*}(c/a = 0) = 0.327$; $F_{\mathrm{I}\theta}^{*}(c/a = 0.57) = 0.224$ and $F_{\mathrm{I}\theta}^{*}(c/a = 0.85) = 0.073$, i.e., $F_{\mathrm{I}\theta}^{*}$ becomes, first, lower by one-third as compared with the case of full slipping and then more than fourfold lower.

In the case of full slipping, the maximum values of the SIF weakly depend on the crack length (within the range $\varepsilon = 0.1\ldots0.5$) and, in the presence of sticking, suffer a significant shifted to the side of small cracks ($\varepsilon = 0.1$).

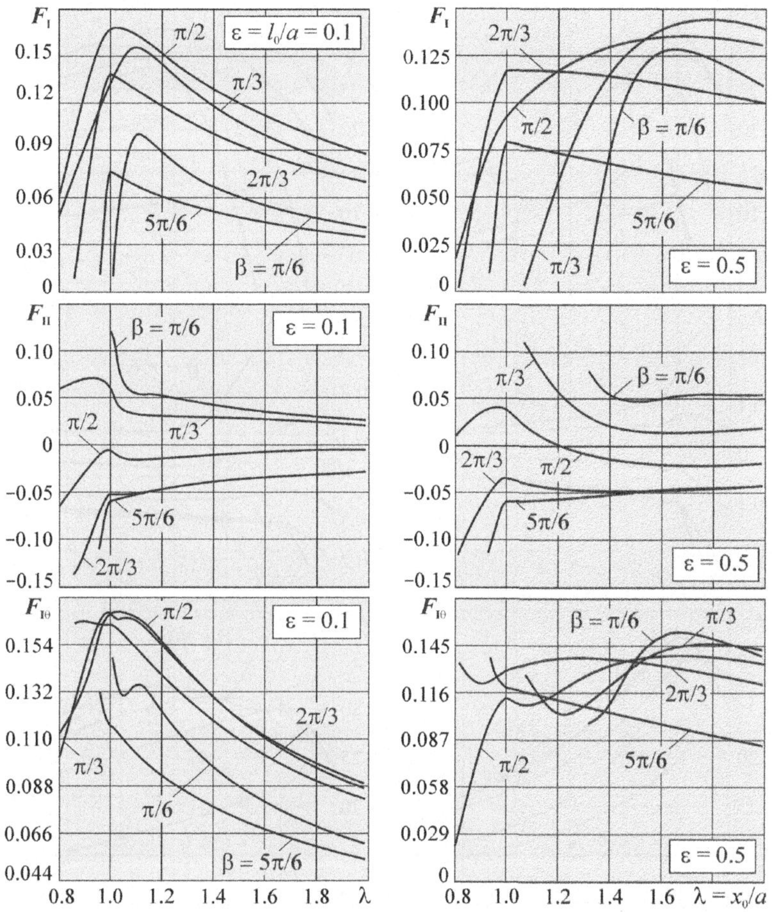

Fig. 5.26 Dependences of the normalized SIF on the location of an edge initially rectilinear arbitrarily oriented crack in the case of full slipping in contact between the bodies ($c/a = 0$); $f = 0.5$

The friction coefficient strongly affects the values of F_{I}, F_{II}, and $F_{\mathrm{I}\theta}$, especially in the absence of sticking in contact ($c/a = 0$): the higher the friction, the greater the SIF.

It is worth noting that the obtained results are in good agreement with the data presented in [21].

In the investigations of convergence of the process of calculations, depending on the choice of the parameter M, i.e., the order of the system of algebraic equations, it was established that if $M = 40$, then we most often obtain four true decimal digits in the values of the SIF; at the same time, for $M = 60$, there are five true digits. Furthermore, the convergence becomes worse as the length of the sticking section (c/a) and the relative crack length increase, and the worst case is realized when

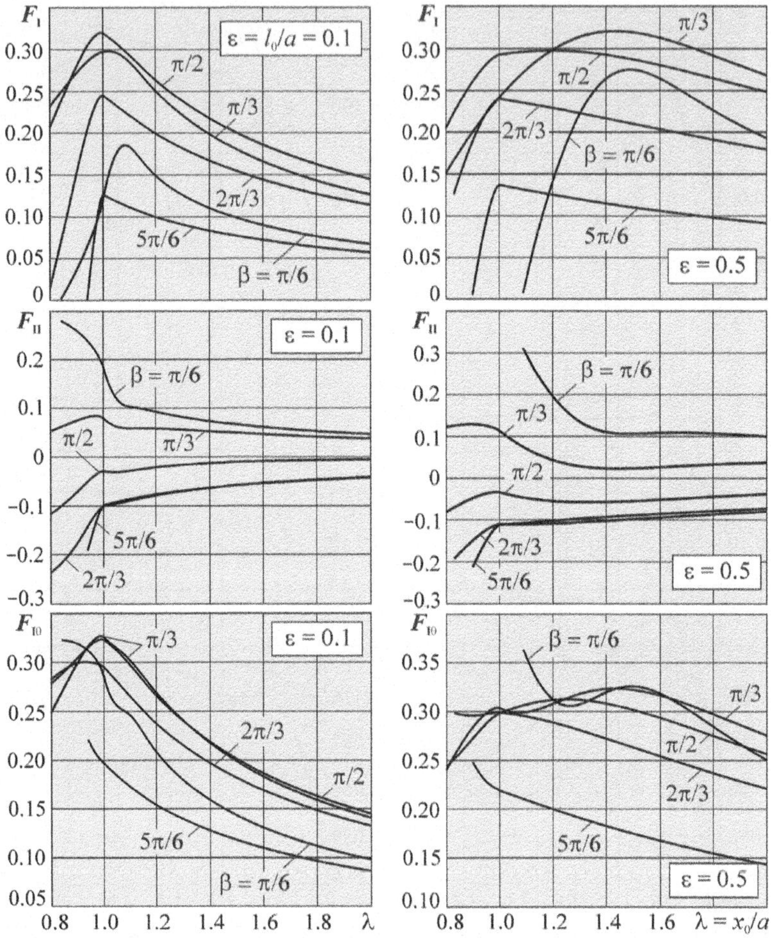

Fig. 5.27 Dependences of the SIF on the location of an edge initially rectilinear arbitrarily oriented crack in the case of full slipping in contact between the bodies ($c/a = 0$); $f = 0.8$

the crack mouth is located at the end of the section of contact loading ($\lambda = 1.0$). All calculations were performed for the order of the system of algebraic equations $M = 60$ with steps $\Delta\lambda = 0.01$.

Crack growth paths (trajectories). In Figs. 5.29, 5.30 and 5.31, we show the dependences of the paths of edge initially rectilinear cracks on the angle β of its inclination to the half-plane boundary and also on its relative initial length ε, relative length of the section of sticking *c*/*a*, and the friction coefficient *f*. These plots were constructed in the coordinate system $x_1O_1y_1$ for $z_{01} = x_0 - a$. The numerical data presented in Table 5.9 supplement these dependences. We also studied the dependence of the crack growth paths on the distance between the crack mouth and the center of the section of contact loading (parameter λ), which is graphically represented in

Table 5.7 Maximal normalized SIF $F^*_{I\theta}$ and the corresponding values of $\lambda^* = x^*_0/a$ for different angles β between the initial crack and the half-plane boundary; $f = 0.5$

β	$\varepsilon = l_0/a = 0.1$				$\varepsilon = 0.5$			
	$F^*_{I\theta}$	λ^*	F_I	F_{II}	$F^*_{I\theta}$	λ^*	F_I	F_{II}
$c/a = 0$ ($Q/fP = 1.00$)								
π/6	0.1479	1.01	0.0100	0.1233	0.1537	1.68	0.1277	0.0523
π/4	0.1695	0.96	0.0552	0.1187	0.1521	1.76	0.1401	0.0350
π/3	0.1686	0.99	0.1294	0.0676	0.1462	1.78	0.1436	0.0159
π/2	0.1688	1.04	0.1678	−0.0104	0.1397	1.66	0.1357	−0.0195
2π/3	0.1645	0.90	0.0559	−0.1139	0.1382	1.30	0.1157	−0.0461
5π/6	0.1319	0.96	0.0111	−0.1089	0.1388	0.93	0.0088	−0.1160
$c/a = 0.57$ ($Q/fP = 0.67$)								
π/6	0.0960	1.12	0.0733	0.0389	0.0849	1.82	0.0722	0.0271
π/4	0.1145	0.97	0.0118	0.0934	0.0790	2.02	0.0730	0.0178
π/3	0.1094	1.11	0.1038	0.0203	0.0739	2.14	0.0724	0.0087
π/2	0.1091	1.08	0.1080	−0.0090	0.0695	2.11	0.0681	−0.0081
2π/3	0.1085	0.90	0.0034	−0.0924	0.0703	1.70	0.0601	−0.0220
5π/6	0.0917	0.97	0.0104	−0.0744	0.0791	0.95	0.0060	−0.0656
$c/a = 0.85$ ($Q/fP = 0.27$)								
π/6	0.0346	1.15	0.0287	0.0118	0.0128	1.76	0.0002	−0.0110
π/4	0.0326	1.19	0.0298	0.0078	–	–	–	–
π/3	0.0303	1.23	0.0294	0.0042	–	–	–	–
π/2	0.0283	1.25	0.0280	−0.0023	–	–	–	–
2π/3	0.0292	1.15	0.0257	−0.0083	0.0096	1.69	0.0002	0.0082
5π/6	0.0307	0.99	0.0123	−0.0202	0.0138	2.01	0.0106	−0.0055

Comment: The dashes correspond to $K_I < 0$

Figs. 5.32 and 5.31 and numerically in Tables 5.10 and 5.11. On the basis of the analysis of these data, we can make the following conclusions:

- As the friction coefficient f decreases and the section of sticking becomes larger, the SIF both for the initial cracks (Fig. 5.28) and along their crack growth paths (Fig. 5.29) noticeably decrease (Table 5.9), and the crack paths become more gently sloping relative to the half-plane boundary and stronger deviate from the contact zone. In particular, in the case of initially longer cracks ($\varepsilon = 0.5$), in the presence of large section of sticking, and for low friction coefficients, the cracks initially directed to the side of the contact zone ($\beta \leq \pi/2$) do not propagate for all ($K_I < 0$) (therefore, their trajectories are not depicted in Fig. 5.29b, unlike the trajectories in Figs. 5.29a, 5.30a, and 5.31a).
- the described trends and the data presented in Table 5.9 are in good agreement with two main approaches to the structural and technological elevation of the fretting

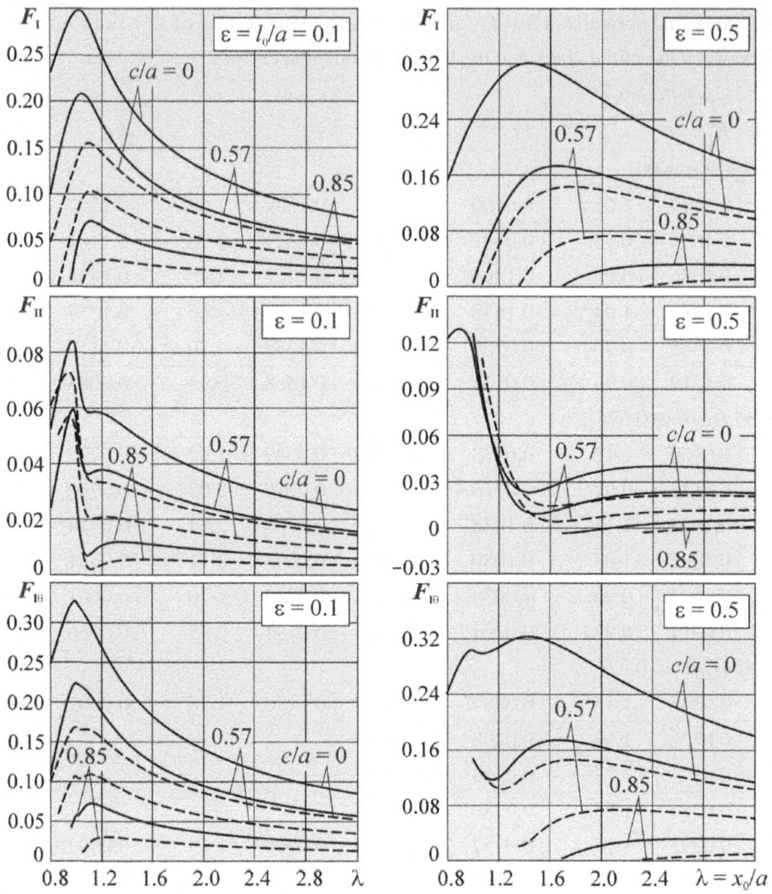

Fig. 5.28 Dependences of the SIF for an edge initially rectilinear crack on its location and the relative length c/a of the sticking section; $\beta = \pi/3$; the solid lines correspond to $f = 0.8$ and the dashed lines correspond to $f = 0.5$

fatigue strength formulated in the monograph by Shevelya and Kalda [57], namely, the increase in strength is realized either by removing the displacements of one surface over the other or, on the contrary, by the facilitation of these displacements, i.e., by guaranteeing that they are realized with the minimum possible friction.

- For initially short cracks ($\varepsilon = 0.1$), the variation of the inclination angle β weakly affects the shape of the trajectories of their propagation (Figs. 5.30a and 5.31a); for longer cracks ($\varepsilon = 0.5$), the influence of the angle β is stronger, especially in the initial stage of crack propagation; in the case of subsequent propagation, independently of the angle of orientation, the crack paths attain their own asymptotes determined by the friction coefficient f for every relative length of the section of sticking (see Figs. 5.30b and 5.31b).

Table 5.8 Maximum SIF $F_{I\theta}^{*}$ and the corresponding values of $\lambda^{*} = x_0^{*}/a$ for different angles β between the crack and the half-plane boundary; $f = 0.8$

	$\varepsilon = l_0/a = 0.1$				$\varepsilon = 0.5$			
β	$F_{I\theta}^{*}$	λ^{*}	$F_{I\theta}^{*}$	λ^{*}	$F_{I\theta}^{*}$	λ^{*}	$F_{I\theta}^{*}$	λ^{*}
$c/a = 0$ $(Q/fP = 1.00)$								
π/6	0.3233	0.84	0.0016	0.2792	0.3634	1.09	0.0077	0.3111
π/4	0.3196	0.97	0.2235	0.1462	0.3566	0.94	0.0929	0.2623
π/3	0.3273	0.99	0.2964	0.0826	0.3235	1.43	0.3210	0.0233
π/2	0.3240	0.99	0.3199	−0.0300	0.3124	1.25	0.2974	−0.0561
2π/3	0.3006	0.93	0.1917	−0.1516	0.3017	0.80	0.0951	−0.2130
5π/6	0.2209	0.94	0.0005	−0.1911	0.2482	0.90	0.0055	−0.2123
$c/a = 0.57$ $(Q/fP = 0.67)$								
π/6	0.2376	0.94	0.0076	0.2022	0.1932	1.59	0.1681	0.0575
π/4	0.2243	0.98	0.1493	0.1085	0.1844	1.65	0.1748	0.0346
π/3	0.2240	1.00	0.2037	0.0556	0.1745	1.67	0.1732	0.0123
π/2	0.2213	1.00	0.2182	−0.0214	0.1649	1.55	0.1581	−0.0274
2π/3	0.2067	0.97	0.1545	−0.0864	0.1615	1.22	0.1324	−0.0567
5π/6	0.1548	0.96	0.0189	−0.1249	0.1520	0.93	0.0098	−0.1269
$c/a = 0.85$ $(Q/fP = 0.27)$								
π/6	0.0759	1.11	0.0631	0.0258	0.0408	2.03	0.0358	0.0118
π/4	0.0756	1.12	0.0697	0.0174	0.0356	2.50	0.0328	0.0082
π/3	0.0726	1.12	0.0713	0.0080	0.0327	2.82	0.0317	0.0046
π/2	0.0690	1.10	0.0673	−0.0088	0.0304	2.93	0.0301	−0.0026
2π/3	0.0687	1.03	0.0580	−0.0224	0.0404	1.00	0.0017	0.0342
5π/6	0.0617	0.98	0.0110	−0.0480	0.0337	1.29	0.0250	−0.0143

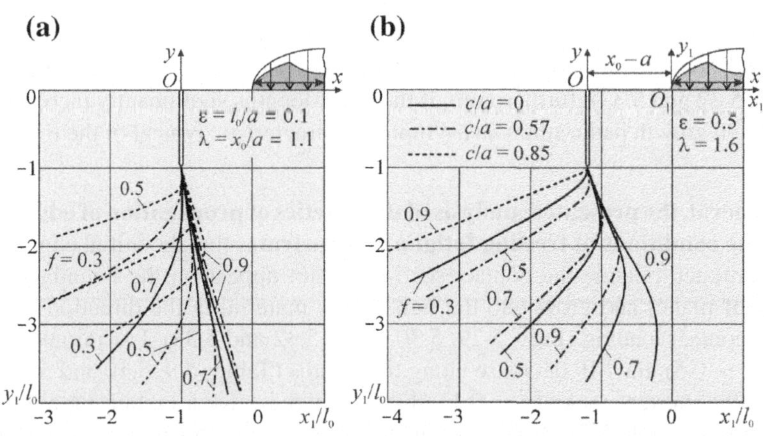

Fig. 5.29 Dependences of the crack growth paths on the friction coefficient f and the length of the section of sticking in contact between the bodies; $\beta = \pi/2$

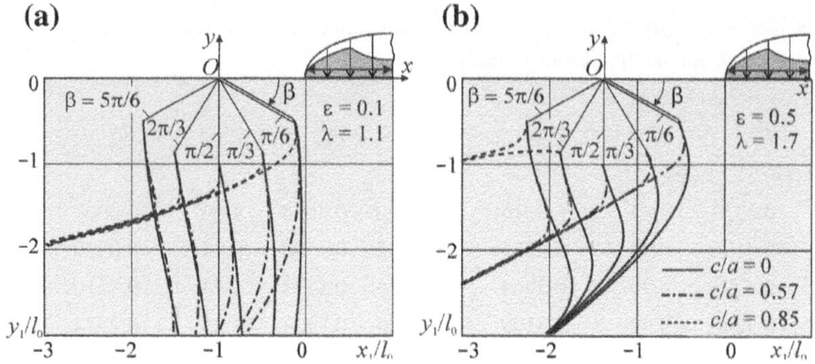

Fig. 5.30 Dependences of the crack growth paths on the initial angle β between the crack and the half-plane boundary; $f = 0.5$

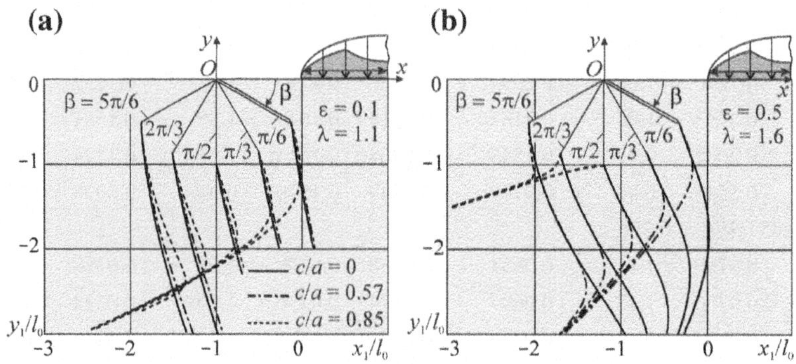

Fig. 5.31 Dependence of the crack growth paths on the initial angle β between the crack and the half-plane boundary; $f = 0.8$

- The shapes of the paths weakly depend on the distance between crack mouth and the boundary of contact loading (for fixed values of the other parameters) (Figs. 5.32 and 5.33); furthermore, if the crack lengths significantly increase, then the crack growth paths attain an asymptote characteristic typical of the fixed value of f.

In general, the presented analysis of the kinetics of propagation of edge cracks under the conditions of fretting fatigue demonstrates that the initial edge macrocracks initiated outside the contact section do not appear on the boundary of the body (half-plane) and grow into the bulk of the material in the direction from the zone of contact loading (Figs. 5.29, 5.30, 5.31, 5.32 and 5.33). For initially longer cracks ($\varepsilon = 0.5$), the SIF decrease along their paths (Tables 5.9, 5.10 and 5.11) and, for initially shorter cracks ($\varepsilon = 0.1$), the SIF increase for a certain time and then also decrease (Table 5.9). In the case of full slipping, we observe similar trends if a constant pressure plays the role of model contact load [9, 22, 47].

Table 5.9 Normalized SIF $F_{I\theta}(l)$ along the crack growth paths depending on the friction coefficient f and the relative length of the sticking section c/a; k is the step number in the process of the path construction

$\beta = \pi/2$, $\varepsilon = 0.1$, $\lambda = 1.1$						
	full slipping ($c/a = 0$)			sticking/slipping; $f = 0.5$		
k	f			c/a		
	0.3	0.7	0.9	0	0.57	0.85
	$F_{I\theta}(l)$					
1	0.0786	0.2532	0.3406	0.1658	0.1087	0.0233
2	0.0784	0.2543	0.3424	0.1663	0.1087	0.0225
3	0.0782	0.2555	0.3444	0.1667	0.1086	0.0217
4	0.0780	0.2568	0.3464	0.1672	0.1085	0.0209
5	0.0777	0.2579	0.3482	0.1676	0.1084	0.0200
30	0.0683	0.2774	0.3839	0.1715	0.0999	0.0151
60	0.0573	0.2837	0.4044	0.1647	0.0822	0.0143
90	0.0511	0.2794	0.4104	0.1526	0.0674	0.0138
120	0.0474	0.2692	0.4078	0.1395	0.0596	0.0134
145	0.0451	0.2583	0.4010	0.1293	0.0560	0.0142

Table 5.10 Normalized SIF $F_{I\theta}(l)$ along the crack growth paths depending on the distance from the crack mouth to the section of contact loading for friction coefficient $f = 0.5$; k is the number of step in the process of construction of the path

k	$\beta = \pi/3$, $\varepsilon = 0.5$, $c/a = 0.57, f = 0.5$		
	$\lambda = 1.84$	$\lambda = \lambda^* = 2.14$	$\lambda = 2.44$
	$F_{I\theta}(l)$ ($F_{I\theta}(l_0, \lambda^*) = \max F_{I\theta}(l_0, \lambda)$)		
1	0.0696	0.0739	0.0720
2	0.0682	0.0733	0.0718
3	0.0666	0.0726	0.0715
4	0.0650	0.0718	0.0712
5	0.0633	0.0709	0.0708
30	0.0376	0.0389	0.0521
60	0.0358	0.0336	0.0296
90	0.0334	0.0321	0.0307
120	0.0314	0.0304	0.0292

Note that all paths studied above were constructed with increments of their growth [47, 53] $h = 0.02l_0$ (see Fig. 5.25) and, in each step, the satisfactory accuracy of calculations was guaranteed for the order of the system of algebraic equations $M = 80$.

Table 5.11 Normalized SIF $F_{I\theta}(l)$ along the crack growth paths depending on the distance between the crack mouth and the section of contact loading for friction coefficient $f = 0.8$; k is the number of step in the process of construction of the path

$\beta = \pi/3$, $\varepsilon = 0.5$, $c/a = 0.57$, $f = 0.8$			
k	$\lambda = 1.37$	$\lambda = \lambda^* = 1.67$	$\lambda = 1.97$
	$F_{I\theta}(l)$ ($F_{I\theta}(l_0, \lambda^*) = \max F_{I\theta}(l_0, \lambda)$)		
1	0.152	0.174	0.165
2	0.149	0.173	0.165
3	0.145	0.171	0.165
4	0.142	0.170	0.165
5	0.139	0.168	0.164
30	0.101	0.111	0.138
60	0.087	0.083	0.082
90	0.078	0.076	0.073
120	0.072	0.070	0.068
150	0.066	0.066	0.064

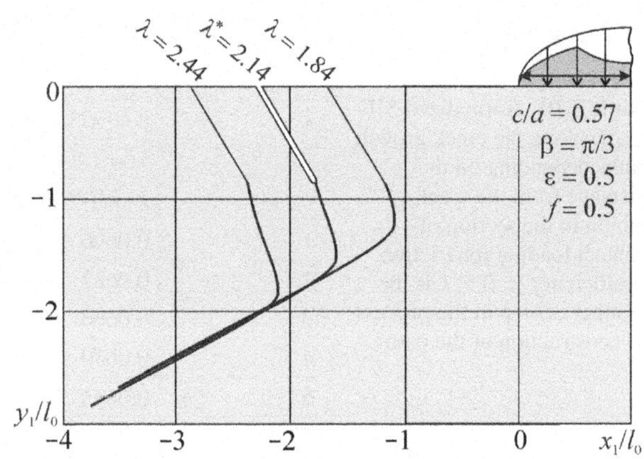

Fig. 5.32 Dependence of the crack growth paths on the distance between the crack mouth and the section of contact loading; $f = 0.5$

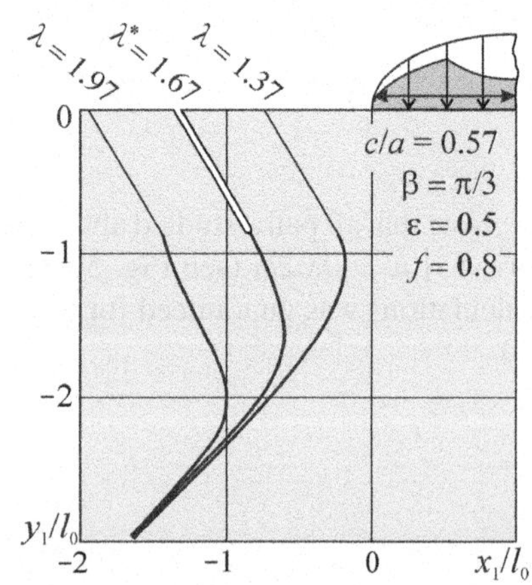

Fig. 5.33 Dependences of the crack growth paths on the distance between the crack mouth and the section of contact loading; $f = 0.8$

5.5 Influence of Rounding of Counterbody Base Edges

The profiles of the elements of numerous nominally immovable connections are rectangular and, in the contact zones of these elements, near corner points (ribs), we observe significant stress concentrations. Cracks are often initiated in these zones and become especially dangerous (quickly propagate) in the case where the joint is subjected to in-service vibration and its elements interact under the conditions of fretting fatigue [66]. Therefore, in practice, the indicated corner points are smoothened by rounding the edges of the bases of contacting bodies (Fig. 5.34). Hence, the problem of influence of the stress concentration and, in particular, of the radius of rounding of the base of counterbody on the stress intensity factors (SIF) at the tip of a crack present in the contact zone is an important problem both for the engineering practice and for the theoretical strength analysis of the contacting systems.

To model the interaction of bodies in the joints and, in particular, the fracture processes in the contact zone, it is customary to use model problems in which a damaged main body is regarded as an elastic half-plane (a cylinder of infinite radius) with cracks and the action of the counterbody is modeled either by the pressure of an elastic or rigid punch with rounded edges (Fig. 5.34) or by the corresponding contact load (Fig. 5.35).

The problems of pressure of a rigid punch with base of any shape onto the elastic half-plane weakened by a system of curvilinear cracks were considered in [45, 46,

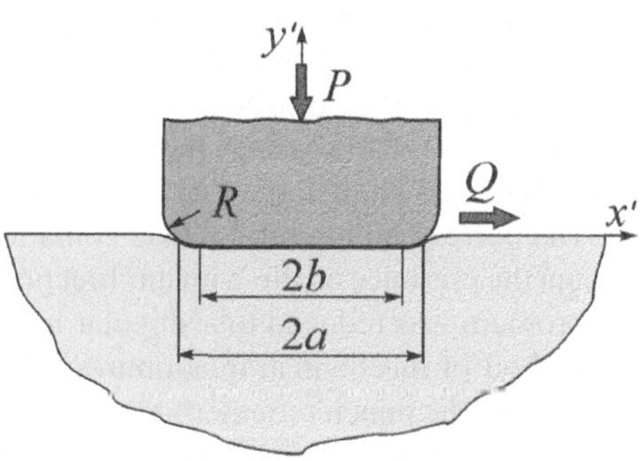

Fig. 5.34 Model scheme of contact interaction

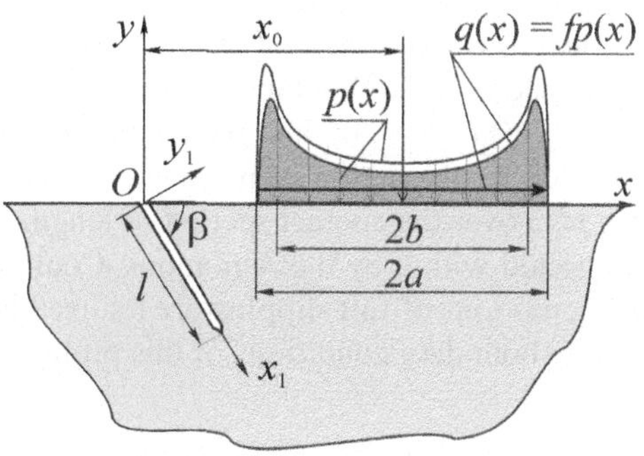

Fig. 5.35 Computational scheme of the problem with model load

54]. Similar problems for a system of rectilinear cracks were studied in [38] with the help of asymptotic approaches. In particular, in the case of pressing of a rigid rectangular punch into the half-plane, the numerical values of the SIF were obtained for an inner vertical crack in [45], for horizontal and vertical cracks in [46], and for an inclined edge crack in [55]. These problems are reduced to singular integral equations. A series of works dealing with the analysis of a half-plane weakened by an inclined edge crack and subjected to the action of a punch one of the edges of which is rounded and the other is pointed on the basis of the method of conformal mappings was published by Hasebe and Qian [32]. One of the first solutions in this direction was obtained by Tonoyan and Minasyan [62] by the method of dual integral equations. A fairly detailed survey of these publications can be found in [46, 54].

As a model load under the conditions of fretting fatigue contact interaction, it is customary to use uniformly distributed pressures, concentrated forces, and elliptic, linear, parabolic, and other distributions. The corresponding investigations were initiated and developed by Rooke and Jones [52] and Edwards [25, 26]. The surveys of these works can be found in [11, 16, 21, 36]. However, in these investigations, the base of the counterbody is, most often regarded as smooth and convex or rectangular without rounding. In other words, among the available works, there are no investigations in which the influence of the radius of rounding of the base of counterbody on the stress intensity factors near the crack tips located in the contact zone is taken into account.

Model contact load. As a continuation of [20, 21], we solve the problem of stress-strain state near the tip of an edge crack in the elastic half-plane subjected to the action of a model load on its boundary. This load characterizes the pressure distribution formed in the course of pressing of a flat elastic punch with horizontal rectilinear base with rounded edges and an arbitrary radius of rounding R (Fig. 5.35). This distribution was established on the basis of the solution of the contact problem of compressing two elastic cylinders along the initial contact strip obtained by Aleksandrov and Romalis [1]. The corresponding model load contained the tangential component taking into account the presence of slip with unidirectional friction in contact between the bodies. The problem was reduced to a singular integral equation and numerically solved by the method of mechanical quadratures. The influence of the radius of rounding of the edges of the punch (counterbody), the level of friction between the punch and the half-plane, and the location, length, and orientation of crack on the stress intensity factors was investigated.

Hence, we assume that one of contacting cylindrical bodies is damaged by an edge macrocrack. We simulate this body by an elastic half-plane with rectilinear crack (cut) (Fig. 5.35) and the contact influence of the other body (counterbody) is modeled by the action of normal static pressure distributed according to a certain law $p(x)$ over the contact section of length $2a$ and alternating tangential forces $q(x)$ connected with it by the Amontons–Coulomb law via the friction coefficient f, i.e., the conditions of full slipping are realized between the analyzed contact bodies.

The boundary conditions of this problem take the form

$$\begin{aligned} \sigma_y(x) \pm i\tau_{xy}(x) &= -p(x)(1 \mp if), \qquad |x - x_0| \le a, \qquad y = 0; \\ \sigma_y(x) - i\tau_{xy}(x) &= 0, \qquad |x - x_0| > a, \qquad y = 0, \end{aligned} \tag{5.15}$$

where x_0 is the abscissa of the middle of the section of external loading in the system xOy. We assume that crack faces are free of loads and do not contact.

We determine the law of distribution of pressure $p(x)$ from the solution of contact problem of compression of two elastic cylinders with an initial contact strip of length $2b$ [1]:

$$\begin{aligned} p(x) = \frac{P(1+\delta)}{b\pi F(\delta)} \Bigg[& (\pi - 2\varphi_0)\cos\varphi + \sin\varphi \ln\left|\frac{\sin(\varphi+\varphi_0)}{\sin(\varphi-\varphi_0)}\right| + \\ & + \sin\varphi_0 \ln\left| tg\frac{\varphi+\varphi_0}{2} tg\frac{\varphi-\varphi_0}{2}\right| \Bigg] = \tilde{p}(x'), \end{aligned} \tag{5.16}$$

$$\varphi = \arcsin\frac{x - x_0}{a}, \quad \varphi_0 = \arcsin\frac{1}{1+\delta}, \quad \delta = (a-b)/b, \quad x' = x - x_{0,},$$

$$F(\delta) = \frac{\pi - 2\varphi_0}{2\sin^2\varphi_0} - \mathrm{ctg}\varphi_0. \tag{5.17}$$

Here, P is the main vector of normal forces in the contact load upon the counterbody (Fig. 5.34), $2b$ is the length of the initial contact line, and $2a$ is the complete length of this line which increasing as a result of contact between the bodies. In this case, the parameter δ characterizes the increment of length of the contact line and is connected with the radius R by the formula

$$R = \frac{b^2 E F(\delta)}{4P(1-\nu^2)}, \tag{5.18}$$

where E and ν are the elastic constants of the materials (we also consider the case where the materials of the contacting bodies are identical). We determine the parameter δ and, hence, the length of the contact section $2a$ (for a given radius of rounding R of edges of the punch, the length of the initial contact line $2b$, and the force of pressing P given by relations (5.17) and (5.18)) from the transcendental equation

$$\frac{\pi - 2\varphi_0}{2\sin^2\varphi_0} - \mathrm{ctg}\varphi_0 = \frac{4PR(1-\nu^2)}{b^2 E}. \tag{5.19}$$

First, we determine the distribution of contact pressure $\tilde{p}(x') = p(x)$ (5.16) under the punch and the length $2a$ of the contact section as functions of the radius of rounding R of the edges of the punch base under the condition of its smooth contact ($f = 0$, $q = 0$) with the boundary of the half-plane along the initial line of length $2b$ under the action of the pressing force P (Fig. 5.34). For this purpose, we use relations (5.16)–(5.19). In particular, the determination of length of the contact section is reduced first to the solution of the transcendental Eq. (5.19) and then to finding

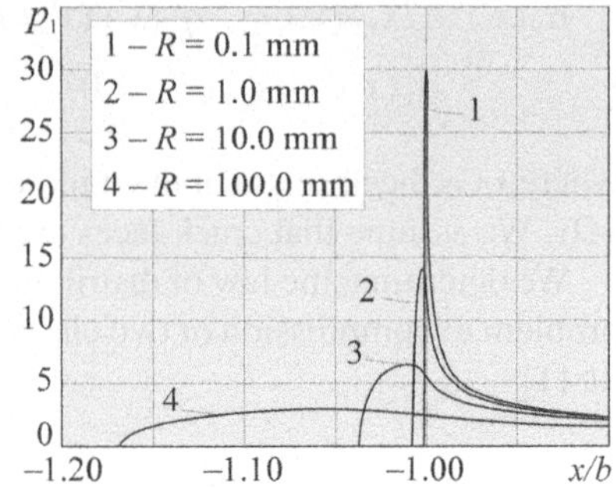

Fig. 5.36 Curves of distribution of the normalized pressure $p_1 = \tilde{p}(x')\pi b/P$ near the left end of the contact section; $P = 2.0$ MN/m; $b = 5$ mm; $\nu_1 = \nu_2 = \nu = 0.3$; $E_1 = E_2 = E = 210$ GPa

the distribution of contact pressure by using relation (5.16). For subsequent investigations, we choose the following rounding radii for the edges of the counterbody: $R = 0.1, 1.0, 10.0, 100.0$ mm. The forces of indentation of the punch are $P = 0.1, 0.5, 1.0, 2.0, 3.0, 4.0$ MN/m. We perform our calculations for the following values of the initial half length of the contact section (half width of the contact strip): $b = 5$ and 10 mm. Note that these values of the force P and the width of the initial strip of the contact section b are chosen with regard for the typical values of pressure of the contact patches onto the base (structural element) in the course of experimental investigations of the fretting fatigue contact interaction and also in the engineering practice [57, 66].

In Fig. 5.36, we present the plots of normalized contact pressure depending on the rounding radius R. The shape of the curves is similar to the curves available from the literature [1]. The numerical data (Fig. 5.36, Table 5.12) show that, as the radius R decreases by an order of magnitude, the maximal values $\tilde{p}_{\max}$ of contact pressure in the half-plane near the edges of counterbody become much (almost twice) higher. In addition, the increase in the radius R leads to the growth of the contact section (significant for $R = 100$ mm). As expected, the level of pressure at the center of the contact section $\tilde{p}(0)$ weakly depends on R (for given P) but strongly varies as a function of the force P. For the values of P (Table 5.12) within the range 0.1…40 MN/m chosen for calculations, the quantity $\tilde{p}(0)$ takes the values 20…200 MPa.

It should be emphasized that the length of initial contact line weakly affects the behavior of contact pressure near the edges of the punch and also the final lengths of the contact section depending on the radius of rounding of the edges. However, for lower values of b (and identical P and R), we obtain larger increments δ and higher values of $\tilde{p}_{\max}$.

Stress intensity factors. The determination of the SIF for the problem with boundary conditions (5.15) is reduced to the solution of the singular integral equation (3.151) with the right-hand side $P(\eta)$, where the functions $\Phi_0(z)$ and $\Psi_0(z)$ can be found from relations (3.122) and (3.176). These relations are the images of the Muskhelishvili complex potentials $\Phi_0(z)$ and $\Psi_0(z)$ [41, 48, 53] in the first main problem of the theory of elasticity for a half-plane without cracks when the pressure $p(x)$ [(5.15), (5.16)] acts upon the half-plane boundary. Thus, we get

Table 5.12 Half length a of the actual (enlarged) contact section, pressure $\tilde{p}(0)$ in its central part, and the maximum pressure $\tilde{p}_{max}$ under the edges of counterbody depending on its loading by the force P and the geometric parameters of the punch base (R, b)

R, mm	$b = 5$ mm			$b = 10$ mm		
	a, mm	$\tilde{p}(0)$, MPa	$\tilde{p}_{max}$, GPa	a, mm	$\tilde{p}(0)$, MPa	$\tilde{p}_{max}$, GPa
$P = 0.1$ MN/m						
0.1	5.0012	6.37	0.44	10.0010	3.18	0.35
1.0	5.0055	6.36	0.20	10.0044	3.18	0.16
10.0	5.0256	6.35	0.11	10.0204	3.18	0.09
100.0	5.1185	6.28	0.05	10.0943	3.17	0.04
$P = 0.5$ MN/m						
0.1	5.0035	31.82	1.28	10.0028	15.91	1.02
1.0	5.0162	31.77	0.63	10.0128	15.90	0.47
10.0	5.0748	31.55	0.33	10.0595	15.86	0.26
100.0	5.3429	30.58	0.15	10.2747	15.66	0.12
$P = 1.0$ MN/m						
0.1	5.0055	63.62	2.03	10.0044	31.82	1.61
1.0	5.0256	63.47	1.12	10.0204	31.79	0.86
10.0	5.1185	62.77	0.52	10.0943	31.65	0.41
100.0	5.5396	59.81	0.24	10.4345	31.02	0.19
$P = 2.0$ MN/m						
0.1	5.0088	127.19	3.22	10.0070	63.64	2.56
1.0	5.0407	126.71	1.79	10.0323	63.54	1.42
10.0	5.1875	124.53	0.83	10.1496	63.10	0.66
100.0	5.8455	115.68	0.38	10.6858	61.15	0.31
$P = 3.0$ MN/m						
0.1	5.0115	190.72	5.06	10.0091	95.44	4.02
1.0	5.0533	189.77	2.35	10.0423	95.25	1.87
10.0	5.2450	185.54	1.09	10.1958	94.39	0.87
100.0	6.0966	168.96	0.48	10.8945	90.65	0.40
$P = 4.0$ MN/m						
0.1	5.0139	254.22	6.13	10.0111	127.24	4.87
1.0	5.0645	252.69	2.85	10.0513	126.93	2.26
10.0	5.2961	245.93	1.31	10.2369	125.54	1.05
100.0	6.3169	220.24	0.57	11.0793	119.62	0.48

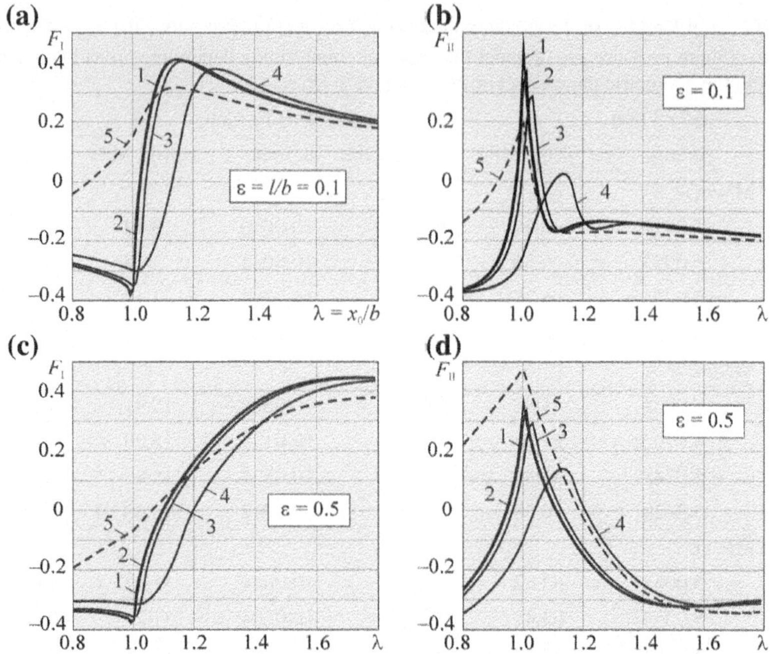

Fig. 5.37 Dependences of the normalized SIF $F_{I,II} = K_{I,II}\sqrt{\pi b}/P$ on the distance from the middle of the contact section to the crack mouth for the angle of its inclination $\beta = \pi/3$ and $f = 0.5$: (1) $r = R/b = 0.02$; (2) $r = 0.20$; (3) $r = 2.00$; (4) $r = 20.00$; (5) uniformly distributed pressure

$$\Phi_0(t) = \frac{1+if}{2\pi i}\int\limits_{x_0-a}^{x_0+a}\frac{p(x)}{x-t}dx;$$
$$\Psi_0(t) = \frac{1}{2\pi i}\int\limits_{x_0-a}^{x_0+a}\left[\frac{(1-if)p(x)}{x-t} - \frac{(1+if)xp(x)}{(x-t)^2}\right]dx. \quad (5.20)$$

For the numerical solution of problem in this case, we apply the Simpson quadrature formula [65] to the right-hand side of the SIE.

The preliminary numerical analysis of the dependences of SIF for a rectilinear crack on the parameters $\varepsilon = l/b$, β, $\lambda = x_0/b$, and f (Fig. 5.35) demonstrates that, in the negative phase of the contact cycle where $q = q^- = -fp(x)$, the crack is located mainly in the zone of compression ($K_I < 0$), whereas in the positive phase, where $q = q^+ = +fp(x)$, it is located in the zone of tension ($K_I > 0$). Therefore, in what follows, we present the SIF for the positive phase.

In Figs. 5.37, 5.38 and 5.39, we present the dependences of the normalized SIF $F_{I,II} = K_{I,II}\sqrt{\pi b}\Big/P$ (curves 1–4) on the relative distance $\lambda = x_0/b$ from the middle of the contact section to the crack mouth for the values of the parameters typical

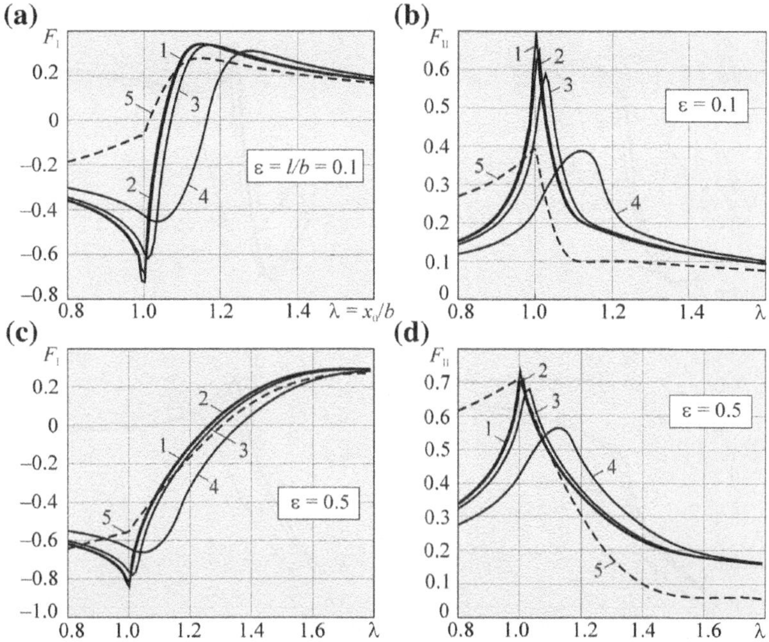

Fig. 5.38 Dependences of the normalized SIF F_I and F_{II} on the distance from middle of the contact section to the crack mouth for the angle of its inclination $\beta = \pi/4$ and $f = 0.5$: (1) $r = R/b = 0.02$; (2) $r = 0.20$; (3) $r = 2.00$; (4) $r = 20.00$; (5) uniformly distributed pressure

mainly of fretting fatigue, namely, the angles made by the crack with the half-plane boundary $\beta = \pi/4$, $\pi/3$, $2\pi/3$, the ratio of the crack length to the half length of the initial contact section $\varepsilon = l/b = 0.1$ and 0.5, and the coefficient of friction between the contacting bodies $f = 0.5$. The ratio of the radius of rounding of the punch base to the half length of the initial contact section $r = R/b = 0.02, 0.20, 2.00, 20.00$. The model contact load is given by relations (5.15)–(5.19). For the sake of comparison, in these figures, we also present the SIF obtained for the case of action of a pressure p uniformly distributed over the segment $2b$ and tangential forces $q = fp$ (curves 5) upon the edge of the half-plane weakened by a crack; in this case, $2pb = P$.

The accumulated results demonstrate that a significant (by three orders of magnitude) decrease in the radius of rounding R of the edges of the counterbody leads to an insignificant increase (by at most 10%) in the maximum values of the SIF K_I (Figs. 5.37a, c, 5.38a,c and 5.39a, c) and to a much more pronounced (almost twofold) growth of the maximal SIF $|K_{II}|$, especially for the acute angles β ($\beta = \pi/3$ and $\pi/4$) and short cracks ($\varepsilon = 0.1$) (Figs. 5.36b, d, 5.37b, d, 5.38b, d and 5.39b, d). However, $\max|K_{II}|$ are attained mainly for $K_I < 0$. Hence, the obtained quantitative estimate is approximate. The exact estimation requires the solution of the problem with regard for the contact of crack faces.

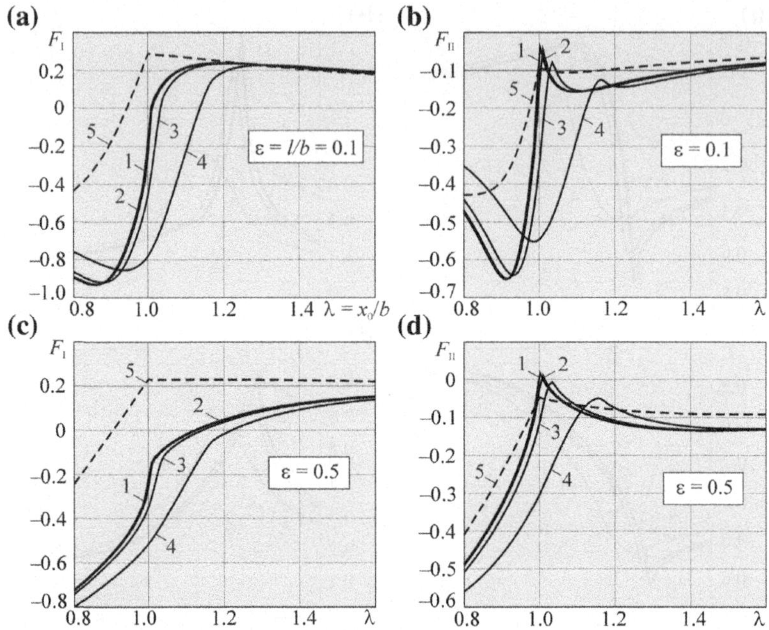

Fig. 5.39 Dependences of the normalized SIF F_{I} and F_{II} on the distance from the middle of the contact section to the crack mouth for the angle of its inclination $\beta = 2\pi/3$ and $f = 0.5$: (1) $r = R/b = 0.02$; (2) $r = 0.20$; (3) $r = 2.00$; (4) $r = 20.00$; (5) uniformly distributed pressure

The functions $F_{\mathrm{I}}(\lambda)$ and $F_{\mathrm{II}}(\lambda)$ for the model load with uniformly distributed pressure (the dashed curves 5 in Figs. 5.37, 5.38 and 5.39) in the analyzed cases are qualitatively and often even quantitatively close to the corresponding main curves, in particular, to curves 1–3 for small radii of rounding R of the counterbody edges. Hence, one can use this distribution to model the contact interaction of fretting fatigue. As expected, the maxima of K_{I} and $|K_{\mathrm{II}}|$ in curve 4 for large radius of rounding are realized for larger λ than in the other cases (curves 1–3, 5), i.e., when the load is located at greater distances from the crack mouth.

The influence of rounding rapidly disappears as the distance from the contact load to the crack mouth independently of the angle of its orientation (Figs. 5.37, 5.38 and 5.39). Even for $\lambda = x_0/b > 1.8$, the difference between the SIF in curves 1–4 does not exceed 1–5%. As the relative crack length ε increases from 0.1 to 0.5, the values of max $|K_{\mathrm{II}}|$, in general, become higher, while the values of max K_{I} decrease (see also [21]). The maxima of the SIF K_{I} are realized for greater λ and the influence of R becomes weaker.

The results of additional calculations show that, for the other values of the friction coefficient ($f = 0.3$ and 0.7), the influence of the rounding radius R on the SIF K_{I} is insignificant, as in the case of $f = 0.5$.

Comparing our results with the data presented in [55] for a plane rectangular rigid punch indented with friction into the half-plane weakened by a crack (Fig. 5.40), we

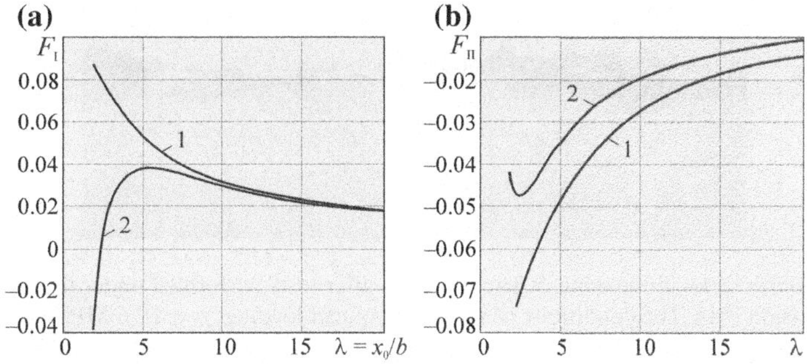

Fig. 5.40 Dependences of the normalized SIF $F_{I,II} = K_{I,II}\sqrt{\pi b}/P$ on the distance from the middle of the contact section to the crack mouth for $\beta = 3\pi/4; f = 0.3; \varepsilon = 1.0$; curves 1 correspond to the results from [55] and curves 2 correspond to our data

conclude that, for K_{II}, our data are close to the results from [55]. At the same time, for K_I, they practically coincide starting from $\lambda = 15$; however, for $\lambda \leq 2$, i.e., when the punch edge or the load are close to the crack mouth, the corresponding data are noticeably different. A similar picture is observed in [26]. The indicated discrepancy is explained by the fact that the curves similar to curves 1 in Fig. 5.40 were obtained in [26] on the basis of the solutions in which the presence of square-root singularity of stresses under the edges of the punch was a priori assumed. However, if we use the model forces $p(x)$ (5.16), then the stresses are finite.

5.6 Shear and Opening Edge Cracks in Contact Zone of Fretting Pair Elements

Edge cracks very often appear in the elements of fretting pairs (couples) both in the contact zone and beyond it (Fig. 5.41; [51]). It is known [59, 66] that these cracks may decrease the fatigue life of the elements of units and machines almost by an order of magnitude. As one of the simplest and widespread model schemes (problems) in the two-dimensional case, which can be used for the investigation the processes of crack propagation in the contact zone of elements of fretting couples, we can mention the model of elastic half-plane with cracks subjected to the action of a model contact load on the boundary. The SIF for an edge crack whose faces do not contact in the contact cycle under the action of model loads were investigated in numerous works (see, e.g., the surveys [16, 21]). However, in the case where the crack faces are in contact in the loading cycle, the problem becomes more complicated and the number of corresponding solutions available from the literature is insignificant [13]. Furthermore, if the contact load in a cycle is alternating, then the problem becomes even more complicated, especially if it is necessary to establish the ranges of the

Fig. 5.41 Microcracks in the zone of friction track after tests performed under the conditions of fretting corrosion [51]. The conditions of vibration contact loading: $p = 19.6$ MPa; slip amplitude $A = 125$ μm; $\nu = 25$ Hz; $T = 293$ K; $N = 5{\cdot}10^5$ cycles. Both elements of the couple are made of VT3-1 (BT-3) titanium alloy (×210)

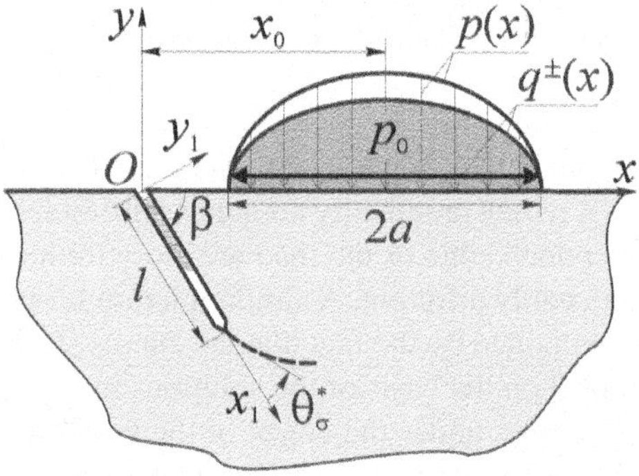

Fig. 5.42 Computational scheme of the problem, that take into account friction between crack faces

SIF ΔK_{II} and $\Delta K_{\mathrm{I}\theta}$ controlling the process of crack growth by the mechanisms of transverse shear and normal opening, respectively.

In what follows, within the framework of the model used to estimate the lifetime of solids under the conditions of their fretting fatigue contact interaction [12, 47], we study the SIFs K_{I} and K_{II}, the mixed-type SIFs $K_{\mathrm{I}\theta}$ and $K_{\mathrm{II}\theta}$, and also their ranges $\Delta K_{\mathrm{I}\theta}$ and $\Delta K_{\mathrm{II}\theta}$ for the edge cracks in the contact zone in the course of mutual displacement of the elements of fretting couples in a single contact cycle (Fig. 5.42). We analyze the dependence of the SIFs on the orientation of the edge crack, the level of friction between its faces, and the location of its mouth relative to the counterbody (model contact load). On the basis of the analysis of values of the parameters ΔK_{II} and $\Delta K_{\mathrm{I}\theta} = \max K_{\mathrm{I}\theta}$, we analyze possible mechanisms of propagation of the edge cracks in the contact zone (transverse shear and normal opening), as well as the most probable their locations and directions of propagation depending on the sets of operating parameters of the fretting couple.

According to the main postulates of the model used for the investigation of fracture and lifetime of the bodies under the conditions of contact interaction of fretting fatigue [12, 47], we model a damaged element of the fretting couple by an elastic half-plane weakened by an inclined edge rectilinear crack (cut) (Fig. 5.42). We refer this half-plane to the main coordinate system xOy. The crack is referred to an auxiliary system $x_1O_1y_1$. The action of the second element (counterbody) is modeled by normal stationary forces $p(x, \lambda)$ and alternating tangential forces $q(x,f, \lambda)$ applied to the half-plane boundary.

We also assume that the normal compression forces are distributed according to the elliptic (Hertzian) law over a part of the half-plane boundary of length $2a$ and the tangential forces in the case of full slipping (without sticking) between the bodies are connected with the normal forces by the Amontons–Coulomb law, i.e.,

$$p(x,\ \lambda) = p_0\sqrt{1 - (x/a - \lambda)^2}, \qquad q^{\pm}(x,\ \lambda) = \pm f p(x,\ \lambda); \qquad |x - x_0| \leq a; \tag{5.21}$$

here, p_0 is the maximum level of model contact pressure and $\lambda = x_0/a$ is a parameter characterizing the distance of from the middle of the contact section to the crack mouth.

In general, depending on the location of the load relative to the crack mouth (parameter λ), its relative length ($\varepsilon = l/a$), the angle of inclination β to the edge of the half-plane, and the friction coefficients f and f_c between the elements of the fretting pair and the crack faces, respectively, the crack in a contact cycle may be either completely open (the faces are not in contact) or closed (partially or completely) with different contact conditions of (sliding with friction or sticking).

In what follows, we consider a somewhat simplified statement of the problem, namely, either the crack faces do not contact and are free of loads, or they contact with friction over the entire crack length. The corresponding boundary conditions take the form [13]

$$N^{\pm}(x_1) + iT^{\pm}(x_1) = 0, \quad 0 \leq x_1 \leq l, \tag{5.22}$$

or

$$v^{+}(x_1) - v^{-}(x_1) = 0, \qquad T^{\pm}(x_1) = f_c \mathrm{sign}\left[T^{\pm}(x_1)\right] \left|N^{\pm}(x_1)\right|, \quad 0 \leq x_1 \leq l; \tag{5.23}$$

here, N^{+} and T^{+} are the normal and tangential components of the forces at the crack faces, respectively, $v^{\pm}$ are normal displacements of the faces in the local coordinate system $x_1O_1y_1$, and f_c is the coefficient of friction between the faces. In the first case, we reduce the problem to the SIE (3.151) and, in the second case, to the system of SIE (3.219) and (3.220) which are solved by the method of mechanical quadratures.

On the basis of the solutions of the SIE, we determine the SIFs K_{I} and K_{II} and, by using the relations of the σ_θ- and τ_θ-criteria of local fracture (2.21) and (2.24), find the mixed-type SIFs $K_{\mathrm{I}\theta}$ and $K_{\mathrm{II}\theta}$. Since the contact interaction of fretting fatigue has a cyclic character, it is also necessary to determine the SIF ranges $\Delta K_{\mathrm{I}\theta}$ and $\Delta K_{\mathrm{II}\theta}$ controlling the process of fatigue crack propagation by the mechanisms of normal opening and transverse shear, respectively. In this case, we assume [12, 47] that, for any distance ($\lambda = x_0/a$) between the model load and the crack mouth, $\Delta K_{\mathrm{II}\theta} = \Delta K_{\mathrm{II}} = |K_{\mathrm{II}}^{+} - K_{\mathrm{II}}^{-}|$, where K_{II}^{+} and K_{II}^{-} are determined by the positive and

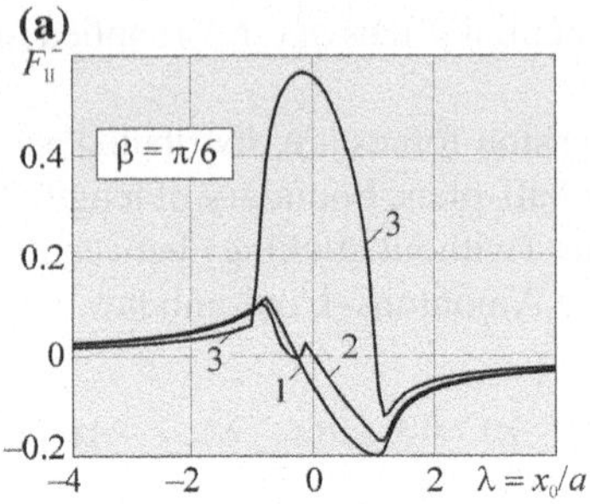

Fig. 5.43 Dependences of the SIF F_{II} on the location of the counterbody in the negative (**a**) and positive (**b**) phases: (1) for smooth contact of the crack faces ($f_c = 0.0$), (2) for the contact of the faces with friction ($f_c = 0.1$), (3) in the absence of the contact of faces; $f = 0.5$

Fig. 5.44 Dependences of the SIFs F_I and $F_{I\theta}$ on the location of the counterbody. The solid lines correspond to the negative phase of a cycle, while the dashed lines correspond to the positive phase; $f = 0.5$

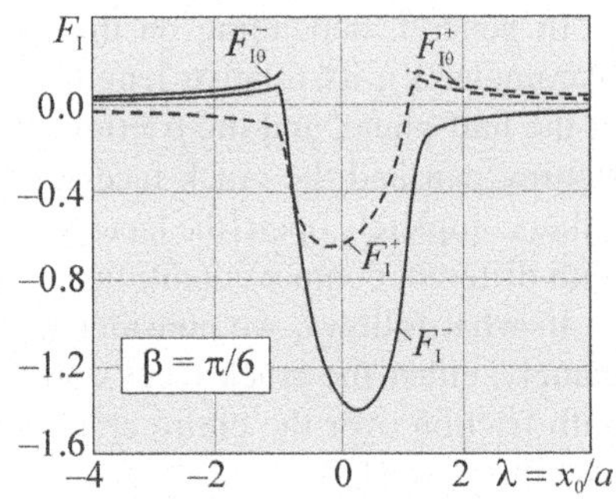

negative phases of the contact cycle, the shear crack in the contact zone propagates along its continuation and the angle of its initial deviation is $\theta_\tau^* = 0$. For the case of opening mechanism, we get $\Delta K_{I\theta} = \max K_{I\theta} = \max\{K_{I\theta}^+, K_{I\theta}^-\}$, and the subsequent process of crack propagation is determined by the angle θ_σ^* (Fig. 5.42) on the basis of the well-known relations of the σ_θ-criterion.

In view of the necessity of investigation of the variations of SIF depending on the location of the model contact load (counterbody) relative to the crack mouth and its orientation, we choose large ranges of the parmeters λ and β, namely, the angle between the crack and the boundary of the half-plane $\beta = 20°\ldots160°$ and the distance $\lambda = x_0/a$ from the middle of the contact section and the crack mouth varied within the range $|\lambda| \le 4.0$. The friction coefficients between the elements of the fretting couple and the crack faces are set equal to $f = 0.5$ and $f_c = 0$ or 0.1, respectively. The relative crack length is equal to $\varepsilon = l/a = 0.2$. The results of evaluation of the quantities K_I and K_{II} are shown in Figs. 5.43, 5.44 and 5.45 and Tables 5.13 and 5.14 in the normalized form, i.e., $F = K\big/\left(p_0\sqrt{\pi a}\right)$.

As follows from Fig. 5.43a, b, the presence of contact between the crack faces enables us to determine the SIF K_{II} much more exactly, while the presence of friction between the crack faces strongly decreases the value of $|K_{II}|$.

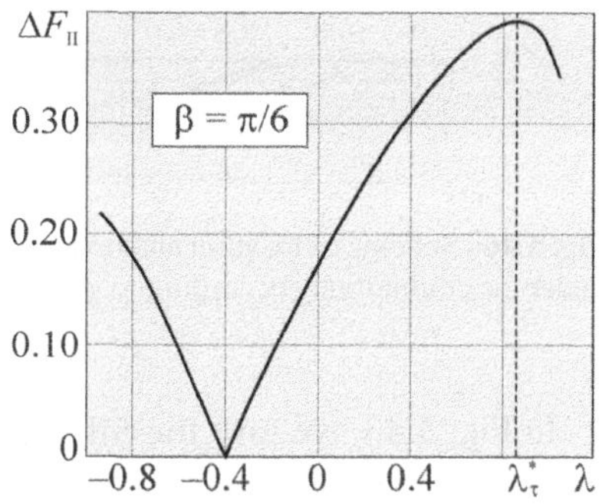

Fig. 5.45 Dependences of the ΔF_{II} range on the location of the counterbody; $\varepsilon = 0.2; f_c = 0; f = 0.5$; $\lambda = x_0/a$

Table 5.13 Dependence of the values of max $\Delta F_{II}(\lambda) = \Delta F_{II}(\lambda_\tau^*)$ and the parameter λ_τ^* of location of the center of the contact section for which this maximum is realized on the orientation of the crack

β, deg.	$f_c = 0.0$		$f_c = 0.1$	
	max ΔF_{II}	λ_τ^*	max ΔF_{II}	λ_τ^*
20	0.4561	0.70	0.3767	0.79
30	0.3922	0.86	0.3303	0.93
40	0.3272	0.95	0.2752	0.98
50	0.3050	−0.79	0.2851	−0.87
60	0.3427	−0.66	0.3027	−0.79
70	0.3704	−0.49	0.3119	−0.60
80	0.3875	−0.26	0.3145	−0.34
90	0.3933	0.00	0.3146	0.00

Table 5.14 Maximum values of the SIFs F_I and $F_{I\theta}$ in a contact cycle for the edge arbitrarily oriented crack, the distance from the crack mouth to the middle of the contact section λ_σ^* for which max $F_{I\theta}$ is attained, and the corresponding angles of initial crack deviation θ_σ^*; $\varepsilon = 0.2$

β, deg.	$f = -0.5$				$f = +0.5$			
	max F_I	max $F_{I\theta}$	λ_σ^*	θ_σ^*, rad.	max F_I	max $F_{I\theta}$	λ_σ^*	θ_σ^*, rad.
20	0.0587	0.1100	−0.97	−1.20	0.0926	0.1324	1.23	−0.80
30	0.0827	0.1452	−0.95	−1.18	0.1188	0.1533	1.24	−0.71
40	0.1058	0.1618	−0.92	−1.20	0.1394	0.1769	1.01	−1.21
50	0.1259	0.1761	−0.88	−1.20	0.1539	0.1812	0.95	−1.15
60	0.1426	0.1763	−0.83	−1.23	0.1635	0.1707	1.23	−0.33
70	0.1547	0.1687	−1.07	−0.48	0.1685	0.1702	1.21	−0.16
80	0.1618	0.1695	−1.12	−0.35	0.1696	0.1696	1.18	0.02
90	0.1672	0.1695	−1.15	−0.19	0.1672	0.1695	1.15	0.19

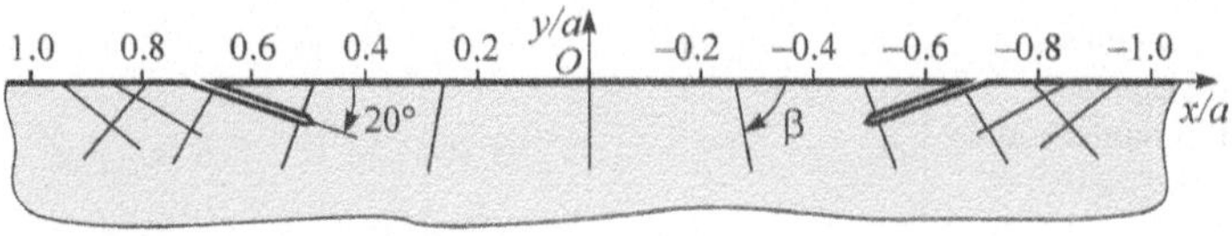

Fig. 5.46 Scheme of location and orientation of shear edge cracks along the contact zone (section) under the counterbody according to data from Table 5.13; $f = 0.5$; $f_c = 0$; $\varepsilon = 0.2$

In Fig. 5.43, we give the SIF F_{II} within the entire range of the parameter λ. At the same time, in finding max $F_{I\theta}$ and max ΔF_{II} presented in the tables, we take into account the fact that these values of SIF are physically meaningful only within the ranges where $F_I > 0$. The quantities with $F_I < 0$ were found from the solution of the problem for an open crack without taking into account the contact of its faces (see Fig. 5.44).

In Fig. 5.45, we illustrate the procedure of finding the maximum value of the range ΔF_{II} for a crack of given orientation (here, $\beta = \pi/6$) and the location of counterbody (the values of parameter $\lambda = \lambda_\tau^*$) for which this maximum is realized. For the angles $\beta = 20°\ldots90°$, these data are presented in Table 5.13. At the same time, for the angles $\beta = 90°\ldots160°$, the corresponding results can be obtained by the mirror imaging.

By using the data from Table 5.13, in Fig. 5.46, we depict the scheme of orientation and location (along the contact section) of shear edge cracks corresponding to the maximum values of the SIF range for K_{II} attained in a single contacting cycle for fixed values of the parameters f, f_c, and ε. This scheme gives us the distances from the middle (boundaries) of the section under loading (counterbody), where the initiation of shear macrocracks is most probable, as well as the directions of their propagation (along their continuation).

The data presented in Table 5.13 enable us to conclude that, in the contact zone (section) under the counterbody, we mainly observe the formation of shear cracks of different orientations closer to the ends of this zone. The simultaneous propagation of these cracks promotes the development of the process of crumbling of the contact surface. However, the process of crack propagation at angles of 20° and 160° proves to be most probable. It is possible to assume that these cracks form the so-called "fretting tongues" in the course of their development (see Fig. 5.46). We also see that the increase in the friction coefficient f_c between the crack faces leads to a decrease in the SIF $|K_{II}|$ and in max ΔK_{II} and the attainment of the indicated maximum value closer to the edges of the counterbody.

The analysis of the data presented in Table 5.14 shows that the highest values of max $F_{I\theta}$ are attained for the cracks oriented at the angles $\beta = 50\ldots60°$ (and, by the mirror imaging, at $\beta = 120\ldots130°$). Hence, these cracks prove to be most dangerous from the viewpoint of formation of mode I cracks. It is also easy to see that these cracks are mainly initiated beyond the contact zone and may propagate toward this zone.

Note that it is possible to predict the residual lifetime of tribojoints under the operating conditions corresponding to fretting fatigue by using the results presented in this chapter on the basis of the available characteristics of cyclic crack growth resistance of the materials of elements of the fretting pair.

5.7 Conclusions

1. We propose a computational model for the investigation of the fracture processes and the evaluation of the residual lifetime (in terms of the crack resistance) of the nearsurface layer of the contact zone of bodies operating under the conditions of fretting fatigue.

For the scheme of contact interaction in which the process of fretting between two bodies pressed to each other is caused, in the analyzed body, by the oscillatory displacements of the counterbody with its full slipping (without sticking), our calculations give the following results:

- For small lengths of the initial edge cracks ($\varepsilon = l_0/a \approx 0.1\ldots0.2$) whose mouth appear on the boundary of the body outside the contact section, the stress intensity factors K_{I} and K_{II} near the crack tips reach their maximal values if the crack mouth is located near the edge of the contact load ($\lambda = x_0/a \approx 1.0$).
- The friction coefficient in contact between the bodies plays an important role in the subsequent propagation of the edge crack. For low friction, the crack propagates into the bulk of the material and deviates from the contact zone. For higher levels of friction, the crack paths are directed toward the contact zone.
- In the course of cracks propagation, the SIF in the vicinity of their tips vanish, and the cracks stop.
- The decrease in the radius of rounding of the edges of counterbody (from 100 to 0.1 mm) insignificantly increases (by at most 10%) the maximum values of the SIF K_{I} and noticeably (sometimes by a factor of two) increases the maximum values of SIF $|K_{\mathrm{II}}|$ at the tip of an edge crack in the main body of the contact couple (pair). This influence is especially pronounced for small relative lengths of the edge cracks ($\varepsilon = l/b = 0.1$) and decreases as their length increases and as the counterbody moves away from the crack mouth (the parameter $\lambda = x_0/b$ increases). The functions $K_{\mathrm{I,II}}(\lambda)$ for a model contact load taking into account the effect of rounding of the edges of counterbody and a model loading with uniformly distributed pressure are qualitatively similar and quantitatively rough. This fact theoretically confirms the possibility of application of a simpler model load (uniform distribution).

2. It is shown that the edge cracks under the counterbody filled either with oil or with fretting products mainly develop to form dish-shaped pittings (cavities). Simultaneously with pitting, the main crack may propagate from the edge crack into the bulk of the material. The formation of pitting near the edge (end) of the contact section or on the boundary proves to be most probable. The obtained results are in good agreement with the experimental data presented in the monograph [66].

Within the framework of the computational model aimed at the analysis of the lifetime and fracture processes in the bodies contacting under conditions of fretting fatigue [14, 15], we simulate and study the formation of pitting in these bodies. It is shown that pitting is initiated and propagates from the surface macrocrack preliminarily formed in the contact section in the course of cyclic filling and wedging of the crack by fretting products. The wedging action of these products is simulated by a uniformly distributed normal pressure in the crack. The action of the counterbody is modeled by a constant pressure in the contact section and alternating (in a contact cycle) tangential forces responsible for the cyclic character of the load and the stress-strain state in the contact zone and near the crack.

3. We investigate the SIF and the paths of edge cracks in elements of the fretting couples in the presence of the sections of sticking and slipping in contact between the bodies.

   We established the following facts:

- As the relative length (c/a) of the section of sticking appears and increases in contact between the bodies, the stress intensity factors near the tip of an edge initially rectilinear crack (initiated in the contact zone) significantly decrease: For $c/a = 0.57$, the maximum SIF become lower, on the average, by one-third and, for $c/a = 0.85$, almost by an order of magnitude as compared with the case of full slipping.
- As the section of sticking enlarges, the crack path deviates from the contact zone as in the case of decrease in the friction coefficient f.
- The crack growth paths weakly depend on the initial angle of their inclination (β) to the boundary of the half-plane edge and the distance between the crack mouth and the boundary of the contact load and have asymptotes determined by the friction coefficient f.
- The stress intensity factors both for the initial cracks and along their trajectories in the fretting zone become lower as the level of friction decreases in the case of full slipping between the contacting bodies and as the sticking section increases for their partial slipping. This feature is in good agreement with the well-known approaches [57] to the structural and technological enhancement of increases the fretting fatigue strength either by removing the displacements of one surface over the other or, on the contrary, by the facilitation of these displacements with an aim to guarantee their realization with minimum possible friction.

4. It is established that, under the conditions of fretting fatigue, a significant influence on the character of the fracture of the contact surface and the sizes of crumbling particles is exerted by the friction coefficient, the distribution of tangential forces in contact between the bodies (in particular, in the presence of the section of sticking), the characteristics of cyclic crack growth resistance of the materials of contacting bodies, and the presence of working products (oil, water, powder-like fretting products, etc.) in the contact zone.

References

1. Aleksandrov VM, Romalis BL (1986) Kontaktnyye zadachi v mashinostroyenii (Contact Problems in Mechanical Engineering). Mashinostroenie, Moscow
2. Balatskii LT (1972) Ustalost' valov v soedineniyakh (Fatigue of Shafts in Joints). Tekhnika, Kiev
3. Balatskii LT (1982) Prochnost' pressovykh soedineniy (Strength of Pressed Joints). Tekhnika, Kiev
4. Botvina LR, Yarema SYa, Grechko VV, Lymar'LV (1982) Kinetics of fatigue fracture of VT3-1 titanium alloy. Sov Mater Sci 17(6):518–524
5. Cattaneo C (1938) Sur contatto di due corpi elatici: distribuzione locate degli sforzi. Rend. dell 'Academia nazionale dei lincei 27(Ser. 6):342, 434, 474
6. Ciavarella M, Demelio G (2001) A review of analytical aspects of fretting fatigue, with extension to damage parameters, and application to dovetail joints. Int J Solids Struct 38(10–13):1791–1811
7. Collins JA (1981) Failure of materials in mechanical design. Wiley, New York
8. Datsishin OP, Marchenko GP (1991) Edge-crack growth. Sov Mater Sci 27(5):465–471
9. Datsyshyn OP (1999) Durability and fracture calculate model for structural materials under fretting fatigue. Naukovyy Visnyk Ukrauinskogo Derzhavnogo Lisotekhnichnogo Universytetu, Issue 9:139–149
10. Datsyshyn OP (1996) Fracture and wear processes simulating under cyclic contact of solid bodies. In: Petit J (ed) ECF-11, vol II. Mechanism and mechanics of damage and failure. EMAS LTD, Warley, pp 1411–1416
11. Datsyshyn OP (2011) Modeling of the initiation of contact fatigue damages and estimation of the durability of elements of tribological conjunctions. Mater Sci 47(2):188–200
12. Datsyshyn OP (2005) Service life and fracture of solid bodies under the conditions of cyclic contact interaction. Mater Sci 41(6):709–733
13. Datsyshyn OP, Hlazov AYu, Levus AB (2014) Specific features of contact of the faces of an edge crack under moving Hertzian loads. Mater Sci 49(5):589–601
14. Datsyshyn OP, Kadyra VM (2006) A fracture mechanics approach to prediction of pitting under fretting fatigue conditions. Int J Fatigue 28(4):375–385
15. Datsyshyn OP, Kalakhan OS, Kadyra VM, Shchur RB (2004) Pitting formation under the conditions of fretting fatigue. Mater Sci 40(2):159–172
16. Datsyshyn OP, Marchenko HP (2008) Stressed state of a half plane with shallow edge crack under Hertzian loading (a survey). Mater Sci 44(1):22–34
17. Datsyshyn OP, Marchenko HP, Hlazov AYu, Levus AB (2004) One approach to the evaluation of durability of solid bodies. Mater Sci 40(4):484–490
18. Datsyshyn OP, Panasyuk VV (1996) Durability and fracture calculational model of solids under their contact interaction. In: Petit J (ed) ECF-11, vol II. Mechanism and mechanics of damage and failure. EMAS LTD, Warley, pp 1381–1386
19. Datsyshyn OP, Panasyuk VV, Glazov AYu (2016) The model of the residual lifetime estimation of trybojoint elements by formation criteria of the typical contact fatigue damages. Int J Fatigue 83(2):300–312
20. Datsyshyn OP, Panasyuk VV, Pryshlyak RE (2014) Effect of rounding the edges in the base of a rider on the stress intensity factors in a body with edge cracks. Mater Sci 50(1):1–13
21. Datsyshyn OP, Pryshlyak RE, Prykhods'ka SV, et al (1998) Influence of the shape of model contact load on the stress intensity factors for an edge crack. Problemy Trybologii 3:3–16
22. Datsyshyn OP, Shchur RB (1998) Development of edge cracks under conditions of fretting fatigue. Problemy Trybologii 2:7–16
23. Dekhovich LA, Makhutov NA (1981) Use of failure mechanics for evaluating fretting fatigue strength. Sov Mater Sci 17(3):280–283
24. Eden EM, Rose WN, Cunningham FL (1911) The endurance of metals. Proc Instn Mech Engrs 4(839): 68–76

25. Edwards PR (1984) Fracture mechanics application to fretting in joints. In: Advanced fracture research, Proceedings 6th international conference on fracture (ICF 6), Oxford e. a.; New Delhi, vol. 6, pp 3813–3836
26. Edwards PR (1981) The application of fracture mechanics to predicting fretting fatigue. In: Waterhouse RB (ed) Fretting fatigue. Elsevier Applied Science, London, pp 67–99
27. Faanes S (1995) Inclined cracks in fretting fatigue. Eng Fract Mech 52(1):1–82
28. Fernando US, Brown MW, Miller KJ (1996) Linear elastic fracture mechanics interpretation of crack growth behaviour in fretting fatigue. In: Advances in fracture resistance in materials (ICF–8 Proceeding), Vol II. Tat McGraw-Hill Publishing, New Delhi, pp 207–215
29. Filimonov GN, Balatskii LT (1973) Fretting v soyedineniyakh sudovykh detaley (Fretting in the Joints of Ship Components). Sudostroenie, Leningrad
30. Fretting fatigue: proceedings international conference of fretting fatigue, Sheffield, UK: The University of Sheffield (1993)
31. Golego NL, Alyab'ev AYa, Shevelya VV (1974) Fretting-koroziya metallov (Fretting Corrosion of Metals). Tekhnika, Kiev
32. Hasebe N, Qian J (1997) Circular rigid punch with one smooth and another sharp ends on a half-plane with edge crack. J App Mech 64:73–79
33. Hills DA, Nowell D (1994) Mechanics of fretting fatigue. Kluwer Academic Publishers, Dordrecht
34. Johnson KL (1985) Contact mechanics. Cambridge University Press, Cambridge
35. Kalakhan OS, Pokhmurs'kyi VI (2000) Influence of incomplete annealing of titanium (α+β)-alloy and its welded joints on fatigue resistance and corrosion-fatigue resistance. Mater Sci 36(2):244–251
36. Keer LM, Bryant MD (1983) A pitting model for rolling contact fatigue. Trans ASME J Lubric Technol 105(2):198–205
37. Kolesnikov Y, Morozov EM (1989) Mekhanika kontaktnogo razrusheniya (Mechanics of Contact Fracture). Nauka, Moscow
38. Kudish II (1986) Contact problem of the theory of elasticity for bodies with cracks. Prikladnaya Matematika i Mekhanika 50(6):1020–1033
39. Lamacq V, Dubourg M-C (1999) Modeling of initial fatigue crack growth and crack branching under fretting conditions. Fatigue Fract Mater Struct 22(6):535–542
40. Maxwell WW, Dudley BR, Cleary AB, Richards J, Shaw J (1967–8) Measures to counter fatigue failures in railway axles. Proc Instn Mech Engrs 182(1):89–108
41. Muskhelishvili NI (1966) Nekotoryye osnovnyye zadachi matematicheskoy teorii uprugosti (Some Basic Problems of the Mathematical Theory of Elasticity). Nauka, Moscow
42. Nix KJ, Lindley TC (1985) The application of fracture mechanics to fretting fatigue. Fatigue Fract Engn Mater Struct 8(2):143–160
43. Nowell D, Dini D, Hills DA (2006) Recent developments in the understanding of fretting fatigue. Eng Fract Mech 73(2):207–222
44. Panasyuk VV (ed) (1988–1990) Mekhanika ruynuvannya ta mitsnist' konstruktsiy (Fracture Mechanics and Strength of Materials): A Handbook. Naukovs Dumka, Kiev
45. Panasyuk VV, Datsyshyn OP, Marchenko HP (1996) Contact problem for a half plane with cracks subjected to the action of a rigid punch on its boundary. Mater Sci 31(6):667–678
46. Panasyuk VV, Datsyshyn OP, Marchenko HP (2000) Stress state of a half-plane with cracks under rigid punch action. Int J Fract 101(4):347–364
47. Panasyuk VV, Datsyshyn OP, Shchur RB (2000) Residual durability of solids contacting under conditions of fretting fatigue. Mater Sci 36(2):153–169
48. Panasyuk VV, Savruk MP, Datsyshyn AP (1976) Raspredeleniye napryazheniy okolo treshchin v plastinakh i obolochkakh (Distribution of Stresses near Cracks in Plates and Shells). Naukova Dumka, Kiev
49. Pokhmurskii VI (1985) Korozionnaya ustalost' metallov (Corrosion Fatigue of Metals). Metallurgiya, Moscow
50. Pokhmurs'kyi VI, Kalakhan OS (1997) Plasma coatings and their ability to protect titanium alloys against corrosion fretting-fatigue fracture. Mater Sci 33(3):331–335

51. Pryshlyak R, Datsyshyn O, Dukhota O, Marchenko H (2014) Shear and tensile edge cracks in the contact zone of elements of the fretting couple. In: Panasyuk VV (ed) Mekhanika ruynuvannya materialiv ta mitsnist' konstruktsiy (Fracture Mechanics of Materials and Strength of Structures): Proceedings of the international Scientific conference. Lviv, pp 125–130
52. Rooke DP, Jones DA (1979) Stress intensity factors in fretting fatigue. J Strain Anal 14(1):1–6
53. Savruk MP (1981) Dvumernyye zadachi uprugosti dlya tel s treshchinsmi (Two-Dimensional Problems of Elasticity for Bodies with Cracks). Naukova Dumka, Kiev
54. Savruk MP, Tomczyk A (2010) Pressure with friction of a perfectly rigid die upon an elastic half-space with cracks. Mater Sci 46(3):283–296
55. Savruk M, Tomczyk A, Yevtushenko A (2007) Płaskie kontaktowe zagadnienie z uwzględnieniem tarcia dla półprzestrzeni ze szczeliną. Acta Mechanica et Automatica 2:41–44
56. Serensen SV (1975) Soprotivleniye materialov ustalostnomu i khrupkomu razrusheniyu (Resistance of Materials to Fatigue and Brittle Fracture). Atomizdat, Moscow
57. Shevelya VV, Kalda GS (1998) Fretting-ustalost' metallov (Fretting Fatigue of Metals). Khmel'nits'kii, Podillya
58. Sosnovskii LA, Makhutov NA, Shurinov VA (1992) Contact-mechanical fatigue: basic regularities, Zavodskaya Laboratoriya 11:44–61
59. Sosnovskii LA, Makhutov NA, Shurinov VA (1992) Fretting fatigue: basic regularities, Zavodskaya Laboratoriya 8:45–62
60. Timoshenko SP, Goodier JN (1951) Theory of elasticity. New York, London et al.: McGraw-Hill, 575 p
61. Tomlinson GA (1927) The rusting of steel surfaces in contact. Proc Roy Soc Lond Ser A 115(1026):472–483
62. Tonoyan VS, Minasyan AF (1975) Asymmetric contact problem for a half plane with vertical finite cut. Doklady Akademii Nauk Armyanskoy SSR 61(5):289–297
63. Troshchenko VT, Tsybanev HV (2002) Fretting fatigue of metallic materials and structural components. In: Troshchenko VT (ed) "Trybofatyka" ("Tribofatigue"): Proceedings of the 4th international symposium on Tribofatigue (ISTF 4), Vol 1. Ternopil', pp 23–28
64. Troshchenko VT, Tsybanev GV, Khotsyanovsky AO (1994) Two-parametr model of fretting fatigue crack growth. Fatigue Fract Eng Mater Struct 17(6):15–23
65. Turchak LI (1987) Osnovy chisel'nykh metodov (Fundamentals of Numerical Methods): A Tutorial. Nauka, Moscow
66. Waterhause RB (1972) Fretting Corrosion. Pergamon Press, Oxford-New York
67. Yarema SYa (1988) Propagation of curvilinear cracks in plates. Izvestiya Akademii Nauk SSSR, Mekhanika Tverdogo Tela 2:156–163

The manufacturer's authorised representative in the EU is Springer Nature Customer Service Centre GmbH, Europaplatz 3, 69115 Heidelberg, Germany. If you have any concerns regarding our products, please contact ProductSafety@springernature.com

Printed and bound by CPI Group (UK) Ltd, Croydon, CR0 4YY
23/07/2026
02175027-0005